OCF Technologies and IoT Programming

OCF技术原理
及物联网程序开发指南

李永华◎编著
Li Yonghua

清華大學出版社
北京

内容简介

本书主要内容包括以下几个方面：物联网的发展概述，主要介绍物联网的产生、架构、技术及发展情况；OCF 技术的基本原理，阐述 OCF 技术框架及核心功能；OCF 资源模型，主要描述 OCF 的资源定义以及资源的操作、功能交互、消息传递方法；OCF 的具体开发方法，包括基于 Mac、Windows、Linux、Android 和 Arduino 开发的方法，主要描述软件工具、编译方法、实例代码和综合实例。本书内容由浅入深，先系统后实践，技术讲解与实践案例相结合，以满足不同层次人员的需求；同时，本书附有实际开发的软件实现代码，供读者自我学习和自我提高使用。

本书可以作为大学信息与通信工程及相关领域的高年级本科生及研究生的教材，也可以作为物联网、OCF 技术开发人员的技术参考书，还可以为物联网方向的创客提供帮助。

图书在版编目(CIP)数据

OCF 技术原理及物联网程序开发指南/李永华编著. —北京：清华大学出版社，2019
(清华开发者书库)
ISBN 978-7-302-51116-8

Ⅰ. ①O…　Ⅱ. ①李…　Ⅲ. ①互联网络—应用 ②智能技术—应用　Ⅳ. ①TP393.4 ②TP18

中国版本图书馆 CIP 数据核字(2018)第 202409 号

责任编辑：盛东亮　张爱华
封面设计：李召霞
责任校对：时翠兰
责任印制：丛怀宇

出版发行：清华大学出版社
　网　　址：http://www.tup.com.cn，http://www.wqbook.com
　地　　址：北京清华大学学研大厦 A 座　　**邮　　编**：100084
　社 总 机：010-62770175　　**邮　　购**：010-62786544
　投稿与读者服务：010-62776969，c-service@tup.tsinghua.edu.cn
　质量反馈：010-62772015，zhiliang@tup.tsinghua.edu.cn
　课件下载：http://www.tup.com.cn，010-62795954
印 装 者：三河市龙大印装有限公司
经　　销：全国新华书店
开　　本：203mm×260mm　　**印　　张**：24　　**字　　数**：661 千字
版　　次：2019 年 1 月第 1 版　　**印　　次**：2019 年 1 月第 1 次印刷
定　　价：99.00 元

产品编号：079340-01

序言

FOREWORD

开放互联基金会(OCF)是物联网行业最大的组织之一,其任务是开发标准规范,促进互操作性,并为物联网(IoT)设备提供测试认证。OCF 联盟拥有超过 400 家会员公司,包括家电行业领导者(海尔、LGE、伊莱克斯、三星电子),芯片制造商(英特尔、高通),软件平台(微软),网络设备(Cisco)和有线服务(CableLabs)等。OCF 联盟还发起一个开源项目——IoTivity,帮助开发者和企业创建解决方案,实现一个开放的物联网规范。

OCF 2.0 版本规范已经发布,业内人士预计,OCF 2.0 版本产品认证将在 2018 年第四季度展开,并在 2019 年继续扩大范围。随着越来越多的 OCF 产品在市场上部署,开发商与学术界对技术理解与实现具有强烈的需求,本书将有助于更好地理解 OCF 技术实现。

向李永华副教授表示衷心的感谢,他抽出个人的时间完成这本有价值的书。毫无疑问,会有很多人阅读这本书,以便更好地开发物联网的产品和服务。

John J. Park

开放互联基金会首席执行官

The Open Connectivity Foundation (OCF) is one of the largest groups in the industry whose mission is to develop specification standards, promote interoperability, and provide a certification program for devices involved in the Internet of Things (IoT). It has over 400 member companies, including industry leaders in home appliances (Haier, LGE, Electrolux, Samsung Electronics), chip manufacturers (Intel, Qualcomm), software platforms (Microsoft), network equipment (Cisco), and cable services (CableLabs). OCF also sponsors an open source project, "IoTivity", to help developers and companies create solutions that map a single, open IoT specification.

As OCF 2.0 Specification is now published, the industry expects that OCF 2.0 certified products will be launched in the fourth quarter of 2018 and continue to expand in 2019. As more OCF products are deployed in the market, there will be strong demand among developers and academia to understand OCF technology for implementation and research. This book will shed light upon OCF technology for those who hope to understand it better.

I would like to express my sincere gratitude to associate professor Youghua Li, who took his personal time to write this valuable book. I have no doubt that this book will be read by many and provide insight on how to develop better IoT products and services.

John J. Park

Open Connectivity Foundation Executive Director

前言
PREFACE

近年来物联网快速发展，各种标准、技术层出不穷，物联网的应用领域不断拓展，国际数据公司预测，到2020年，全世界联网装置将超过2120亿个，市值将达到万亿美元。技术多样化发展的同时，也为物联网互联互通带来隐患，业界制定统一标准的呼声也越来越高。2016年10月，OCF（Open Connectivity Foundation，开放互联基金会）成立，探索建立物联网统一标准，真正开发一套通用的物联网互联架构，为物联网未来的发展提供了新的思路。

OCF由Linux基金会负责运营，目前包括各种各样的会员，涉及芯片、模块、产品、安全、家电、系统、集成等多方面的物联网厂商，这种跨领域组成的OCF，有助于开发一套通用的物联网互联架构，为未来物联网更加广泛的应用提供了技术保障。

OCF最初的框架来源OIC和AllJoyn开源项目，采用Apache和BSD许可协议。无论是终端产品、应用、服务，通过OCF技术就可以互相通信。OCF是由开放、统一的框架和核心资源组成，让开发者通过其软件开发框架开发各种应用，以便使邻近的系统、应用或设备得以互联互通、控制及共享资源。OCF最终希望打造一个跨平台、接入方式、编程语言的开放软件架构，可以让不同的设备（例如电视、路由器、冰箱、洗衣机、智能照明系统）和其他设备无缝地连接起来，并跨越iOS、Android、Windows、Linux或Mac等不同的操作系统。

本书以当前物联网的发展为背景，总结OCF技术的原理及应用方法。从物联网技术开发方法出发，系统地介绍如何利用OCF技术进行不同系统下的产品研发，继而进行相应的应用。因此，本书面向未来的物联网工业创新与发展，通过OCF软件架构，紧紧跟随技术的发展，为物联网技术的发展提供创新型人才。同时，本书总结了实际科研中的应用技术，不仅包括处理能力较强的各种标准客户端系统应用，也包括能力相对较弱的瘦客户端系统应用，希望对教育教学及工业界有所帮助，起到抛砖引玉的作用。

本书的主要内容包括如下几个方面：物联网的发展概述，主要介绍物联网的产生、架构、技术及发展情况；OCF技术的基本原理，阐述OCF技术框架及核心功能；OCF资源模型，主要描述OCF的资源定义以及资源的操作、功能交互和消息传递方法；OCF的具体开发方法，包括基于Mac、Windows、Linux、Android和Arduino开发的方法，主要描述软件工具、编译方法、实例代码和综合实例。

本书的内容和素材主要来自OCF的官方网站（www.openconnectivity.org）。首先，本书是作者近几年承担的科研成果和教育成果的总结，在此特别感谢林家儒教授的鼎力支持和悉心指导；其次，本书是作者指导的研究生在物联网和智能硬件方面的研究工作及成果的总结，在此特别感谢万昊、谭扬、黄旭新、陈佳丰、王玥等同学的大力协助；再次，OCF联盟为本书提供了第一手资料，在此向联盟的鼎力支持表示感谢；最后，父母妻儿在精神上给予我极大的支持与鼓励，才使得此书得以问世，向他们表示感谢！

本书由北京市教育科学“十二五”规划重点课题（优先关注）、北京市职业教育产教融合专业建设模式研究(ADA15159)资助；同时，本书也由北京邮电大学教育教学改革项目(2017JY04)资助，在此一并表示感谢！

本书内容由浅入深，先系统后实践，技术讲解与实践案例相结合，以满足不同层次人员的需求；同时，本书附有实际开发的软件实现代码，供读者自我学习和自我提高使用。本书可作为大学信息与通信工程及相关领域的高年级本科生及研究生的教材，也可以作为物联网、OCF技术开发人员的技术参考书，还可以为物联网方向的创客提供帮助。

本书主要由李永华编著。此外，李昕烨、陈河泉、李和禹、陈向梅、张秋彤、张国利也参与了部分内容的编写。

由于作者的水平有限，书中难免存在疏漏之处，衷心地希望各位读者多提宝贵意见及具体的整改措施，以便作者进一步修改和完善。

李永华于北京邮电大学

2018年4月

目录
CONTENTS

第1章

CHAPTER 1

物联网技术概述

互联网从产生到现在，不断地发展和变化。在互联网之上，创造了许多新的业务应用和商业模式。互联网不断变化的同时，各种技术的发展也在改变着IT行业的发展方向。当前，有线和无线的宽带接入无处不在，而且价格不断地下降；智能硬件设备可以搭载多种传感器，变得更加强大，向微型化发展；多种设备之间的通信互联，导致互联网的进一步发展，这就是物联网。当接入设备能够获取更多网络数据时，物理实体通过相关数据，可以为网络用户提供更多智能服务，成为物联网发展的主要驱动力量。

物联网描述了一个所有物体全部成为互联网中的元素，所有物体都拥有独有的特征，并且通过互联网访问可以获取它的位置与状态、可以添加各种服务进行智能扩展，融合了数字与物理世界，极大地影响着个人和社会环境。因此，通过物联网，世间万物正朝着"永远在线"的方向发展，例如，绿色IT、能源效率、智能家居、车联网、可穿戴设备、智能医疗等各种领域，不胜枚举。物联网的发展同时伴随着挑战：一方面，不同技术之间如何打破壁垒，真正实现万物互联，包括不同接入技术之间、不同操作系统之间、不同编程语言之间的互联互通；另一方面，在信息安全方面，为确保向社会各领域提供一个公平而且可信任、开放的物联网，标准化和监管是必不可少的。

物联网的发展在国际上引起了高度的重视。例如，美国提出智慧地球；中国政府提出感知中国；欧盟提交《物联网——欧洲行动计划》公告，通过构建新型物联网管理框架来引领世界物联网的发展，这些都是对未来物联网发展的战略性构想。在企业层面的发展也是如火如荼，例如，苹果公司发布新一代操作系统，实现更流畅的苹果设备无缝对接、个人健康信息管理应用、智能家居应用等。谷歌公司2014年并购的物联网公司包括Nest Labs(智能恒温器制造商)、MyEnergy(在线家庭水、电、天然气使用管理方案提供商)、DropCam(家居安防摄像头制造商)、Revolv(智能家居自动化控制中枢制造商)等。Intel公司通过高性能、连接性、安全性为物联网实现了更加智能的嵌入式解决方案。Microsoft公司为企业提供一种独特的集成方法，使其能够通过收集、存储和处理数据来利用物联网的优势。该方法通过将各种产品组合进行扩展，包括一系列个人计算机、平板电脑、企业网络边缘的行业设备、开发工具、后台系统服务以及多样化的合作伙伴生态系统。

国内的企业中，海尔公司提出了U+平台(由智能数字家电产业技术创新战略联盟推出)，这是一个成熟的商业生态系统，具有行业性质的开放平台，芯片、模组、电控厂商、开发者、投资者、电子商务、云服务平台和跨平台合作等所有的参与者都能从中受益。百度公司的语音技术通过免费、开放的策略，打造周边信息查询、导航、公交线路、到站提醒、盲人路线自定义，以及丰富的旅游、餐饮、购物等生活服务语音模块，并进入智能手机、车载、教育等多个服务领域。阿里巴巴与多个家电厂商签署战略合作协议，将共同构建基于阿里云的物联网开放平台，实现家电产品的连接对话和远程控制，在未来形成统一的物联网产品应用和通信标准，实现系列家电产品的无缝接入和统一控制。腾讯公司从手机QQ、QQ空间等移动社交平台转移到以应用宝为核心的分发渠道，随着开放平台及智能硬件的发展，开放连接的另一端开始指向硬件领域，腾讯社交智能硬件开放平台，QQ物联正式切入物联网领域。

总体而言，目前世界上的物联网发展呈现了多元的趋势，大型的科技企业准备制定自己的物联网标准，也有一些企业联合起来成立了联盟，制定共同的物联网标准。在全球物联网加速发展并推动产业变革之际，企业进一步升级设备应用，逐渐向智能生活移动终端过渡，以争夺物联网领域的主导权，多个国际物联网联盟高度重视物联网领域并展开实际的研发。因此，未来一定会出现实用化、智能化的物联网标准，这对于未来社会的发展必将产生深远的影响。

1.1 物联网基本架构

物联网(Internet of Things，IoT)是通过互联网、传统电信网等信息载体，让所有独立寻址的普通物理对象实现互联互通的网络。物联网一般为无线网，由于每个人周围的设备可以达到数千个，所以物联网将包含数万亿个物体。在物联网上，每个人都可以应用智能感知将真实的物体联网，查找出它们的具体位置。通过物联网可以用中心计算机对机器、设备、人员进行集中管理、控制，也可以实现对设备控制、位置搜寻、物品防盗以及智能推荐等的各种应用。物联网将现实世界数字化，应用范围十分广泛，具有十分广阔的市场和应用前景。根据国际上一般的分类方法，物联网的应用领域主要包括运输和物流领域、健康医疗领域、智能环境领域、个人和社会领域等。

1.1.1 物联网的由来

互联网是一个不断发展进化的实体，在人类社会的发展中越来越重要，通过扩展不断创造新的价值。互联网开始于"计算机的互联网"，是世界级的网络服务，万维网则建立了初始的顶层平台。而近几年，互联网却开始向"人的互联网"方向转变，创造了如Web 2.0的概念并使用其内容。

技术的发展扩展了互联网的边界，宽带网络连接变得更加普及，无论是在发达国家，还是在发展中国家，带宽变得更加廉价。例如，非洲某些地区，由于光纤网络的发展带来了社会各个方面的显著进步。一方面，设备硬件的处理能力与存储空间正在飞速增长，技术发展让这些设备变得越来越小；设备的变化不仅让人们可以更好地使用互联网，而且创造了一系列新的发展机会。整个人类社会正在经历一场以个人计算机领域为主到以移动电子设备为主的巨变，包括智能手机、笔记本电脑及平板电脑。另一方面，这些设备由于传感器和探测器的发展，而使其性能大幅度提升；多个方面的结合创造了一个任何位置的设备均可以连接至网络中的环境，也就有了设备感知及计算，通过信息传输，传感器成为互联网的一部分。除此以外，物理设备也可以搭载智能硬件而被其他设备感知，把物理世界与虚拟世界通过智能设备联系到一起，把互联网扩展成物联网。

物联网的实践最早可以追溯到1990年施乐公司的网络可乐贩售机。1991年，美国麻省理工学院的Kevin Ashton教授首次提出物联网的概念。1995年，比尔·盖茨在《未来之路》一书中也曾提及物联网，但未引起广泛重视。1999年，美国麻省理工学院建立了自动识别中心，提出"万物皆可通过网络互联"，阐明了物联网的基本含义。

随着技术和应用的发展，物联网的内涵已经发生了较大变化。2003年，美国《技术评论》提出传感网络技术将是未来改变人们生活的十大技术之首。2005年11月17日，在突尼斯举行的信息社会世界峰会上，国际电信联盟(ITU)发布《ITU互联网报告2005：物联网》，引用了"物联网"的概念。物联网的定义和范围已经发生了变化，覆盖范围有了较大拓展，不再只是基于RFID(Radio Frequency

Identification，射频识别)。为了促进科技发展，寻找新的经济增长点，各国政府开始重视下一代的技术规划，将目光放在了物联网上。2008 年 11 月，在中国北京大学举行的第二届中国移动政务研讨会上，提出“知识社会与创新 2.0”。移动技术、物联网技术的发展代表着新一代信息技术的形成，并带动经济社会形态、创新形态的变革，推动面向知识社会、以用户体验为核心的创新形态的形成。创新与发展更加关注用户、注重以人为本，而创新 2.0 的形成，又进一步推动了新一代信息技术的健康发展。

1.1.2　物联网的结构

在物联网概念中，“物”的定义是非常广的，包含各种不同的物理元素，其中有人们每天使用的个人产品，如智能手机、平板电脑和数码相机等；也包括环境中的元素，使其可以通过网关与人们相连接。基于以上观点的“物”，将有数量庞大的设备与物体接入到互联网中，每一个点都提供了数据与信息，甚至服务。物联网思维让连接从原来的“任何时间、任何地点、任何人”变成了“任何时间、任何地点、任何物”。这些事物加入网络，使得通过智能处理和服务支持经济发展、环境保护和人类健康成为可能。物联网的各个元素的示意图如图 1-1 所示。

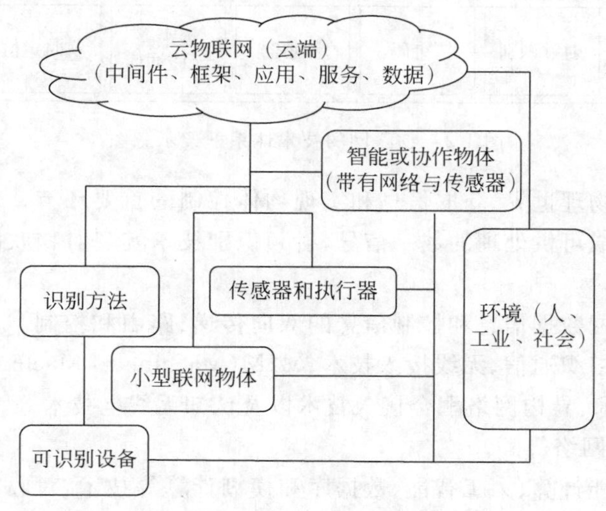

图 1-1　物联网的各个元素的示意图

图 1-1 是抽象意义上的物联网生态系统示意图，其中设备本身需要能够被相应的技术识别。通过多种手段辨识出设备的属性当然也包括相关位置信息。相应地，这些使用传感器的联网物体开始变得小型化，并且融入人们的日常生活，传感器和执行器通过网络环境可以做出相关反应，并且可以根据具体状况和时间做出更高级的服务。智能或协作物体感知到相关的活动与状态，并且将它们连接到物联网中。中间件和框架允许其他设备接收所需的数据，运行相关的应用和服务，例如，云端可以提供容量，使这些应用和服务的质量更高，从而使物联网按照相关的设想来改变环境。

从这个意义上说，几乎所有的事物都能连接网络，即使是个人的随身物品也有迹可循，它的状态和位置信息将会在高层次的服务中被实时获取。在这种背景下，确定物联网的作用范围是非常重要的。几十亿物体连入网络，每一个物体都提供了数据，而它们中的大部分都能影响所处的环境。当处理这些庞大的数据时，需要智能设备进行判断决策。因此，可以利用今天已经熟知的互联网技术，逐渐向物联网方面进化，以这种形式奠定物联网的基础。当物联网领域出现连续的技术更新并趋近成熟时，物联网进化的核心驱动力便是各种各样的应用。

随着物联网的发展，人们提出了物联网的技术体系框架，从可实现的角度对物联网的发展进行了总结，如图 1-2 所示。

图 1-2　物联网的技术体系框架示意图

感知层是物联网感知物理世界、获取信息和实现物体控制的首要环节。传感器将物理世界中的物理量、化学量、生物量转换成可供处理的数字信号，射频识别技术实现对物联网中物体的标识和信息的获取。

传输层主要实现物联网数据信息和控制信息的双向传递、路由和控制。重点包括低速近距离无线通信技术、低功耗路由、自组织通信、无线接入技术、M2M(Machine to Machine，机器对机器)通信增强、IP 承载技术、网络传送技术、异构网络融合接入技术以及认知无线电技术等。例如，移动网络、互联网、无线网络、卫星和 Post-IP 网络等。

支撑层综合运用高性能计算、人工智能、数据库和模糊计算等技术，对收集的感知数据进行通用处理，重点涉及数据存储、并行计算、数据挖掘、平台服务、信息呈现等。例如，智能处理、分布式并行计算、云计算技术、海量存储与数据挖掘、管理系统及数据库以及综合设计验证等。

应用层是一种松耦合的软件组件技术。它将应用程序的不同功能模块化，并通过标准化的接口和调用方式联系起来，实现快速可重用的系统开发和部署；可提高物联网框架的扩展性，提升应用开发效率，充分整合和复用信息资源。例如，运营平台、信息中心、内容服务以及专家系统等。

1.2　物联网相关技术

当前，全球制造业的发展越来越呈现数字化、网络化和智能化的新特征。

从国家层面看，美国提出“工业互联网”战略、德国提出“工业 4.0”战略、中国提出“互联网＋”战略，主要意图都是抢占智能制造这一未来产业竞争的制高点。

从企业层面看，应把实施互联网战略作为企业提升竞争力的关键环节。互联网不再只是企业生产的工具和手段，而是已成为支撑企业成长的关键要素和支撑平台。基于互联网思维下的企业转型升级，推动众多新业态的崛起，已经成为当前制造业发展的新亮点，更是当前经济形势下难得的发展机遇。

从技术层面看，物联网是互联网概念的延伸。如果说互联网是"软"的，那么物联网则是"软（云平台）＋硬（智能硬件）"的模式。这也是为什么互联网巨头转型到物联网时，需要做硬件，而传统的硬件公司转型到物联网就要做云平台，这是物联网软硬件的互补性需求。物联网必须"软硬"结合，硬件是基础设施，连接是必要条件。通过硬件获取数据及软件的处理，才是物联网的核心价值。因此，物联网带来的价值绝大多数将来自云平台上的数据。也就是说，未来物联网出现的地方，大数据、云平台就像孪生兄弟一样相伴而行。如今市场上缺少的不是硬件个体，而是如何让硬件之间互联互通，同时提供多样的云平台服务，如数据的存储、分析、处理和推送，这样才能对人类社会的进步起到推动作用。

物联网本身可以是任意大小的网络。例如，物联网系统可以是一种智能家庭局域网，为了方便人们在室内的各种活动而形成一种多样化网络架构，通常由灯光照明、电器控制、遮阳控制、节能控制、远程抄表、应用软件、互联规划和共享网络等子系统组成，主要利用综合布线技术、网络通信技术、智能家居系统设计方案、安全防范技术、自动控制技术和音视频技术将与家居生活有关的设施集成，构建高效的住宅设施与家庭日程事务的管理系统，提升家居安全性、便利性、舒适性和艺术性，并实现环保节能的居住环境。也就是说，如今物联网发展所取得的成果，是建立在技术进步基础上的。下面将简单介绍物联网的相关技术。

1.2.1 接入技术

在智能家居领域，有线组网技术是最先兴起的，这种组网方式具有稳定性高、通信速度快等优点，但存在着网络庞大、布线复杂、灵活性差等诸多缺点。总的来说，经过多年的发展，有线组网技术普及率较高，市场出现了许多成熟的技术。有线方式主要包括电力线载波、电话线方式、以太网方式的 IEEE 802.3 以及专用总线方式的 IEEE 1394 等。物联网中使用的传统布线方案，不仅布线复杂且成本高昂，因此，基于无线技术的物联网互联方案渐渐获得市场的青睐。下面对目前无线和有线的接入方式做简单的介绍。

1. ZigBee 技术

ZigBee 技术作为一种低速短距离传输的无线网络协议，在物联网领域获得了广泛的使用，替代了传统的有线布线。基于 ZigBee 的智能家居能源管理系统，简化了家庭布线的复杂度，实现了家电的无线互联，为用户提供了舒适的家庭环境、方便易用的家电管理以及远程查看和管理家电的功能。ZigBee 协议以低功耗、低成本、低速率、低复杂度和易组网等特点已经在医疗监护、环境监测、智能交通、智能电网和智能家居等方面得到了广泛研究与应用。

为了方便管理家电的工作模式，使用基于 ZigBee 技术的智能插座和红外遥控器等技术实现对家电的无线控制。其中，智能插座可以对所有家电执行简单的开关控制；而红外遥控器则可以对空调、电视机等多控制状态的家电执行控制，为家庭用户提供了方便易用的家电控制，美中不足是无法满足远距离家庭用户管理家电的需求。针对这个问题，有文章提出使用基于 Web 的动态网页远程管理家电的智能家居系统，使用嵌入式操作系统，以 ARM Cortex-M3 作为系统控制器，使用 ZigBee 无线传感器网络实现家电控制以及环境信息采集，提供动态网页供用户远程访问家电环境信息以及控制家电设备。

2. 红外通信技术

红外通信技术是以红外线作为通信介质的特定应用，通常用在移动电话、笔记本电脑和掌上电脑中。红外通信是被设计用于短距离、低功率、无需许可证的通信。但是红外通信由于技术原因，导致传输距离受到很大限制。例如，红外线不能穿越阻挡信号的物体、传输角度不能过大等。因此，在早期的智能家居应用中使用红外技术的效果并不理想。

在智能家居系统中，常常利用 ZigBee 技术来控制红外遥控器，从而间接地控制基于红外通信的家用电器。因为角度的偏离会影响设备对红外线的接收，可将红外遥控器固定在室内合适的位置，这样人们就可以通过 ZigBee 技术向红外遥控器发送指令，这些指令经过遥控器识别后转换成相应的红外通信指令，并发送到相应的红外控制家电上。使用基于 ZigBee 技术的红外遥控器，可以免去对家电的更新，只对遥控器进行重新设计，实施起来较为方便，但是这样仅仅实现了对红外家电的简单控制，没有任何家电信息的收集与交互。

3. 蓝牙技术

蓝牙技术是无线数据和语音传输的开放式标准，它将各种通信设备、计算机及其终端设备、各种数字数据系统、家用电器采用无线方式连接起来。该技术用于替代便携、固定电子设备上所使用的电缆连线，是短距离无线连接技术，工作在 2.4GHz 的开放频段上，采用 1600 次/s 的扩频技术，发射功率为三类，即 1mW、10mW 和 100mW，通信距离为 10～100m，传输速率约为 3Mb/s。在传输数据信息的时候，还可传输一路话音信息，这也是蓝牙技术的重要特点之一。

蓝牙技术适用于在短距离（大约 10m）范围内替代电缆，如果增大发射功率，它的传输距离可达 100m，而家庭中各家电之间的相隔距离一般不会超过此距离。蓝牙传输速率完全可以满足家庭网络中各家电间的数据传输；而且，蓝牙的抗干扰能力强，它的快速跳频使系统更加稳定，前向纠错能力可以限制噪声的影响，这样家庭中的各种蓝牙家电可以互不干扰而正常工作；蓝牙系统具有连接的普遍性、标准的开放性以及强大的扩展性，可以满足家庭网络中的多种需要；蓝牙芯片的成本相对较低，因而可大大降低网络家电的成本；中国无线电管理委员会已对蓝牙技术开放了相应的频段，这起到了推广的作用。

基于蓝牙技术的智能家居控制方法，在需要控制的家电中嵌入蓝牙模块，从而组成蓝牙控制网络。利用嵌入式模块构成智能家居控制器、个人计算机模拟家庭主网关、单片机加蓝牙模块模拟信息家电，把三者结合组成模拟的家庭控制子网，实现家庭控制子网对家电设备的基本操作；采用以智能家居控制器为中心的完全星形组网方式，所有子网设备直接与智能家居控制器进行通信以完成子网的控制和管理功能，组网简单，实现即插即用。

4. WiFi

WiFi 是一种可以将个人计算机、智能移动设备等终端以无线方式互相连接的技术，可以方便地与现有的有线以太网络整合，组网成本低，通常对应以个人计算机共享上网为主要应用模式的家庭网络服务。

在嵌入式处理器和操作系统基础上，利用 WiFi 模块互联各个家电终端，组建家庭无线网络。系统采用典型的客户机与服务器架构，充分发挥客户端个人计算机的处理优势，实现在 WiFi 热点区域接入无线通信终端，对家电设备远程控制。也可以基于 ZigBee 网络与 WiFi 视频监控网络构成智能家居监控系统，利用 ZigBee 传输标量数据和控制信号、WiFi 传输音视频数据，并为下层的 ZigBee 网络提供互联网接入的网关功能，发挥两种无线技术的优势。

5. 蜂窝移动通信网

蜂窝移动通信采用蜂窝无线组网方式，在终端和网络设备之间通过无线通道连接起来，进而实现用户在活动中可相互通信。其主要特征是终端的移动性，并具有越区切换和跨本地网自动漫游的功能。蜂窝移动通信是指由基站子系统和移动交换子系统等设备组成蜂窝移动通信网，并提供话音、数据、视频图像等业务。移动通信网主要包括第二代移动通信系统 GSM、CDMA，第三代移动通信系统

WCDMA、CDMA2000 以及 TD-SCDMA,第四代移动通信系统,第五代移动通信系统,等等。

6. 电力线

网络信息传输介质用电力线有很大优势。目前,大多数家居中已经铺设电力线,电缆不需要另外布设,虽然这样降低了施工难度,但是家庭中所使用的手持移动设备不能采用电力线接入网络。另外,电力线的缺点是传输速率较慢,仅有 300kb/s,不能满足数字信号音频和视频信号的传输,保密性差,标准未统一,接入设备昂贵等。目前,国际上采用电力线作为联网传输介质推出的解决方案有 X-10、CEBus 等。

7. 电话线

HomePNA(Home Phoneline Networking Alliance,家庭电话线网络联盟)是一个非营利性组织,致力于协调采用统一的标准、统一电话线网络的工业标准。该联盟在 1998 年由 11 个公司共同建立,旨在以家庭电话线为连接介质构造家庭网络,使用频分复用技术,在同一条电话线上同时传送话音和数据信号,当用户上网时,打电话和收发传真都不受影响。HomePNA 先后推出两个版本,分别是 HomePNA 1.0 和 2.0,其中 1.0 版本支持 1Mb/s 的速率,2.0 版本支持 10Mb/s 以上速率。韩国三星公司曾推出基于此项技术的家用智能产品。这种方式是通过在电话线上加载高频载波信号来实现信息传递的,可以同时满足电话业务、XDSL 和家庭内部数据传输,且互不干扰。利用墙上预留的电话线插座,能够避免重新布线,但是在家庭中电话线插座不可能随处安装,在扩充新节点时还是会面临重新布线的问题。

8. 以太网

以太网指的是由 Xerox 公司创建并由 Xerox、Intel 和 DEC 公司联合开发的基带局域网规范。该技术基于铜介质的双绞线和同轴电缆实现信号的双向传输,数据传输率很高,可以达到 10Mb/s、100Mb/s 和 1000Mb/s,能够传输数据、电话、视频以及家电控制信息,主要用于有线局域网和高速互联网。现阶段,以太网技术在目前的家庭设备互连中是最简单也是最普及的,成本低,但专门布线费用高,安装维护比较困难,几乎不具备移动性,家庭中的用户宁可使用已经铺设好的电话线或电缆,也不愿意再安装以太网线。因此,以太网方式可能是家庭网络发展初期的解决方案,但不是家庭网络的最终方案。

9. 专用总线

通过采用专用总线的形式来实现家庭控制网络组建,并完成小区的智能相连,如 RS-485 总线解决方案。它的优点是抗干扰能力比较强,技术相对成熟;缺点是需要重新铺设线路,给用户带来麻烦。

10. RFID 技术

RFID 技术是一种通信技术,可通过无线电信号识别特定目标并读写相关数据,而无须在识别系统与特定目标之间建立机械或光学接触。常用的有低频、高频、超高频和微波等技术。RFID 读写器分为移动式和固定式,目前 RFID 技术应用很广,例如应用在图书馆、门禁系统和食品安全溯源等方面。

从概念上来讲,RFID 类似于条码扫描。对于条码技术而言,它是将已编码的条形码附着于目标物并使用专用的扫描读写器,利用光信号将信息传送到扫描读写器;而 RFID 则使用专用的 RFID 读写器及专门的、可附着于目标物的 RFID 标签,利用频率信号将信息由 RFID 标签传送至 RFID 读写器。从结构上来讲,RFID 是一种简单的无线系统,用于控制、检测和跟踪物体。该系统只有两个基本器件,由一个询问器和很多应答器组成。RFID 技术的飞速发展对于物联网领域的应用具有重要意义。

11. NFC 技术

NFC(Near Field Communication,近场通信)技术是一种短距离高频的无线电技术,在 13.56MHz 频率运行于 20cm 距离内。NFC 技术由 RFID 技术演变而来,由飞利浦半导体、诺基亚和索尼公司共同研制开发，其基础是 RFID 及互联技术。其传输速率有 106kb/s、212kb/s 和 424kb/s 三种。目前 NFC 技术已成为 ISO/IEC IS 18092 国际标准、ECMA-340 标准与 ETSI TS 102 190 标准。NFC 技术采用主动和被动两种读取模式；在单一芯片上结合感应式读卡器、感应式卡片和点对点的功能，能在短距离内与兼容设备进行识别和数据交换，但是使用这种手机支付方案的用户必须更换特制的手机。目前，这项技术已被广泛应用，如果配置了这种支付功能的手机，就可以行遍全国，手机可以用作机场登机验证、大厦的门禁钥匙、交通一卡通、信用卡和支付卡等。

1.2.2 基于网络的信息管理技术

无线通信技术提供了家电互联的可能性，而 Web 技术的融入将会使家电信息的管理变得更加方便。越来越多的传感器正在接入互联网，包括便携式传感器，如手机中自带的各种传感器、环境感应传感器组成的传感器网络、智能家居中的传感器、各种商品货物上的标识传感器，它们共同组成了物联网。

传感器的目的是监测数据，形成一个庞大的监测网络，供用户参考。为了更好地管理传感器网络和利用传感器网络所监测的数据，Web 技术脱颖而出，它将传感器网络中的各种数据资源、监测设备、应用系统和计算资源都融合在互联网上，形成了另外一种概念上的物联网，即物联 Web。它能够让人们实时、全面、智能、可靠地监测和管理现实世界的对象，用户能够方便地获得感兴趣的传感器或传感器的观测值，从而根据这些观测值来决定对设备的下一步动作。

1. 家电传感器检索技术

用户要方便地获得智能家居传感器或传感器对家电的观测值，需要用到检索技术。当前的检索方式可以分为两类：语义检索和相似性检索。其中，语义检索分为简易文本检索和详细文本检索。简易文本检索支持检索描述传感器的文本元数据，例如传感器的类型、位置和测量单位等，但是实际应用性不强，因为人们描述同一概念的方式不同，简易文本的绑定容易造成检索误差；详细文本检索将描述传感器多个相关的文本元数据标记在传感器上，提供传感器描述表及详细的传感器描述，这在一定程度上降低了用户检索的误差，但是执行复杂，相关的文本元数据确定不易。相似性检索技术在一定程度上降低了语义检索的文本输入准确度要求，以图片或者数据检索与其相似的传感器或传感器数据，提高了易用性。检索技术可以用于智能家居的客户端，使用相似性检索有利于客户端在不同地方使用，不用进行客户端与智能家居的单独匹配，适合多种平台。

2. 嵌入式 Web 服务器

智能家居控制器是一个由软件和硬件共同组成的、具有计算机能力的功能实体，实现它的一种方法就是采用个人计算机。个人计算机不仅具有足够的能力实现控制与网关功能，而且可以很方便地实现控制与网关功能，但是作为控制中心也有它的不足，如体积大、功耗大和成本高，所以使用个人计算机作为智能家居控制器并不合适。为了克服个人计算机的这些不足，可以采用嵌入式系统来实现基于以太网的智能家居控制器。嵌入式系统以应用为中心，针对家居控制中心的要求，在软硬件上可定制设计，使之在功能、可靠性、成本、体积和功耗等方面符合家庭控制中心的要求。

基于嵌入式 Web 服务器系统的网络智能家居控制器，不仅符合物联网的发展趋势，而且具有很好的技术研究和应用价值，其中一些用于控制电灯、电冰箱、空调等家用电器的运行状态。例如，物联网控制器采用 ARM 处理器作为核心处理单元，硬件平台由处理器和以太网控制器等组成，软件平台主要以精简 TCP/IP 协议栈为核心，软硬件共同组建了嵌入式物联网 Web 服务器。精简 TCP/IP 协议栈包括以太网控制器驱动程序、ARP 协议模块、IP 协议模块、ICMP 协议模块、TCP 协议模块和 HTTP 协议模块。在 HTTP 协议的基础上，建立嵌入式 Web 服务器 TCP/IP 应用。用户可以使用任意浏览器对家居中的设备和环境进行检测和控制。

采用ZigBee技术与嵌入式Web服务器相结合的方式可以控制物联网。使用ZigBee技术可以组建家居无线控制网络,并且实现所有节点之间的数据收发,组网灵活,加入或退出节点都极为方便;加入嵌入式Web服务器,不仅提供了用户通过浏览器查询家居温度、控制家居中灯节点的功能,还可以设计一个较为友好的访问界面。

总体来说,当今主流的物联网无线接入技术,可以很好地实现物联网之间的互联,而Web技术的引入,极大地方便了家电信息的查询管理。对于物联网的管理,除了嵌入式Web服务器以外,还有基于Android和iOS平台的应用程序,通过相应的软件来对物联网进行管理。这些物联网能够启动的前提是控制端与各个接入家庭网的智能家电,在系统上必须保证匹配。也就是说,所有家电必须要有统一的标准,这样才能使控制端较好地对其进行操作。而当今物联网的控制标准尚缺,平台匹配较差,物联网需要定制,大大影响了用户对家电的选择性,因此出台相应的物联网标准和适用于智能家电系统的平台极为迫切。

3. 云计算技术

云计算是由谷歌公司提出的一种网络应用模式。狭义云计算是指IT基础设施的交付和使用模式,指通过网络以按需、易扩展的方式获得所需资源;广义云计算是指服务的交付和使用模式,指通过网络以按需、易扩展的方式获得所需服务。这种服务可以是和IT、软件、互联网相关的服务,也可以是任意其他的服务,具有超大规模、虚拟化、安全可靠等独特功能。

云计算是把一些相关网络技术和计算机融合在一起的产物,对存储系统和计算机进行调整,将分布式计算的信息和数据利用互联网完成信息传输,使资源的利用效率更高,目的是把各种任务进行低成本处理并融合为功能完整的实体。云计算是以加强改善其处理能力为重点,用户终端的负担相应降低,I/O设备能够简化,还可以对它的计算功能进行合理的分配应用。例如,百度等搜索功能就是它的应用之一。

随着物联网产业的深入发展,物联网发展到一定规模后,物理资源与云计算结合水到渠成。对于一部分物联网行业应用,如智能电网、地震台网监测等,终端数量的规模化导致物联网应用对物理资源产生了大规模需求,一个是接入终端的数量可能是海量的,另一个是采集的数据可能是海量的。云计算在物联网中的应用主要有三种:IaaS(Infrastructure-as-a-Service,基础设施即服务)模式、SaaS(Software-as-a-Service,软件即服务)模式和PaaS(Platform-as-a-Service,平台即服务)模式。

(1) IaaS模式在物联网中的应用。无论是横向的、通用的支撑平台,还是纵向的、特定的物联网应用平台,都可以在IaaS技术虚拟化的基础上实现物理资源的共享,实现业务处理能力的动态扩展。IaaS技术在对主机、存储和网络资源的集成与抽象的基础上,具有可扩展性和统计复用能力,允许用户按需使用。除网络资源外,其他资源均可通过虚拟化提供成熟的技术实现,为解决物联网应用的海量终端接入和数据处理提供了有效途径。同时,IaaS对各类内部异构的物理资源环境提供了统一的服务界面,即为资源定制、出让和高效利用提供了统一界面,也有利于实现物联网应用的软系统与硬系统之间某种程度的松耦合关系。目前,国内建设与物联网相关的云计算中心、云计算平台,主要是IaaS模式在物联网领域的应用。

(2) SaaS模式在物联网中的应用。SaaS模式的存在由来已久,被云计算概念重新包装后,除了可以利用云计算的技术(如IaaS技术)外,其没有本质上的变化。通过SaaS模式,实现物联网应用提供的服务被多个客户共享使用。这为各类行业应用和信息共享提供了有效途径,也为高效利用基础设施资源、实现高性价比的大量数据处理提供了可能。在物联网范畴内出现的一些变化是,SaaS应用在感知层进行了拓展，它们依赖感知层的各种设备采集了大量数据，并以这些数据为基础进行关联分析和处理，向最终用户提供业务功能和服务。

(3) PaaS模式在物联网中的应用。研究机构Gartner把PaaS分成两类：APaaS(Application PaaS，应用部署和运行平台)和IPaaS(Integration PaaS，集成平台)。APaaS主要为应用提供运行环境和数据存储；IPaaS主要用于集成和构建复合应用。人们常说的PaaS平台大都指APaaS，如Force.com和GoogleAppEngine。在物联网范畴内，由于构建者本身价值取向和实现目标的不同，因此PaaS模式存在不同的应用模式和应用方向。

从目前来看，物联网与云计算的结合是必然趋势，但是物联网与云计算的结合也需要水到渠成。不管是PaaS模式还是SaaS模式，物联网的应用都需要在特定的环境中才能发挥应有的作用。

4. 大数据技术

对于大数据，研究机构Gartner给出了这样的定义：大数据是海量、高增长率和多样化的信息资产，它需要新处理模式才能获得更强的决策力、洞察发现力和流程优化能力。大数据技术的战略意义不在于掌握庞大的数据信息，而在于对这些有意义的数据进行专业化处理。换言之，如果把大数据比作一种产业，那么这种产业实现盈利的关键在于提高对数据的加工能力，通过加工实现数据的增值。

从技术上看，大数据与云计算的关系就像一枚硬币的正反面一样密不可分。大数据必然无法用单台的计算机进行处理，必须采用分布式架构。它的特色在于对海量数据进行分布式数据挖掘，但它必须依托云计算的分布式处理、分布式数据库、云存储和虚拟化技术。随着云时代的来临，大数据也引起了越来越多的关注。著云台的分析师团队认为，大数据通常用来形容一个公司创造的大量非结构化数据和半结构化数据，这些数据下载到关系数据库用于分析时会花费过多时间和金钱。大数据分析常和云计算联系在一起，因为实时的大型数据集分析需要像MapReduce一样的框架来向数十、数百甚至数千的计算机分配工作。

大数据需要特殊的技术，以有效地处理某段时间内的数据。适用于大数据的技术包括大规模并行处理数据库、数据挖掘、分布式文件系统、分布式数据库、云计算平台、互联网和可扩展的存储系统。物联网通过智能硬件为云计算提供大量的数据。物联网、云计算与大数据技术之间的关系如图1-3所示。物联网依靠智能硬件接入，云计算包括云存储虚拟化、分布式处理以及分布式数据库，大数据技术包括分布式数据挖掘。

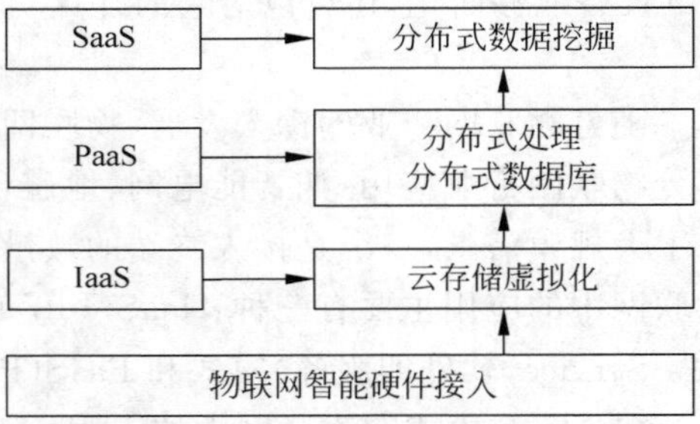

图1-3　物联网、云计算与大数据技术之间的关系

1.2.3　物联网语义

物联网系统分为四个层面：感知层、传输层、支撑层和应用层。感知层主要是对物体进行识别或数据采集；传输层是通过现有的通信网络将信息进行可靠传输；支撑层和应用层则是对采集的数据进行智能处理或展示。基于物理、化学、生物等技术发明的传感器标准已经有多项专利。而传输层的各种通信标准也已基本成熟，建立新的物联网通信标准难度较大，可行性较小。因此，物联网标准的关键和亟待统一的是关于应用层的标准，而其中尤以数据表达、交换和处理标准为核心。

目前，针对物联网应用层的数据交换标准主要有PML(Physical Markup Language，实体标记语言)、EDDL(Electronic Device Description Language，电子设备描述语言)、M2M XML(M2M eXtensible Markup Language，M2M可扩展标记语言)和NGTP(Next Generation Telematics Protocol，下一代车

载智能通信协议)等。其中,PML是电子产品代码在物联网中交换信息的共同语言,用来描述人及机器都可以使用的自然物体的标准。EDDL可以描述现场设备中的数据,用于工程、调试、监视运行和诊断。M2M XML是一种用于终端设备间的通信协议,它包含一个用于分析协议的与语义无关的Java API。NGTP是宝马公司推出的开放式车载智能通信协议架构平台,它使用统一、开放的接口来区分车载智能通信服务供应链的各个环节。此外,各种行业标准层出不穷。

可以看出,现有的物联网应用层的数据交换标准大多是针对某一特定领域或行业业务提出的,有一定的局限性,所以当前物联网缺少的是一个统一的物联网数据交换标准体系。欧盟有关机构正在进行数据交换标准融合的研究,目标是综合考虑相关领域已有的基于XML的数据交换标准,以便为那些在不同的标准中语义上具有等价性的数据元素(尽管它们可能有不同的名字)提供全球唯一的交叉引用方式和标识结构,从而提炼出一个基础的元数据标准,把这个标准作为物联网数据交换的核心,那么,对于不同的行业应用,就可以基于元数据扩展出相应的行业数据交换标准。

总体来说,物联网的标准化工作已经得到了业界的普遍重视,但对于应用层的标准化工作来说,需要客观分析物联网标准的整体需求,从国际标准、国家标准、行业标准、地区标准等多个层次进行统筹设计;还需要协调各个标准的推进策略,优化资源配置。

1. 物联网数据交换标准的语义基础是本体

本体起源于哲学,被Neches等人引入计算机科学领域后,在人工智能、语义Web、软件工程以及信息架构等领域得到了广泛应用。本体最流行的定义是Gruber在1993年给出的,即本体是概念模型明确的规范说明。Studer在对前人的定义进行概括后提出本体的概念包括四个方面:一是概念模型,它是客观世界现象的抽象模型,其表示的含义独立于具体的环境状态;二是明确,所使用的概念及使用这些概念的约束都有明确的定义;三是形式化,本体的表示是形式化的,可以被计算机处理;四是共享,本体中体现的是共同认可的知识,反映的是相关领域中公认的概念集,它所针对的是团体而不是个体。

本体的目标是获取相关的领域知识,提供对该领域知识的共同理解,确定该领域内共同认可的词汇,并从不同层次的形式化模式上,给出这些词汇(术语)和词汇间相互关系的明确定义。所以,本体是具有不同知识表示Web应用系统之间进行数据或知识交换共享的基础结构。通过定义共享和公共的领域知识,本体可帮助在机器之间或机器与人之间进行更加精确的交流,实现相互之间的语义交换,而不只是语法级的交互。

按照领域依赖程度,Guarino将本体划分为四类:一是顶级本体,用于描述通用的概念和概念之间的关系,如时间、空间、物质、对象、事件、动作等,顶级本体独立于特定的问题和领域,与具体的应用无关;二是领域本体,用于描述特殊领域(如教育或金融)中的概念,即陈述性知识;三是任务本体,用于描述特定任务或活动(如入学或取款)中的概念,即过程性知识;四是应用本体,可通过进一步特殊化领域本体和任务本体,将其用于描述既依赖于特定领域又依赖于特定任务的概念,这些概念通常对应于领域个体执行特定活动(如学生入学或客户取款)时所扮演的角色。Daniel等人利用现有的本体创建工具,构建了通用的语义传感器本体,实现了大型的语义传感器网络基础设施。

本体从底层向上分为顶级本体、领域本体、任务本体以及应用本体,这些不同层次的本体可提供整个世界的共性描述,而物联网正是要将世界连接起来。

首先,物联网所连接的各种物体都处在同一个世界中,它们都具有某些共同的特点,即人们对于这个世界的基本认识,如时空、物质、事件和行为等,所以物联网数据交换标准体系的基础是顶级本体标准。其次,物联网各个垂直的应用领域都有特殊性。具体到每一个领域,都有可能、有必要发展一套依托于领域本体的标准。但是,很多类型的业务词汇和流程是可以跨越多个垂直应用领域而公用的,所以,还有必要发展跨领域的物联网任务本体标准,即某个领域的本体标准可能构建于多个任务本体标准之上,而某个任务本体也有可能被多个领域本体所引用。最后,具体到每个企业、组织甚至个人,它们针对自身的物品、行为、过程等,也可以建立起基于顶级本体、领域本体和任务本体的应用本体标准,以供其他个体在与自身发生信息交换时共享这些事先定义好的内容。

构建本体时要确定本领域内公认的词汇,建立对某个领域知识的共同理解和相关描述,并能够给出领域词汇和词汇之间的相互关系在不同层次的形式、模式上的明确定义,从而能够完整地提取领域知识。本体层首先需要对基本的类/属性进行描述,同时还必须对本体以及本体之间的关系进行描述,是语义网的核心层。本体层的专用描述语言规范也出现许多,得到大家认可的有 DAML(DARPA Agent Markup Language,DARPA 代理标记语言)、SHOL(Simple HTML Ontology Language,简单 HTML 本体语言)、OIL(Ontology Inference Language,本体推理语言)以及 DAML+OIL。目前,学术界使用最多的本体描述语言是由 W3C 组织推荐的 OWL(Ontology Web Language,本体 Web 语言)。

2. 物联网数据交换标准的语法基础是 XML

ASN.1 是 ISO 和 ITU-T 的联合标准,ASN.1 本身只定义了表示信息的抽象句法,但是没有限定其编码的方法。各种 ASN.1 编码规则提供了由 ASN.1 描述其抽象句法数据值的传送语法(具体表达)。标准的 ASN.1 编码规则有 BER(Basic Encoding Rules,基本编码规则)、CER(Canonical Encoding Rules,规范编码规则)、DER(Distinguished Encoding Rules,唯一编码规则)、PER(Packed Encoding Rules,压缩编码规则)和 XER(XML Encoding Rules,XML 编码规则)。

XML 是 W3C 组织于 1998 年推出的一种用于数据描述的元标记语言标准。作为 SGML(Standard Generalized Markup Language,标准通用标识语言)的一个简化子集,它结合了 SGML 丰富的功能和 HTML 的简单易用,同时具有可扩展性、自描述性、开放性、互操作性、可支持多国语言等特点,因而得到了广泛的支持与应用。对于作为物联网数据交换标准的格式来说,XML 具有以下显著优点。

(1) 可定义行业或领域标记语言。XML 可以用 Schema 来定义,一份遵循 Schema 定义的 XML 文档才是有效的。因此,XML 可以针对不同应用建立相关的标准语言,例如化学标记语言、数学标记语言、语音标记语言等,包括目前物联网中很多已经存在的标准,都是基于 XML 定义的。

(2) 具有结构化的通用数据格式。XML 使用树形目录结构形式,可以自行定义文字标签并指定元素间的关系,同时它也是 W3C 公开的一种数据格式,没有版权的使用限制,因而十分适合作为不同应用程序之间的信息交换格式。

(3) 可提供整套方案。XML 拥有一整套技术体系,如可扩展样式表语言(XSL)、XML 数据查询技术(xQuery)以及文档对象模型(DOM)等。

(4) 在语法上结构化信息表达能力和本体在语义上透明性之间的优势互补,为物联网数据交换标准的建立提供了很好的解决思路。

基于上述内容,物联网数据交换标准应以 XML 为语法格式,以标准化的本体为语义共识。按照本体的分类,物联网数据交换标准体系应以顶级本体为基础,以纵向的领域本体和横向的任务本体为支撑,建立起各种不同的应用本体标准,其整个物联网数据交换标准体系示意如图 1-4 所示。

3. 物联网数据互操作是 RDF

RDF(Resource Description Framework,资源描述框架)模型和 RDF Schema 是语义网体系结构框架的互操作层。

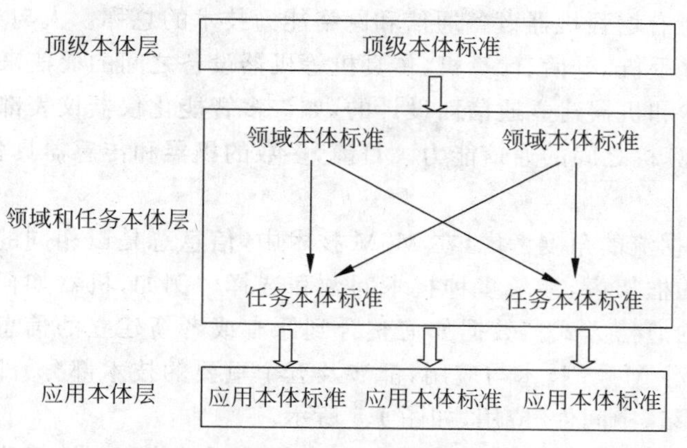

图 1-4 物联网数据交换标准体系示意图

语义数据的定义和互操作由这层来完成。W3C 组织研究开发了用来描述资源及其之间关系的 RDF 规范。RDF 通常采用三元组来表示互联网上的各种信息资源、属性及其值，具体表示为 RDF 的声明，即某个资源的某个属性(谓词)的值是客体(某个资源或者原生值)。

RDF Schema 是在 RDF 的基础上，引入了描述类和属性的能力，它定义了属性的定义域与值域类以及属性之间的关系等，就像是一本大词典，定义了计算机可以理解的词汇，计算机在分析执行程序的时候，直接在词典中查询这些定义就可以知道数据所包含的语义。

物联网中的数据语义不同，虽然可以通过 XML、本体等技术来构建一套物联网数据交换标准体系，并成立相关标准组织来进行管理，但这只能在一定程度上解决一定范围内的数据交换问题，而不可能也没必要建立一整套全面的数据交换标准，并要求所有参与者都要符合这个标准。所以，整个物联网数据交换标准体系应该是由少数几个顶级本体标准、大多数领域本体标准与任务本体标准，以及数量众多的应用本体标准组成。正因为如此，还需要在物联网的各个终端的必要位置上，设置恰当的转换器或者接口，从而实现针对同一对象、应用或业务而语义不同的标准之间的转换。

1.2.4 M2M 技术

随着技术的发展，越来越多的设备具有了通信和联网能力，这将使得所有物体联网逐步变为现实。人与人之间的通信需要更加直观、精美的界面和更丰富的多媒体内容，而 M2M 的通信更需要建立一个统一、规范的通信接口和标准化的传输内容。另外，通信网络技术的出现和发展给社会生活带来了极大的变化。人与人之间可以更加快捷地沟通，信息的交流更顺畅。但是，目前仅仅是计算机和其他一些 IT 类设备具备这种通信和网络能力。众多的普通机器设备几乎不具备联网和通信能力，如家电、车辆、自动售货机、工厂设备等。

ETSI 是国际上较早系统地展开 M2M 相关研究的标准化组织，2009 年初成立了专门的工作组负责统筹 M2M 的研究，旨在制定一个水平化的、不针对特定 M2M 应用的端到端解决方案的标准。其研究范围可以分为两个层面：第一个层面是针对 M2M 应用用例的收集和分析；第二个层面是在用例研究的基础上，开展与应用无关的 M2M 业务需求分析、网络体系架构定义、数据模型、接口和过程设计等工作。

M2M 技术的目标是使所有机器设备都具备联网和通信能力，其核心理念是所有物体连接网络。M2M 技术具有非常重要的意义，有广阔的市场和应用前景，推动着社会生产和生活方式新一轮的变革。

M2M是一种理念，也是所有增强机器设备通信和网络能力技术的总称。人与人之间的沟通很多也是通过机器实现的，例如，通过手机、电话、计算机、传真机等机器设备之间的通信来实现人与人之间的沟通。另外一类技术是专为机器和机器建立通信而设计的，如许多智能化仪器仪表都带有通用接口，增强了仪器与仪器之间、仪器与计算机之间的通信能力。目前，多数的机器和传感器具备本地或者远程的通信与联网能力。

M2M系统框架从数据流的角度考虑，在M2M技术中，信息总是以相同的顺序流动。在这个基本的框架内，涉及多种技术问题和选择。例如，机器如何连接网络？使用什么样的通信方式？数据如何整合到原有或者新建立的信息系统中？但是，无论哪一种M2M技术与应用，都涉及五个重要的技术部分，即机器、M2M硬件、通信网络、中间件、应用，如图1-5所示。

应用
中间件
通信网络
M2M硬件
机器

图1-5 M2M系统框架示意图

智能化机器使机器"开口说话"，让机器具备信息感知、信息加工（计算能力）、无线通信能力。M2M硬件进行信息的提取，从各种机器/设备那里获取数据，并传送到通信网络。通信网络将信息传送到目的地。中间件在通信网络和IT系统间起桥接作用。应用对获得数据进行加工分析，为决策和控制提供依据。

因此，M2M不是简单的数据在机器和机器之间的传输，更重要的是，它是机器和机器之间的一种智能化、交互式的通信。也就是说，即使人们没有实时发出信号，机器也会根据既定程序主动进行通信，并根据所得到的数据智能化地做出选择，对相关设备发出正确的指令。可以说，智能化、交互式成为M2M有别于其他应用的典型特征，这一特征下的机器也被赋予了更多的"思想"和"智慧"。

M2M是将数据从一台终端传送到另一台终端，也就是机器与机器的对话。但是，从广义上M2M可代表机器对机器、人对机器、机器对人、移动网络对机器之间的连接与通信，它涵盖了所有实现在人、机器、系统之间建立通信连接的技术和手段。

M2M产品主要由三部分构成：第一，无线终端，可以是特殊的行业应用终端，也可以是通常的手机或笔记本电脑；第二，传输通道，从无线终端到用户端的行业应用中心之间的通道；第三，行业应用中心，也就是终端上传数据的汇聚点，对分散的行业终端进行监控。其特点是行业特征强，用户自行管理，而且可位于企业端或者可以托管。

M2M应用市场正在全球范围内快速增长，随着通信设备、管理软件等相关技术的深化，M2M产品成本的下降，M2M业务将逐渐走向成熟。目前，中国、美国和加拿大等国已经实现M2M产品在安全监测、机械服务、维修业务、自动售货机、公共交通系统、车队管理、工业流程自动化、电动机械、城市信息化等领域的应用。

1.3 物联网的发展

1.3.1 两化融合及互联网+

两化融合是指电子信息技术广泛应用到工业生产的各个环节，信息化成为工业企业经营管理的常规手段。信息化进程和工业化进程不再相互独立进行，不再是单方的带动和促进关系，而是两者在技术、产品、管理等各个层面相互交融，彼此不可分割，并催生工业电子、工业软件、工业信息服务等新产业。两化融合是工业化和信息化发展到一定阶段的必然产物。信息化与工业化主要在技术、产品、业务、产业四个方面进行融合。也就是说，两化融合包括技术融合、产品融合、业务融合、产业衍生四个方面。物联网在制造业的两化融合可以从以下四个角度理解：生产自动化，将物联网技术融入制造业生产，如工业控制技术、柔性制造、数字化工艺生产线等；产品智能化，在制造业产品中采用物联网技术提高产品技术含量，如智能家电、工业机器人、数控机床等；管理精细化，在企业经营管理活动中采用物联网技术，如制造执行系统、产品追溯、安全生产的应用；产业先进化，制造业产业和物联网技术融合优化产业结构，促进产业升级。

2015 年 3 月 5 日第十二届全国人民代表大会第三次会议上，李克强总理在政府工作报告中首次提出“互联网＋”行动计划。李克强总理所提的“互联网＋”与较早相关互联网企业讨论聚焦的“互联网改造传统产业”相比已经有了进一步的深入和发展。李克强总理在政府工作报告中首次提出的“互联网＋”实际上是创新 2.0 下互联网发展新形态、新业态，是知识社会创新 2.0 推动下的互联网形态演进。

伴随知识社会的来临，驱动当今社会变革的不仅仅是无所不在的网络，还有无所不在的计算、无所不在的数据、无所不在的知识。“互联网＋”不仅仅是互联网移动了、泛在了、应用于某个传统行业了，更是无所不在的计算、数据、知识造就了无所不在的创新，推动了知识社会以用户创新、开放创新、大众创新、协同创新为特点的创新 2.0，改变了我们的生产、工作、生活方式，也引领了创新驱动发展的新常态。

新一代信息技术催生了创新 2.0，而创新 2.0 又反过来作用于新一代信息技术形态的形成与发展，重塑了物联网、云计算、大数据等新一代信息技术的新形态。新一代信息技术的发展又推动了创新 2.0 模式的发展和演变，Living Lab（生活实验室、体验实验区）、Fab Lab（个人制造实验室、创客）、AIP（“三验”应用创新园区）、WiKi（维基模式）、Prosumer（产消者）、Crowdsourcing（众包）等典型创新 2.0 模式不断涌现。新一代信息技术与创新 2.0 的互动与演进推动了“互联网＋”的发展。关于知识社会环境下新一代信息技术与创新 2.0 的互动演进可参阅《创新 2.0 研究十大热点》。互联网随着信息通信技术的深入应用带来了创新形态演变，其本身也在变化并与行业新形态相互作用共同演化，如同以工业 4.0 为代表的新工业革命以及以 Fab Lab 与创客为代表的个人设计、个人制造、群体创造。可以说“互联网＋”是新常态下创新驱动发展的重要组成部分，是物联网发展的重大驱动力量。

“互联网＋”是创新 2.0 下的互联网发展新形态、新业态，是知识社会创新 2.0 推动下的互联网形态演进。“互联网＋”代表一种新的经济形态，即充分发挥互联网在生产要素配置中的优化和集成作用，将互联网的创新成果深度融合于经济社会各领域之中，提升实体经济的创新力和生产力，形成更广泛的以互联网为基础设施和实现工具的经济发展新形态。“互联网＋”行动计划将重点促进以云计算、物联网和大数据为代表的新一代信息技术与现代制造业、生产性服务业的融合创新，发展壮大新兴业态，打造新的产业增长点，为大众创业、万众创新提供环境，为产业智能化提供支撑，增强新的经济发展动力，促进国民经济提质增效升级。

1.3.2 物联网联盟

2014 年，很多具有强大影响力的国内外企业组建了物联网联盟，为未来物联网领域的标准规范奠定了基础；另外，联盟之间的许多工作有重复之处，并且一个企业参与了多个联盟，如何进一步融合是未来联盟之间发展的重点。下面简单介绍目前物联网行业技术标准的重要联盟。

1. 开放互联联盟/开放互联基金会

2014 年 7 月成立了开放互联联盟（Open Interconnect Consortium，OIC），该联盟最初有戴尔、惠普、英特尔、联想和三星等诸多会员。2016 年 10 月，OIC 和 AllSeen 联盟同时宣布双方正式合并，成立开放互联基金会（Open Connectivity Foundation，OCF），合并后双方将遵循 OCF 的名称和运作规则。OCF

成为全球物联网行业最大的标准组织,并有望成为国际物联网标准的实际制定者,推动物联网产业形成统一的产品标准,打造更好的用户体验。目前,OCF 正组织撰写一系列开源标准。在这些标准的帮助下,各类联网设备将能寻找、隔离和确认彼此,进行沟通、相互影响、完成数据交换。该联盟在 2015 年末发布了面向开发者的首个开源代码。

2. IEEE P2413 项目

技术标准的传统权威 IEEE 发起了整顿 IoT 领域的 P2413 项目,力图统一物联网技术标准,解决业内在标准制定上的重复浪费工作。2014 年 7 月,IEEE P2413 项目召集 23 家供应商和其他相关方举行了首次会议,希望制定一整套明确的物联网书面标准。但是,以物联网的快速发展势头,IEEE 这样的预订日程实在太过漫长。可能等到 IEEE 的标准解决方案面世时,制造商与供应商推出的事实标准或已被业界默认接受。目前,该工作组的供应商和组织机构包括思科、华为、通用电气、甲骨文、高通和 ZigBee 联盟等。

3. 工业互联网联盟

这家机构于 2014 年 3 月正式成立,其原始成员为 AT&T、思科、通用电气、IBM 和英特尔。该联盟现有逾百名成员,华为、微软、三星等业内知名企业均在其中。其着重于各家企业的物联网建设及策略,并未着力制定行业标准,而是与认证机构合作,以求确保各商业领域的物联网技术融会贯通。其宗旨是让多家正在开发 IoT 与 M2M 技术的企业实现共同协作、相互影响,这涉及界定基本标准要求、参考架构和概念证明的问题。

4. AllSeen 技术联盟

AllSeen 技术联盟诞生于 2013 年 12 月,目前拥有近 200 家会员企业,既有高通、思科、TP-LINK、Silicon Image、Technicolor、Microsoft 这类软硬件厂商,也有海尔、LG、松下、夏普、Electrolux 等消费类电子产品厂商。AllSeen 技术联盟的创始会员高通公司对其最为重要,该协会的开源软件框架 AllJoyn,是基于高通的代码和技术平台创建的。AllSeen 技术联盟的目标是让配置不同操作系统和通信网络协议的家用、商务设备实现协作互助。2015 年 3 月,Microsoft 公司宣布 Windows 10 全面支持 AllJoyn 技术并推出适用于 AllJoyn 的工具包。该联盟目前已经与 OIC 合并,共同致力于物联网互联互通协议的构建。

5. Thread 联盟

Thread 联盟诞生于 2014 年 7 月,谷歌公司旗下智能家居公司 Nest 和三星等 50 家机构都是该联盟的成员,中国家电企业美的集团也在其中。Thread 是一种基于 IP 的安全网络协议,用来连接智能产品。该联盟因此得到先发优势,其协议支持已上市的芯片,并能给所有设备都分配一个 IPv6 地址。由于 Thread 仅定义联网,因此为支持 AllSeen 和 OIC 这类更高层面的标准奠定了基础,该联盟已经进行产品认证。

此外,过去一年多还有许多机构和行业协会在物联网领域跃跃欲试,例如,通信标准化协会、国际自动化协会等。今年还会有新的联盟、企业及企业联合体"杀"入物联网。IoT 领域已有许多标准认证主体,有的机构致力于完善补充现有方案标准,有的机构致力于技术标准冲突解决。然而,很多认证标准的冲突与重复都出自同一联盟内部的企业成员之间,这无疑加大了标准制定的难度。因此,在 IoT 行业之初,标准统一的可能性很小,必然存在多个标准并行的局面,以保障各方的利益。

1.3.3 OCF 技术

2016 年底,全球物联网工业界组成 OCF,为物联网发展迈出一大步。开发基于 OCF 技术的物联网通信控制管理系统，是整个行业发展的希望。

OCF 认为，要使得数十亿的连接设备（设备、电话、计算机和传感器）能够相互沟通，涉及制造商、操作系统、芯片组及底层物理传输。OCF 构建了一套协议规范，赞助了一个开源项目 Iotivity，使得实现这些构想成为可能。作为一个开放的协议规范，OCF 将为物联网市场带来巨大的机遇，加速行业创新，帮助开发者和企业找到解决方案。对于消费者、商业和工业界来说，OCF 将确保互操作性。

OCF 将完成定义规范、认证产品和品牌，通过一个开放源码 Iotivity 项目可以实现可靠的互操作性，这个开放的协议规范允许任何人来实现，开发者很容易使用。它还包括可预见认证设备的品牌保护和服务级别的互操作性，实现了应用程序开发者和设备制造商的产品在不同操作系统上的互操作，如 Android、iOS、Windows、Linux 和 Tizen 等。

Iotivity 是一个开放源代码的软件框架，实现设备到设备的无缝连接，解决物联网的各种需求。作为一个全新的开源软件系统，Iotivity 能为跨越不同设备的分布式应用，提供一个具有移动性、安全性和动态配置的环境。它不仅能解决异构分布式系统的固有问题，还能解决伴随移动性系统产生的随时接入问题，使开发人员能集中注意力解决应用程序的核心问题。因此，OCF 技术将在今后的物联网推广中占有重要地位。

1.4 RESTful

REST（Representational State Transfer，表述性状态转移）是一种软件架构风格、设计风格，而不是标准，它只是提供了一组设计原则和约束条件，主要用于客户端和服务器端交互类的软件。基于这个风格设计的软件可以更简洁、更有层次、更易于实现缓存等机制，其实现的位置如图 1-6 所示。

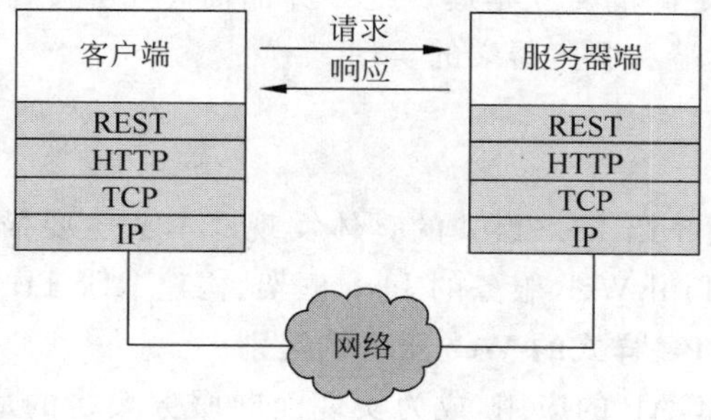

图 1-6 REST 位置示意图

1.4.1 概述

REST 描述了一个架构样式的网络系统，例如，Web 应用程序首次出现在 2000 年 Roy Fielding（HTTP 规范的主要编写者之一）的博士论文中。在目前主流的三种 Web 服务交互方案中，REST 相比于 SOAP（Simple Object Access Protocol，简单对象访问协议）以及 XML-RPC（XML Remote Procedure Call，XML 远程过程调用）更加简单明了，无论是对 URL（Uniform Resource Locator，统一资源定位符）的处理，还是对载荷的编码，REST 都倾向于用更加简单的方法设计和实现。值得注意的是，REST 并没有一个明确的标准，而更像是一种设计的风格，交互风格如图 1-7 所示。

REST 指的是一组架构约束条件和原则，满足这些约束条件和原则的应用程序或设计就是 RESTful。

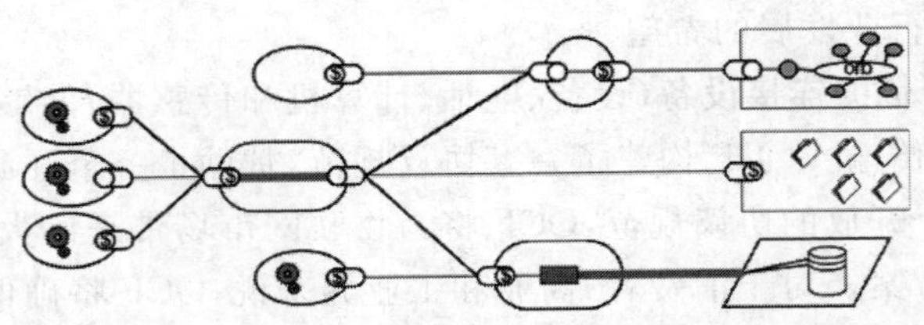

图 1-7 REST 交互风格

Web 应用程序最重要的 REST 原则是，客户端和服务器端之间的交互请求是无状态的。从客户端到服务器端的每个请求都必须包含理解请求所必需的信息。如果服务器端请求在任何时间点重启，客户端不会得到通知。此外，无状态请求可以由任何可用服务器端回答，这十分适合云计算之类的环境，客户端可以缓存数据以改进性能。

在服务器端，应用程序状态和功能可以分为各种资源。资源是一个有趣的概念实体，它向客户端公开。资源是应用程序对象、数据库记录、算法等。每个资源都使用 URI(Universal Resource Identifier，统一资源标识符)得到一个唯一的地址。所有资源都共享统一的接口，以便在客户端和服务器端之间传输状态。它使用的是标准的 HTTP 方法，如 GET、PUT、POST 和 DELETE。超媒体是应用程序状态的引擎，资源表示通过超链接互联。

另一个重要的 REST 原则是分层系统，也就是说，无法了解与之交互的中间层以外的组件。通过将系统知识限制在单个层，可以限制整个系统的复杂性，促进了底层的独立性。

当 REST 架构的约束条件作为一个整体应用时，将生成一个可以扩展到大量客户端的应用程序。它还降低了客户端和服务器端之间的交互延迟。统一界面简化了整个系统架构，改进了子系统之间交互的可见性，REST 简化了客户端和服务器端的实现。

1.4.2 实现

了解了什么是 REST，下面介绍 RESTful 的具体实现。本节主要描述 RESTful Web 服务与 RPC 样式的 Web 服务的区别、RESTful Web 服务的 Java 框架、构建 RESTful Web 服务的多层架构。

1. RESTful Web 服务与 RPC 样式的 Web 服务的区别

使用 RPC 样式构建基于 SOAP 的应用，成为实现面向服务架构的最常用方法。RPC 样式的 Web 客户端将一个装满数据的数据包(包括方法和参数信息)通过 HTTP 发送到服务器端。服务器端打开数据包，使用传入参数执行指定的方法，将结果打包作为响应发回客户端，客户端收到响应并打开数据包，每个对象都有自己独特的方法以及仅公开一个 URI 的 RPC 样式 Web 服务，URI 表示单个端点。这种方法忽略 HTTP 的大部分特性且仅支持 POST 方法。

由于轻量级以及通过 HTTP 直接传输数据的特性，Web 服务的 RESTful 方法已经成为最常见的替代方法。可以使用各种语言(如 Java、Perl、Ruby、Python、PHP、JavaScript 和 Ajax)实现客户端。RESTful Web 服务通常可以通过自动客户端或代表用户的应用程序访问。但是，这种服务的简便性让用户能够与之直接交互，使用它们的 Web 浏览器构建一个 GET URL 并读取返回的内容。

在 RESTful 样式的 Web 服务中，每个资源都有一个地址。资源本身都是方法调用的目标，方法列表对于所有资源都是标准方法，包括 HTTP 的 GET、POST、PUT 和 DELETE，还可能包括 HEADER 和 OPTIONS。

在 RPC 样式的架构中，关注点在于方法；而在 RESTful 样式的架构中，关注点在于资源，它将使用标准方法检索并操作信息字段（使用表示的形式），资源表示使用超链接形式。

Leonard Richardson 和 Sam Ruby 在他们的著作 *RESTful Web Services* 中引入了术语 REST-RPC 混合架构。REST-RPC 混合 Web 服务不使用打包封装方法、参数和数据，而是直接通过 HTTP 传输数据，这与 RESTful 样式的 Web 服务是类似的。但是它不使用标准的 HTTP 方法操作资源，它在 HTTP 请求的 URI 部分存储方法信息。知名的 Web 服务（如 Yahoo 的 Flickr API 和 Delicious API）都使用这种混合架构。

2. RESTful Web 服务的 Java 框架

有两个 Java 框架可以帮助构建 RESTful Web 服务。Erome Louvel 和 Dave Pawson 开发的 Restlet 是轻量级的。它实现针对各种 RESTful 系统的资源、表示、连接器和媒体类型之类的概念，包括 Web 服务。在 Restlet 框架中，客户端和服务器端都是组件。组件通过连接器互相通信。该框架最重要的类是抽象类 Uniform 及其具体的子类 Restlet，Uniform 的子类是专用类，如 Application、Filter、Finder、Router 和 Route。这些子类能够一起处理验证、过滤、安全、数据转换以及将传入请求路由到相应资源等操作。Resource 类生成客户端的表示形式。

JSR-311 是 Sun Microsystems 的规范，可以为开发 RESTful Web 服务定义一组 Java API。Jersey 是对 JSR-311 的参考实现。JSR-311 提供一组注解，相关类和接口都可以用来将 Java 对象作为 Web 资源展示。该规范假定 HTTP 是底层网络协议，它使用注释提供 URI 和相应资源类之间的清晰映射，以及 HTTP 方法与 Java 对象方法之间的映射。API 支持广泛的 HTTP 实体内容类型，包括 HTML、XML、JSON、GIF 和 JPG 等。它还将提供所需的插件功能，以允许使用标准方法，通过应用程序添加其他类型。

3. 构建 RESTful Web 服务的多层架构

RESTful Web 服务和动态 Web 应用程序在许多方面都是类似的。有时它们提供相同或非常类似的数据和函数，尽管客户端的种类不同。例如，在线电子商务分类网站为用户提供一个浏览器界面，用于搜索、查看和订购产品。如果为公司、零售商甚至个人能够自动订购产品提供 Web 服务，它将非常有用。与大部分动态 Web 应用程序一样，Web 服务可以从多层架构的关注点分离中受益。业务逻辑和数据可以由自动客户端和图形用户界面客户端共享，不同点在于客户端的本质和中间的表示层。此外，从数据访问中分离业务逻辑可实现数据库独立性，并为各种类型的数据存储提供插件能力。

图 1-8 是自动化客户端架构，包括 Java 和各种语言编写的脚本，这些语言包括 Perl、Ruby、PHP、Python 或命令行工具（如 Curl）。在浏览器中运行且作为 RESTful Web 服务消费者运行的 Ajax、Flash、JavaFX、GWT、博客和 WiKi 都属于此列，因为它们都代表用户以自动化方式运行。自动化 Web 服务客户端在 Web 层向资源请求处理程序发送 HTTP 响应。客户端的无状态请求在头部包含方法信息，即 POST、GET、PUT 和 DELETE，这又将映射到资源请求处理程序中的相应操作。每个请求都包含所有必需的信息，包括资源请求处理程序用来处理请求的依据。

从 Web 服务客户端收到请求之后，资源请求处理程序从业务逻辑层请求服务。资源请求处理程序确定所有概念性的实体，系统将这些实体作为资源公开，并为每个资源分配唯一的 URI。但是，概念性的实体在该层是不存在的，它们存在于业务逻辑层。可以使用 Jersey 或其他框架（如 Restlet）实现资源请求处理程序，它应该是轻量级的，将大量工作委托给业务逻辑层。

Ajax 和 RESTful Web 服务本质上是互为补充的。它们都可以利用大量 Web 技术和标准，例如

HTML、JavaScript、浏览器对象、XML/JSON 和 HTTP。当然也不需要购买、安装或配置任何主要组件支持 Ajax 前端和 RESTful Web 服务之间的交互。RESTful Web 服务为 Ajax 提供了非常简单的API,以处理服务器端资源之间的交互。

图 1-8 中的浏览器客户端作为图形用户界面的前端,使用表示层中的浏览器请求处理程序生成 HTML 提供显示功能。浏览器请求处理程序可以使用模型-视图-控制器模型(Struts、JSF 或 Spring 都是 Java 的示例)。它从浏览器接收请求,从业务逻辑层请求服务,生成表示并对浏览器做出响应。表示供用户在浏览器中显示使用。表示不仅包含内容,还包含显示的属性,如 HTML 和 CSS(Cascading Style Sheets,串联样式表)。

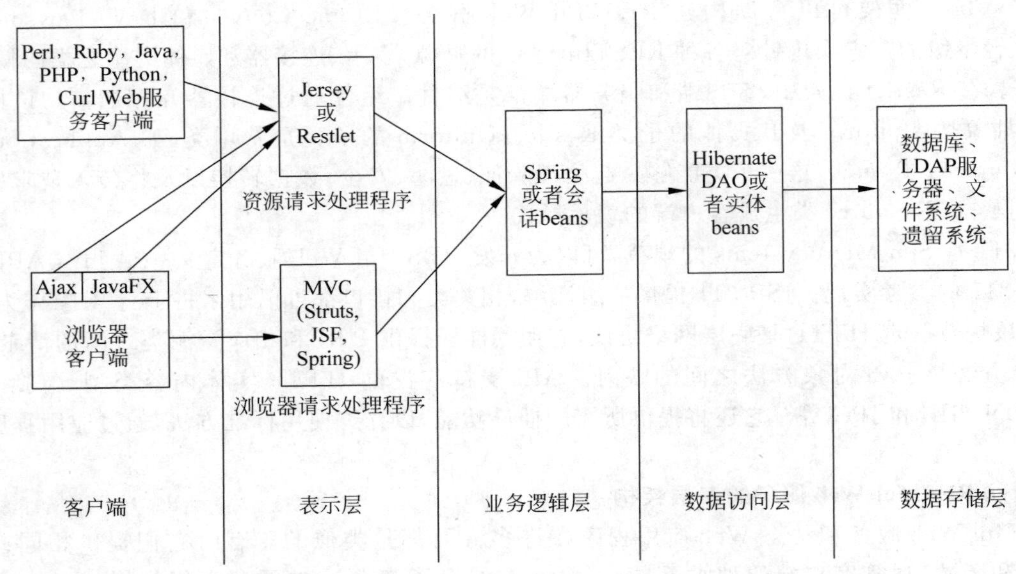

图 1-8 自动化客户端架构

业务规则可以集中到业务逻辑层,该层充当表示层和数据访问层之间的数据交换中间层。数据以域对象或值对象的形式提供给表示层。从业务逻辑层中解耦浏览器请求处理程序和资源请求处理程序有助于促进代码重用,并能实现灵活和可扩展的架构。此外,由于将来可以使用新的 REST 和模型-视图-控制器框架,因此实现它们会变得更加容易,无须重写业务逻辑层。

数据访问层提供与数据存储层的交互,可以使用 DAO 设计模式或者对象-关系映射解决方案(如 Hibernate、OJB 或 iBatis/Mybatis)实现。作为替代方案,业务逻辑层和数据访问层中的组件可以实现为 EJB 组件,并取得 EJB 容器的支持,该容器可以为组件生命周期提供便利,管理持久性、事务和资源配置。但是,这需要一个遵从 Java EE 的应用服务器(如 JBoss),并且可能无法处理 Tomcat。该层的作用在于针对不同的数据存储技术,从业务逻辑层中分离数据访问代码。数据访问层可以作为连接其他系统的集成点,还可以成为其他 Web 服务的客户端。

数据存储层包括数据库、LDAP 服务器、文件系统和遗留系统。使用该架构,可以看到 RESTful Web 服务的力量,它可以灵活地成为任何企业数据存储的统一 API,从而向以用户为中心的 Web 应用程序公开垂直数据,并自动化批量报告脚本。

REST 描述了一个架构样式的互联系统(如 Web 应用程序)。REST 约束条件作为一个整体应用时，将生成一个可扩展、简单、有效、安全和可靠的架构。由于具有简便、轻量级以及通过 HTTP 直接传输数据的特性，RESTful Web 服务成为基于 SOAP 服务的一个最有前途的替代方案。用于 Web 服务和动态 Web 应用程序的多层架构，可以实现重用性、简单性、扩展性和组件响应性的清晰分离。开发人员可以轻松使用 Ajax 和 RESTful Web 服务一起创建丰富的界面。

RESTful 的关键是定义可表示流程元素/资源的对象。在 REST 中，每一个对象都是通过 URL 来表示的，对象用户负责将状态信息打包进每一条消息内，以便对象的处理总是无状态的。RESTful 是组合管理及流程绑定，适应企业等级的发现、绑定的灵活性和复杂性。

OCF 使用 RAML(RESTful API Modeling Language，RESTful API 建模语言）描述，RAML 是对 RESTful API 的一种简单和直接的描述。它是一种让人们易于阅读，并且能让机器对特定的文档解析的语言。RAML 基于 YAML(Yet Another Markup Language，又一种标识语言)，能帮助设计 RESTful API 和鼓励对 API 的发掘和重用，依靠标准和最佳实践，从而编写更高质量的 API。通过 RAML 定义，机器能够看得懂，所以可以衍生出一些附加的功能服务，例如，解析并自动生成对应的客户端调用代码、服务器端代码结构及 API 的说明文档。

RAML 本质上可以理解为一种文档的书写格式，这种格式是特别针对 API 的，就像 Markdown 是针对 HTML 的一样。并且，RAML 也同样具备了像 Markdown 那样的可读性，即使看 RAML 源码也能很快明白其意图。另外，基于 RAML 语言，有不少辅助 API 开发的工具，如 RAML2HTML 和 API-Console，它们都可以将 RAML 源文件转换成精美的文档页面，其中，RAML2HTML 转换结果是静态的，API-Console 则可以产生可直接交互的 API 文档页面，类似 1.6 节介绍的 Swagger。

1.5 Swagger

Swagger 与 RAML 相比，RAML 解决的是设计阶段的问题，而 Swagger 则是侧重解决现有 API 的文档问题，它们最大的不同是 RAML 需要单独维护一套文档，而 Swagger 则是通过一套反射机制从代码中生成文档，并且借助 Ajax 直接在文档中对 API 进行交互。因为代码与文档是捆绑的，所以在迭代代码的时候，就能方便地将文档也更新了，不会出现随着项目推移代码与文档不匹配的问题，另外，Swagger 是基于 JSON 进行文档定义的。

Swagger API 框架用于管理项目中的 API 接口，是当前流行的 API 接口管理工具。Swagger 是一个开源框架(Web 框架)，功能强大，UI 界面友好，支持在线测试。

Swagger 的目标是对 REST API 定义一个标准的与语言无关的接口，可让人和计算机无须访问源码、文档或网络流量监测，就可以发现和理解服务。通过 Swagger 进行正确定义，用户可以理解远程服务，并使用最少实现逻辑与远程服务进行交互。与底层编程所实现的接口类似，Swagger 消除了调用服务时可能存在的猜测。

Swagger 是一组开源项目，主要项目如下。

Swagger-Tools：提供各种与 Swagger 进行集成和交互的工具。例如，模式检验、Swagger 1.2 版本的文档转换成 Swagger 2.0 版本的文档等功能。

Swagger-Core：用于 Java/Scala 的 Swagger 实现。与 JAX-RS(Jersey、Resteasy、CXF)、Servlets 和 Play 框架进行集成。

Swagger-JS：用于 JavaScript 的 Swagger 实现。

Swagger-Node-Express：Swagger 模块，用于 Node.js 的 Express Web 应用框架。

Swagger-UI：一个不依赖 HTML、JavaScript 和 CSS 的集合，可以为 Swagger 兼容 API 动态生成优雅文档。

Swagger-Codegen：一个模板驱动引擎，通过分析用户 Swagger 资源声明，以各种语言生成客户端代码。

Swagger-Editor：可让使用者在浏览器里以 YAML 格式编辑 Swagger API 规范，并实时预览文档；可以生成有效的 Swagger JSON 描述，并用于所有 Swagger 工具（代码生成、文档等）中。

第 2 章 CHAPTER 2 OCF 技术基础

物联网即万物互联，就是要将所有的人和设备联系在一起，根据目前的互联网应用，发挥更加多样的功能，从而丰富我们的生活。因而，物联网的产出之一便是智能生活，在不久的将来，与人们的生活息息相关的各种设备都通过网络互联实现联动，为人们提供更加丰富的生活情景，如智能家居、智能办公室、车联网以及各种智能硬件等应用。

在物联网中，将会有数以亿计的设备，物联网无法像互联网一样将所有的设备都注册在一个控制中心，但交互最多的还是网络互联的邻近设备。因此，业界迫切需要有一个服务框架，能自动识别出邻近物联网中存在的设备和服务，而且随着设备暴露出越来越多的连接和控制接口，安全问题也日益突出。

如果将不同生活场景看作一个邻近的物联网络，那么，今后人们生活的情景有可能如图 2-1 所示。

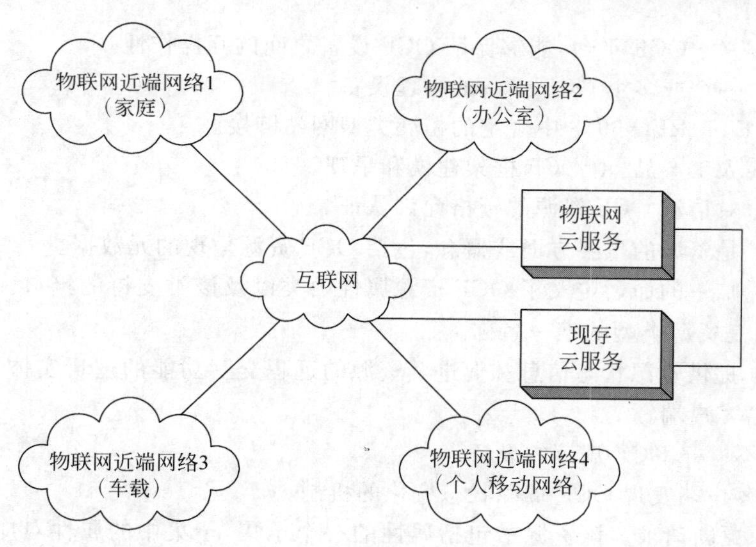

图 2-1　物联网的应用场景

在同一个物联网络中的智能设备能自动发现其他设备和服务的存在，并与之进行端对端的通信。对于那些需要通过网络地址转换的设备，可以通过云端的发现服务去寻找自己感兴趣的设备，当然这些云端的服务也可以让不同邻近区域的设备实现通信，除此之外，云端的服务还可以为整个物联网提供某些特定的功能。

从图 2-1 可以看出，在物联网中，邻近物联网设备之间的相互交互是实现物联网丰富功能的关键所在。整个物联网最关键的便是设备广播和发现、网络的动态移动管理、安全性和隐私、不同操作系统的交互和拓展性。因此，本章将详细介绍目前全球最受关注的开源项目——OCF 技术，它是目前最主要的物联网框架之一。

2.1 OCF 术语和定义

OCF 核心资源：在 OCF 规范中定义的 OCF 资源。

配置源：在云平台或服务网络的一个实体或一个本地只读文件，包括提供与 OCF 设备的相关配置信息。

实体：物质世界的一个要素，由 OCF 设备实现。

观察：该行为通过发送检索请求来监测 OCF 资源，检索请求由托管 OCF 资源的 OCF 服务器端缓存，并且每次 OCF 资源发生改变时，检索请求都会被重新处理。

OCF 客户端：在 OCF 服务器端连接 OCF 资源的逻辑实体。

OCF 集合：包括零或多个 OCF 链接的 OCF 资源。

OCF 设备：呈现一个或多个 OCF 角色(OCF 客户端或 OCF 服务器端)的逻辑实体。注意，一个物理平台上可以有多个 OCF 设备。

OCF 功能：在任何 OCF 设备中包含的基础或核心功能。

OCF 框架：一套在规范中定义通用的功能和相互作用，能够实现众多网络设备之间的互操作性，包括 IoT。

OCF 基础设施网关：OCF 平台，能够保证 OCF 设备之间的互操作性。

OCF 平台：包含一个或多个 OCF 设备的物理设备。

OCF 链接：在 IETF RFC 5988 中规定的拓展类型网站链接。

OCF 资源：代表人工产品，由 OCF 框架建模和呈现。

OCF 资源接口：对请求 OCF 资源的资格允许认证。

OCF 资源属性：是资源的重要方面或概念，包括 OCF 资源呈现的元数据。

OCF 资源类型：唯一的命名定义了 OCF 资源属性的类以及该类支持的接口。每一个资源都有一个 rt 属性，该属性值是资源类型的唯一名称。

OCF 服务器端：提供资源状态信息和促进其资源的远程交互功能的逻辑实体，可以实现服务器端将非设备资源暴露给客户端。

非 OCF 设备：不遵守 OCF 规范的设备。

通知：使 OCF 客户端发现 OCF 资源状态变化的机制。

部分更新：资源更新请求。该资源中包括属性的一个子集，子集中的属性对应用于资源类型的接口是可见的。

远程接入端点客户端：支持 XMPP(Extensible Messaging and Presence Protocol，可扩展消息处理现场协议)功能的 OCF 客户端。

远程接入端点服务器端：支持 XMPP 的 OCF 服务器端，并且可以将其资源发布给 XMPP 服务器端，从而可以实现远程寻址及接入。

资源目录：掌握其他 OCF 服务器端上资源描述的 OCF 设备，且允许其查找这些资源。该功能的使用可以通过服务器端睡眠/服务器端不监听响应多播请求的方式来实现。

场景：场景值中所列的值。一个场景是一组资源的一个规定设置，其中每一个资源都有对于该属性预先设置的值，这些值可以进行修改。

场景集合：包含可能的场景值列表以及当前场景值的OCF资源。该资源是一个有附加数据的集合资源，场景资源的成员值是场景成员。

场景值：呈现OCF资源可能状态的场景列表。

场景成员：一种OCF资源，包含场景值对应资源的属性值映射。

规则：该OCF资源包含一个条件，当该条件为真时则会在OCF服务器端启动一个脚本。

规则条件：是一个表达式，描述如何对资源特性的一个值评估。规则条件是以扩展巴科斯范式的格式表达的，并且使用特定服务器端资源属性的引用。

规则成员：OCF资源，其中包含当规则条件为真时设置的资源属性值。

脚本：当规则条件为真时被执行的规则成员。

默认接口：当在一个请求中省略了一个接口时，使用该接口产生响应。

参数：如果资源被一个链接的目标URI所引用，则提供关于该资源的元数据。

2.2 OCF技术简介

本节主要介绍OCF系统的架构、OCF框架功能和OCF角色的示例场景。

1. OCF的系统架构

OCF的宗旨：定义一个通用的通信框架，实现不同操作系统、不同设备之间的互联互通。它包含的核心功能主要有标识和寻址、设备和资源发现、资源模型、CRUDN(Create/Retrieve/Update/Delete/Notify，创建/检索/更新/删除/通知)操作、报文传输、设备管理和安全等。

OCF采用资源表示方法，运用RESTful的操作接口来使用资源，所有OCF设备由包含的资源及相应的URI表示。每个提供资源表示的OCF设备都能充当服务器端的角色。相应地，主动发起资源操作请求的设备充当客户端的角色。资源操作包括创建、检索、更新、删除和通知。

OCF的系统架构实现了物联网的物理设备或应用之间基于资源的交互。OCF的系统架构充分利用现存的行业标准和技术，并且提供了建立连接(无线或有线)的解决办法，为各种场景、操作系统提供管理设备之间的信息流动。

OCF的系统架构提供了如下可能：多个细分市场(消费者、企业、工业、汽车和医疗等)的通信和互操作框架、操作系统、平台、通信方式、传输和使用情况；一种描述环境促成信息和语义互操作的通用、一致的模型；用于发现和连接的通用通信协议；通用的安全和鉴定机制；创新和产品差异化的机会；表明不同设备能力可拓展的解决方案，适用于智能设备、最小连接以及可穿戴设备。

OCF的系统架构基于面向资源的设计原则，在OCF的系统架构中物理世界的实体，如温度传感器、电灯或电器被呈现为资源。和一个实体的相互作用是通过其资源表示实现的，使用遵循表述性状态转移架构风格的操作，如RESTful交互。OCF的系统架构定义了OCF框架整体结构为一个信息系统以及构成OCF的实体间的交互；实体呈现为OCF资源，携带着它们独特的标识符以及支持OCF资源的RESTful操作接口。

每个RESTful操作都有一个操作的发起者(客户端)以及一个操作的应答者(服务器端)。在OCF框架中，客户端和服务器端概念是由OCF角色实现的。任何OCF设备都可以作为OCF客户端或服务器端，在OCF设备上初始化RESTful操作。同样，任何将实体呈现为OCF资源的OCF设备，也可以作为OCF服务器端使用。遵循REST架构风格，每个在OCF的RESTful操作中，包括所有能够理解交互内容的必要信息，并且用一小部分通用操作就可以驱动，如 CRUDN 操作，这些也包括了对 OCF 资源的表示。图 2-2 描述了 OCF 的系统架构。

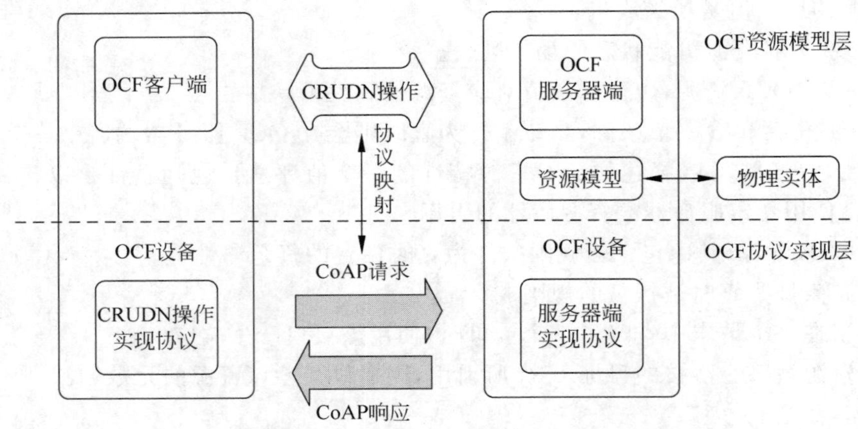

图 2-2 OCF 的系统架构

架构概念分为三个主要方面：资源模型、RESTful 操作和抽象化。

资源模型：提供了逻辑模型所需的抽象和概念，并在逻辑上操作其应用程序及环境。资源模型和应用领域无关，如智能家居、工业或汽车。资源模型定义了抽象化实体的 OCF 资源，且 OCF 资源表示映射为实体的状态。其他模型概念可以用来模拟行为模型等。

RESTful 操作：一般的操作被定义为用 RESTful 范例来模拟协议以及 OCF 资源的交互。具体的通信或信息传递是协议抽象的一部分，并且 OCF 资源和特定协议的映射将在本书中讲述。

抽象化：在资源模型中的抽象和 RESTful 操作被抽象原语映射到具体元素。实体处理程序用于映射到 OCF 资源以及连接抽象原语，这些抽象原语用来映射逻辑 RESTful 操作到数据连接协议或技术，实体处理程序也会被映射到不是本地 OCF 支持的实体。

OCF 功能结构如图 2-3 所示，分为 L2 连接层、网络层、传输层、OCF 框架层和应用配置层。

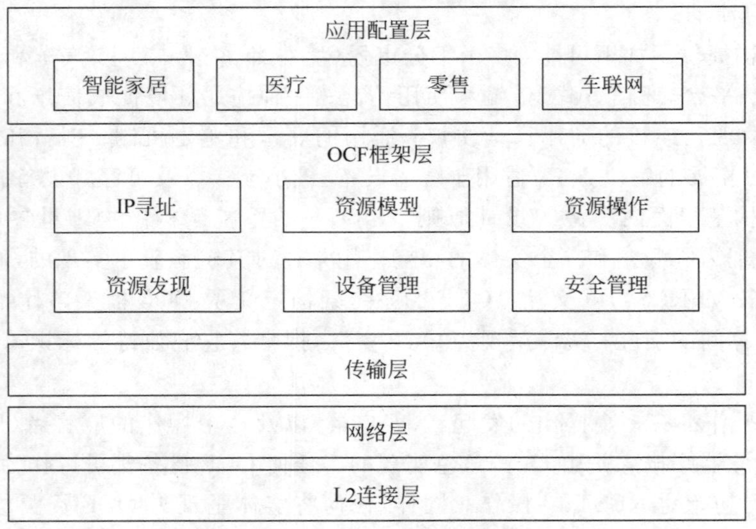

图 2-3 OCF 功能结构

L2 连接层：建立物理层和数据链路层连接（如 WiFi、蓝牙）。

网络层：提供 OCF 设备所需要的数据交换网络功能（如互联网）。

传输层：提供端到端的传输路径，如 TCP、UDP 及新的 IETF 传输协议。

OCF 框架层：提供 IP 寻址、资源发现、资源操作、设备管理和安全管理等核心功能。

应用配置层：提供对应于某一领域的数据模型及功能，如智能家居和车联网等。

当两个 OCF 设备互相交流时，一个 OCF 设备的每个功能模块和另一个 OCF 设备的对应模块以图 2-4 所示的方式进行交互。

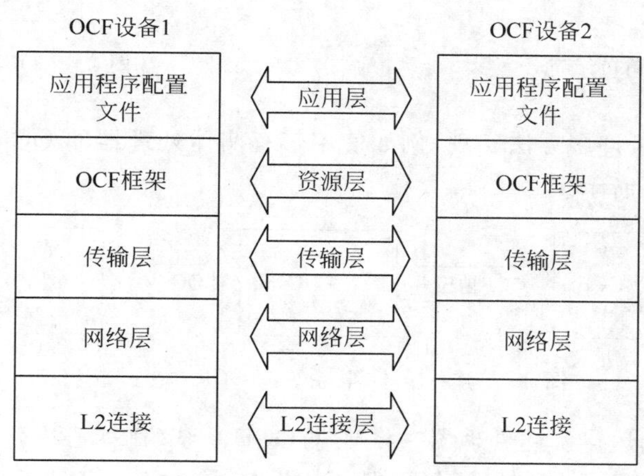

图 2-4　OCF 交流分层模型

2. OCF 框架层功能

OCF 框架层由提供 OCF 操作核心的功能组成，主要包括如下几点。

(1) IP 寻址：定义标识符和寻址能力。

(2) 资源发现：定义发现可用的 OCF 设备和 OCF 资源的过程。

(3) 资源模型：根据资源表示实体的能力，并且定义操作资源的机制。

(4) 资源操作：提供用于 OCF 客户端和 OCF 服务器端之间互动的通用方案。消息是提供 RESTful 操作的具体消息传递协议，如 CRUDN 操作、CoAP 消息和传递协议。

(5) 设备管理：指明管理 OCF 设备能力的原则，并且包括设备配置、初始化以及设备监测与诊断。

(6) 安全管理：包括对实体的安全访问所需的身份验证、授权和访问控制机制。

3. OCF 角色的示例场景

OCF 的交互定义于逻辑实体之间，该逻辑实体被称为 OCF 角色。OCF 定义了三种角色：OCF 客户端、OCF 服务器端和 OCF 中介。

图 2-5 说明了在一个场景中 OCF 角色的例子，该场景是一个智能手机给恒温器发请求消息，原始消息通过 HTTP 发送，但是在传输中被网关转换为 CoAP 请求消息，然后再传递给恒温器。在这个例子中，智能手机承担了 OCF 客户端的角色，网关承担了 OCF 中介角色，恒温器承担了 OCF 服务器端的角色。

如果连接非 OCF 生态系统，如图 2-6 所示，监测心率传感器（心率监测仪）的腕表设备，该传感器由非 OCF 的协议实现。提供了一个详细的逻辑观点。

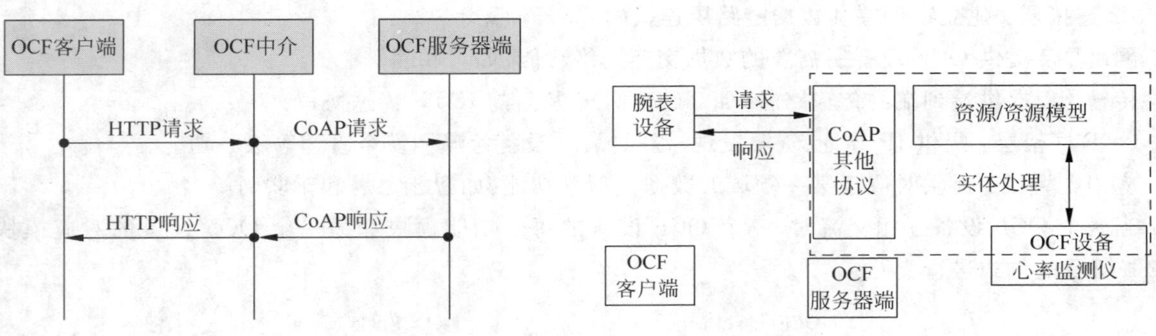

图 2-5 OCF 角色示意

图 2-6 OCF 结构细节

OCF 结构细节可以通过许多方法实现，例如使用带有实体处理器的 OCF 服务器端，该处理器直接连接非 OCF 设备，如图 2-7 所示。

图 2-7 连接非 OCF 系统的 OCF 服务器端

启动时，OCF 服务器端运行实体处理器来发现非 OCF 系统（例如心率传感器设备），并且为每个发现的设备或功能创建资源。实体处理器给每个发现的设备或功能创建资源，并且将其 OCF 资源绑定，通过 OCF 服务器端，使得这些资源可被发现。

一旦资源被创建并且可发现，之后显示设备可以发现这些资源，并且使用规范中描述的机制去操作它们。对 OCF 服务器端资源的请求会被实体处理器翻译，并且通过非 OCF 设备支持的协议传递给非 OCF 设备，非 OCF 设备返回的信息会被映射成该资源相应的应答。

2.3 OCF 标识与寻址

为了有效促进 OCF 架构元素间的互动，需要一种方法去标识元素、命名元素以及对这些元素寻址。

对于 OCF 系统中的标识符，应该唯一地表示环境或域中的元素。环境、域根据使用情况或者应用程序决定。元素的标识符在其生命周期应该是不可变的，并且在该环境或域中是独一无二的。

与其他系统一样，OCF 地址是用来定义一个位置、访问元素的方法，以便与其进行交互，地址在环境中是可变的。

名字是用来分辨架构中的元素，名字在该元素生命周期中可以改变。可以通过已知其他的一个或多个条件决定另一个名字或地址（如用地址决定名字或名字决定地址）。

在多个环境中，可以分别定义每个方面（如一个环境可以是一层）。所以，地址可以是寻址资源的 URL 以及连续层寻址的 IP 地址。在某些情况下，这两种资源都需要。例如，对一个特定的资源呈现做检索操作，客户端需要知道目标资源的地址以及暴露该资源的服务器地址。

在使用的环境或者域中，名字或地址可以被用作标识符，反之亦然。例如，URL 可以被用作资源的标识符并且可被制定为 URI。

本节的其余部分从资源模型和它所支持交互的角度讨论标识符、地址以及命名。交互是基于

RESTful 的,例如,对资源的操作、传输协议的映射,可以使用 CoAP 协议和 HTTP 协议。

标识符应该在环境或使用域内是唯一的。目前有许多方案可以产生具有所需属性的标识符。标识符是特定背景的,即在该环境或域中是独一无二的。标识符背景还可以是独立的,即这些标识符在所有环境和域中都是独一无二的。特定背景标识符可以被定义为像单调枚举的简单方案或者定义为重载的地址或名称。例如,一个 IP 地址可以是一个在智能网关后的私人域汇总的标识符。另外,背景独立标识符需要更强的方案去派生全局独立标识符,例如 UUID(Universally Unique Identifier,全局唯一标识符)中的任何一个。背景独立标识符可通过域的分层来生成,即层次的根由一个 UUID 来定义,子域通过连接这个域内特定背景标识符和其父域的背景独立标识符来产生背景独立标识符。

资源可以使用 URI 标识,如果该 URI 是一个 URL,则可以用同一个 URI 去寻址该资源。在一些情况下,资源需要和 URI 不一样的标识符,在这种情况下,资源属性的值是标识符。当 URI 是 URL 的格式时,该 URI 可用来寻址该资源。一个 OCF 的 URI 是基于在 IETF RFC 3986 中定义的,URI 的一般格式为< scheme >://< Authority >/< Path >? < Query >。对于 OCF,URI 的格式为 oic://< Authority >/< Path >? < Query >。

每个使用的部分说明如下:URI 的方案是"oic"。"oic"方案表示了语义,在 OCF 规范定义中使用。如果一个 URI 的"//"前面的部分省略,那么就默认其使用 OCF 方案。

在请求者通过网络发送消息之前,每个传输绑定需要指明一个 OCF 的 URI 是如何转换为传输协议的 URI。在接收端也是相似的,在切换到接收者的资源模型层之前,每个传输绑定需要指明 OCF 是如何从一个传输协议 URI 转换为 OCF 的 URI。

如果权限是局部 OCF 设备,那么"oic"被用于权限。权限的一般形式为< host >:< port >,其中,< host >是名字或者终端网络地址,< port >是网络端口号。< host >提供的内容如下:对于 IP 网络,< host >是< authority >的主机名或 IP 地址;对于非 IP 网络,< host >是名字或者合适的标识符;如果< authority >是支持资源的 OCF 设备,那么关键词 oic 可能会用于< host >。

路径必须是唯一的字符串,是唯一在该 OCF 服务器环境中标识或指定的一个资源。在此版本中,一个路径不应该包括非 ASCII 字符或 NUL 字符。路径应该以"/"(斜线)开始。为了方便人们阅读,路径可能用"/"分隔。在 OCF 环境中,"/"分离段被视为一个字符串,直接引用资源(即扁平结构)而不是解析为一个层次架构。在该 OCF 服务器中,路径或路径中的子字符串会通过哈希算法或其他方案来进行压缩,当然要假设产生的引用在该主机环境内是唯一的。

一旦产生路径,该资源的客户端或者 URI 的接收方会使用该路径为不透明字符串,并且无法解析其结构、组织或者语义。

一个查询字符串应该包括"< name >=< value >"字段对的列表,每对由";"分隔。查询字符串将被映射到相关协议的语法,用于消息传递,如 CoAP。

URI 的产生:URI 可以是绝对的(完全限定),也可以是相对的。URI 可以由创造该资源的 OCF 客户端直接定义,这样的 URI 可以是相对的或者是绝对的。一个相对的 URI 应该是相对于其归属的 OCF 设备。另外,URI 可以由该资源的 OCF 服务器端,自动根据该资源预先定义的公约或者组织,依据一个接口、一些原则、不同的根或基础来生成。

URI 的使用:引用一个 URI 的绝对路径被认为是一个不透明字符串,客户端无法通过其推断出任何准确的或隐藏 URI 的结构,即 URI 仅仅是一个地址。拥有资源的 OCF 设备将每个资源的 URI 看作是一个只用来寻址该资源的不透明字符串。例如,URI 的"/a"和"/a/b"认为是明确的地址,而且资源 b 不是资源 a 的子资源。

命名空间：相对的 URI 前缀“/oic/”被保留为一个 OCF 规范中定义的、用于命令 URI 的空间，不能用于非 OCF 定义规范中的 URI。

网络寻址：OCF 规范中使用 IP 地址，即当设备使用 IP 配置的接口时，使用 IP 地址。当该装置只有其对等端的身份信息，需要相关的机制将标识符映射到相应地址。

2.4 OCF 数据类型

表 2-1 包含了 OCF 数据类型的定义。数据类型来源于欧洲计算机制造联合会中定义的 JSON 值。但是，OCF 资源可以重载 JSON 定义的值，来指定 JSON 值的特定子集。OCF 数据类型可以用于特定的用途，例如，因为特定的原因改变一个字符串的长度。

表 2-1 OCF 数据类型的定义

命　名	JSON 值	描　述
boolean	布尔型	二进制值 0,1
BSV	字符串	一个以空白(例如空格)分隔的、由字符串编码的值组成的列表。在 BSV 中值的类型由使用 BSV 的特性来描述，如 BSV 整数
CSV	字符串	一个以逗号分隔的、由字符串编码的值组成的列表。在 CSV 中值的类型由使用 CSV 的特性来描述，如 CSV 整数
date	字符串	在 ISO 8601 中定义，格式是[yyyy]-[mm]-[dd]Z
datetime	字符串	在 ISO 8601 中定义，级联的 data 与 time，并且 data 与 time 中间用 T 分隔。格式是[yyyy]-[mm]-[dd]T[hh]:[mm]:[ss]Z
enum	枚举	枚举类型
float	数值	有符号的 IEEE 754 单精度浮点值
integer	数值	有符号的 32 位整数
JSON	对象/数组	JSON 元素，可以是欧洲计算机制造联合会定义的一个对象或者数列表示的数据。这个 JSON 对象或者数列可以用 JSON 图标来描述
string	字符串	字母字符串，不可以超过最大长度(即 64 字节)
time	字符串	在 ISO 8601 中定义但被 UTC 限制，且尾部加 Z。格式是[hh]:[mm]:[ss]Z
URI	字符串	用于标识一个资源的字符串，其值不应超过最大长度(即 256 字节)
UUID	字符串	一个根据 IETF RFC 4122 标识符的格式

第3章 OCF的资源模型

CHAPTER 3

3.1 基本概念

资源模型定义了一些概念和机制，这些概念和机制提供了 OCF 生态系统中设备之间的一致性和核心互操作性。在此基础上，资源模型的概念和机制被映射到传输协议上，使得两个设备之间可以进行通信，即每次传输都提供了通信协议的互操作性。因此，资源模型允许独立于传输来定义互操作性。

此外，资源模型中的概念支持基本设备及彼此之间关系的建模，并在一个环境中捕获互操作性所需的语义信息。通过这种方式，OCF 超越了简单的协议互操作性，在可穿戴设备和物联网生态系统中，可以获取真正互操作性所需的丰富语义。

在 OCF 资源模型中，基本概念有实体、资源、统一资源标识符、资源类型、属性、表示、接口、集合和链接。此外，通用的操作有创建、检索、更新、删除和通知，也就是 CRUDN。这些概念和操作以各种各样的方式组合，可以为 OCF 框架的各种应用场景定义所需的语义和互操作。

OCF 资源模型框架中，在一个给定应用中的软硬件人工产品、值、用例和上下文被称为一个实体。当一个实体需要可视化、交互或者被操作时，它就会由一个抽象概念表示，这个抽象概念称为资源。一个资源的压缩和表示是一个实体的最重要部分，通过使用 URI 可以被标识、寻址和命名。

“属性”就是“键值对”，表示资源的主要部分。这些属性的一个“快照”就是资源的“表示”。“表示”是一个特定的视图，且可应用于该视图中的机制就是“接口”。与一个资源进行交互是由“请求”和“响应”完成的，其中就包含“表示”。

一个资源实例源于一个资源类型。一个资源和另一个资源之间的单向关系被定义为一个“链”。一个有“属性”和“链路”的资源就是一个“集合”。

一组“属性”可以用来定义一个资源的一个状态。通过使用合适的“表示”，使用该资源的响应或到该资源的请求，该资源的状态可以被检索或更新。

一个资源（和资源类型）可以表示，并用于暴露一种能力。通过与该资源的交互可以训练或使用这种能力。这样的能力可以用于定义类似发现、管理和广播这样的过程。例如，“在一个设备上发现资源”可以定义为一个特定资源表示的检索、该资源的属性值描述或者引用设备上的资源。

具有请求或响应的表示信息可以“在线”传输，这可以通过使用传输协议序列化或者封装到传输协议负载上的方式，具体的方法由请求或响应到传输协议的映射规范来决定。

在本书中，以 RAML 定义标准规范。同时，也可以使用 JSON 模式。有关 OCF 规范中定义的资源类型，将在第 9 和 10 章中介绍。

3.2 OCF 资源

一个资源可能会被定义为一种或多种资源类型，请求实例化/创建一个资源应该指明定义该资源的一个或多个资源类型。一个资源将被托管在一个设备中，如上所述，一个资源应该有一个 URI，该 URI 可以在创建或实例化资源时由管理机构分配，也可以通过资源类型的规范预定义，如图 3-1 所示。

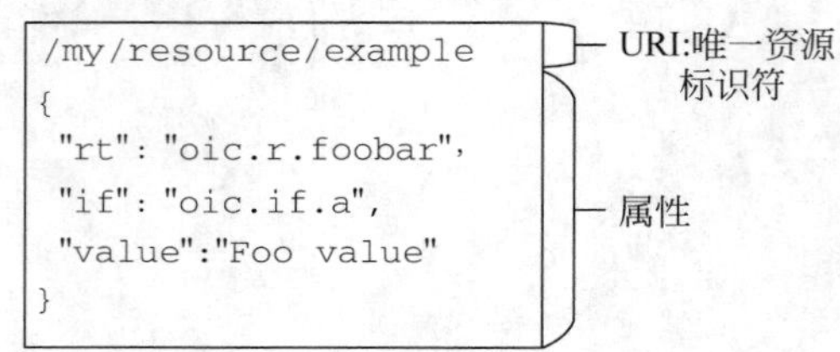

图 3-1 一个资源例子

核心资源是 OCF 规范中定义的、用于实现功能交互（如发现、设备管理等）的资源。其中，功能交互在第 6 章中定义。在核心资源中，"/oic/res""/oic/p"和"oic/d"是所有的设备上都应该支持的；同时，设备可能根据支持的交互功能，支持其他的核心功能。

3.3 资源属性

一个属性描述了资源所暴露的一个方面或概念，包括与该资源相关的元信息。一个属性应该有一个名称（即属性名）和一个值（即属性值）。属性是以键值对的形式存在，"键"就是属性名，"值"就是属性值，键值对就类似于<属性名>=<属性值>。例如，如果"温度"属性有一个属性名"temp"和一个属性值"30F"，则该属性就表达为"temp=30F"。属性的特定形式是由编码模式决定的。例如，在 JSON 中，属性表示为"键"：值（如"temp"：30）。除此之外，属性的定义包括以下内容。

（1）值类型：定义了属性值可能采用的值。值类型可以是一个简单数据类型（如 string、boolean 等），如上文所定义，也可以是一个用特定模式定义的复杂数据类型。值类型可以定义值规则，即定义一些规则，属性值会采用这些规则，并将其用于属性值中。这些规则可以定义值的范围、最大、最小值、公式、枚举值集合、阵列、条件值，甚至是对其他属性值的依赖关系。这些规则可以用于验证属性值中的特定值，并且标记错误。

（2）强制性：指明对于一个给定的资源类型，该属性是否是强制的。

（3）访问模式：指明该属性是否可以被读写。更新等同于写入。"r"表示读，"w"表示写，二者都是可以被指定的；同时，写操作并不会意味着就会自动进行读操作。

一个属性的定义可能包括下列附加信息，这些项可以提供有用的信息。

（1）属性名：指明一个属性人性化的名称，通常不会在线发送。

（2）描述：一个描述性的文本，定义该属性的目的和预期的使用。

一个属性可能被用在一个 URI 的查询部分，作为选择一个特定资源的标准。该过程可以通过将属性声明为查询的一个字段来完成（即<属性名>=<想要的属性值>）。在 OIC 1.1 版本的规范中，查询过滤器中只允许 ASCII 字符串，不允许 NUL 字符串。这意味着，只有使用 ASCII 字符的属性值，才可以在查询过滤器中被匹配。当查询中声明的属性与目标资源的完整表示中对应的属性相匹配时，该资源就被选择。完整表示目标所有资源类型的属性集合，是这个集合的快照。如果在查询的"过滤"字段声明了一个属性，那么，声明的属性就会与接口定义表示相匹配，以隔离表示的特定部分。

通常来说，一个属性只有在与其相关的资源内才是有意义的。然而，有一些属性可能被所有的资源支持，称为公共属性，这些属性在资源之间保持资源完整性，即它们的"键＝值"对在所有的资源中的含义都相同。下面介绍公共属性的字段。

公共属性可以在所有资源中定义。以下几种属性被定义为公共属性：资源类型、资源接口、名称和资源标识。

一个公共属性的名称应该是唯一的，不应该被其他属性使用。当定义一个新的资源类型时，如果是非公共属性，则不应该使用已经存在的公共属性名称（如"rt""if""p"和"id"）。如果定义一个新的公共属性，应该确定其名称不应该被其他属性所使用。一个新公共属性名称的唯一性，可以通过检查所有已存在、OCF 所定义的资源类型属性来核实。然而，随着资源类型的增长，这可能越来越难以处理。为了阻止未来出现这样的名称冲突，OCF 为公共属性保留了一个特定的名称空间。

潜在的方法有：①可能会分配一个特定的前缀（例如"oic"），接在该前缀后面的名称前（如"oic. psize"）仅用于公共属性；②包含一个或两个字母的名称被保留用于公共属性，其他属性名称的长度都要大于两个字母；③公共属性可能嵌套在特定的对象下，以与它们自身区分。

所有资源属性总结如下。

(1) 资源类型（"rt"）：该属性用于声明资源类型。由于一个资源可以被定义多个资源类型，因此，资源类型属性值可以用于声明多个资源类型。例如，"rt"：["oic. wk. d"，"oic. d. airConditioner"]表明，包含该属性的资源被定义为"oic. wk. d"或"oic. d. airConditioner"资源类型。

(2) 接口（"if"）：该属性声明了资源支持的接口。接口属性值可以有多个，列出资源支持的所有接口。

(3) 名称（"n"）：该属性声明了分配给资源"人类可读"的名称。

(4) 资源标识符（"id"）：对于一个资源的特定实例，该属性值是唯一的实例标识符（在整个主机服务器的范围内唯一），该标识符的编码依赖于具体设备。

在 OCF 规范中使用的属性名称和属性值如下。

(1) 属性名称："键值对"中的"键"。属性名称是区分大小写的，并且其数据类型是字符串类型，但是只允许 ASCII 字符串，不允许嵌入 NUL 字符。

(2) 属性值："键值对"中的"值"。当数据类型是字符串时，属性值是区分大小写的。枚举类型的值只能是 ASCII 字符。

名称是资源的人性化名称，即一个特定的资源实例名称（如 MyLivingRoomLight），名称属性在表 3-1 中进行了定义。

表 3-1　名称属性的定义

属性	属性名	属性值类型	属性值规则	单位	访问模式	是否强制	描　述
名称	n	字符串			读	否	可理解的资源名称；可以被本地设置或远程设置（例如，由一个用户进行设置）

对于一个资源的特定实例来说，资源标识符属性应该是唯一的实例标识符（在整个主机服务器端的范围内）。该标识符的编码是依赖于设备和实现的，资源标识符属性的定义如表 3-2 所示。

表 3-2 资源标识符属性的定义

属　性	属性名	属性值类型	属性值规则	单位	访问模式	是否强制	描　述
资源标识符	ID	字符串	依赖实现		写	否	资源的唯一标识符(在设备的所有资源中)

3.4 资源类型

资源类型是指一类资源,而一个资源是资源类型的具体实例。也就是说,资源类型定义了一类资源。一个资源的资源类型是通过使用资源类型公共属性来声明的,或者是在使用了资源类型参数的链接中声明的。

资源类型可以是预定义的(例如,在 OCF 规范中的核心资源类型或垂直域规范中的资源类型),也可以是自定义的,由制造商、终端用户或者设备开发人员定义(例如,供应商定义的资源类型)。资源类型及其定义细节可以通过外部(例如,通过文档)传输,也可以通过使用 API、应用程序下载和使用的元数据来进行明确定义。OCF 推荐使用 RAML 或 JSON 模式明确定义资源类型。

每一个资源类型都会有一个 ID 来标识,资源类型 ID 是一个分段的小写字符串,各段之间用“.”隔开。整个字符串表示资源类型 ID。当定义 ID 时,每一段可能会表示适用于该资源类型的任何语义。例如,每段都可以表示一个命名空间。一旦 ID 被定义,ID 应该被非透明地使用,而不是从每个段中推断出任何信息。当字符串"oic"用作资源类型 ID 定义的第一段时,该字符串被保留为 OCF 定义的资源类型。资源类型 ID 可以用于寻找资源类型的定义,可以作为权威的引用。属性名称和属性值都是区分大小写的,并且“rt”属性中的资源类型 ID,应该以区分大小写的方式进行比较。

3.4.1 资源类型属性

当一个资源被实例化或创建时,资源应该具有一个或多个资源类型,这些资源类型是作为资源模板的。资源所遵守的资源类型应该使用资源的“rt”公共属性进行声明。对于用作标准的资源类型,“rt”公共属性的属性值应该是一个资源类型 ID 的列表(即"rt"=<资源类型 ID 列表>)。资源类型公共属性定义如表 3-3 所示。资源类型可以在用户(即客户端)和资源主机(即服务器端)之间被明确地发现或隐式地共享。

表 3-3 资源类型公共属性定义

属　性	属性名	属性值类型	属性值规则	单位	访问模式	是否强制	描　述
资源类型	rt	字符串	属性类型的 ID		读	是	属性名“rt”是在 IETF RFC 6690 中具体定义的

3.4.2 资源类型定义

在 OCF 规范中,资源类型声明如下。

(1) 预定义的 URI(可选):在一个 OCF 规范中,可以为一个特定的资源类型指定一个预定义的 URI。当一个资源类型有一个预定义的 URI 时,该资源类型的所有实例将只使用这个预定义的 URI,其他资源类型的实例不会使用该预定义的 URI。

(2) 资源类型名称(可选):指明资源类型的人性化名称。

(3) 资源类型 ID："rt"属性的值，指明了资源类型(如"oic. wk. p")，由多个段组成的小写字母字符串，各段之间用"."隔开；每段可能表示一个命名空间，后一段(从左到右)表示的就是前一段的子命名空间；在实现的过程中会不透明地使用这些段，并且使用区分大小写的字符串匹配。

(4) 资源接口：该资源类型可能支持的一些接口。

(5) 资源属性：应用于一个资源类型的所有属性的定义。资源类型的定义中，应该规定一个属性是强制的、在某种条件下是强制的还是可选的。

(6) 相关资源类型(可选)：可以作为该资源类型一部分，是被引用的其他资源类型规范，也适用于集合。

(7) Mime 类型(可选)：包括序列化资源支持的 Mime 类型(如 application/cbor、application/json 和 application/xml)。

表 3-4、表 3-5 将一个 foobar 资源类型及其相关属性，作为一个例子进行了描述。

表 3-4 foobar 资源类型示例

预定义的 URI	资源类型名称	资源类型 ID ("rt"属性的值)	资源接口	描述	相关的功能交互	强制/条件/可选
无	foobar	oic. r. foobar	oic. if. a	foobar 资源示例	驱动	可选

表 3-5 foobar 属性示例

属性	属性名	属性值类型	属性值规则	单位	访问模式	是否强制	描述
资源类型	rt	字符串			读	是	资源类型
资源接口	if	字符串			读	是	接口
资源值	value	字符串			读	是	属性

foobar 资源类型的实例如下所示。

```
{
"rt": ["oic.r.foobar"],
"if": ["oic.if.a"],
"value": "foo value"
}
```

foobar 资源类型的示例 Schema 如下所示。

```
{
"$schema": "http://json-schema.org/draft-04/schema",
"type": "object",
"properties": {
"rt": {"type": "string"},
"if": {"type": "string"},
"value": {"type": "string"}
},
"required": ["rt", "if", "value"]
}
```

3.4.3 多"rt"值资源

多"rt"值资源是指具有多种资源类型。这样的资源与多个资源类型相关联，因此它的"rt"属性值具有多个资源类型 ID(如"rt"：["oic. r. switch. binary","oic. r. light. brightness"])。"rt"属性值中资源类型 ID 的顺序是无意义的。例如,"rt":["oic. r. switch. binary","oic. r. light. brightness"]和"rt":["oic. r. light. brightness","oic. r. switch binary"]具有相同的含义。

资源类型为多"rt"值的资源应满足以下条件。

(1) 属性名称除了"公共属性"以外,每个资源类型的属性名称应该是唯一的(在多"rt"值资源的范围内),否则会出现冲突的属性语义。如果两个资源类型具有相同名称的属性,则多"rt"值资源不应由这些资源类型组成。

(2) 多"rt"值资源满足每种资源类型的所有要求,并符合每个组件资源类型的 RAML/JSON 定义。因此,多"rt"值资源的强制属性应为每种资源类型的所有强制属性的并集。例如,在资源"rt"为["oic. r. switch. binary","oic. r. light. brightness"]中的强制属性是"值"和"亮度",前者属于"oic. r. switch. binary",后者属于"oic. r. light. brightness"。

(3) 多"rt"值资源接口集合应是组件资源类型中接口集合的并集。响应于接口上的 CRUDN 操作资源表示应为该接口定义模式的并集。多"rt"值资源的默认接口应为基准接口("oic. if. baseline"),因为这是资源类型之间唯一保证的通用接口。

为了清楚起见,如果每个资源类型支持相同的一组接口,则所得到的多"rt"值资源具有相同的接口集,默认接口为"oic. if. baseline"。

对多"rt"值资源的默认接口"oic. if. a""oic. if. s""oic. if. r"和"oic. if. rw"等"rt"查询是通用"rt"查询的扩展。当服务器端收到带有"rt"查询的多"rt"值资源的检索请求(即 GET /ResExample? rt=oic. r. foo)时,服务器端应当在查询目标为多"rt"值资源某一项"rt"属性中进行响应,且应仅返回与查询相关联的属性。例如,当目标多"rt"值资源"rt":["oic. r. switch. binary","oic. r. light. brightness"]在接收到使用"rt"查询"GET /ResExample? rt =oic. r. switch. binary:"时,服务器端只响应"oic. r. switch. binary"的属性。

3.5 设备类型及资源接口

设备类型是设备的类。定义的每个设备类型将包括该设备类型实现的最小资源类型列表。设备可能会暴露超出最小列表的额外标准和供应商定义的资源类型。设备类型用于资源发现。像资源类型一样,设备类型可以使用在资源类型公共属性中或资源类型参数的链接中。设备类型可以由制造商、最终用户、设备的开发人员进行预定义或自定义,设备类型及其定义的详细信息可能用其他方式交互(如文档中)。

对于 OCF 规范的接口,它首先提供了资源内部的一个视图,然后根据资源的视图定义允许的请求和响应,是由接口提供的视图定义了对这个资源请求和响应的上下文。因此,对于一个资源的相同请求,针对不同的接口时会产生不同的响应。

资源的接口可以由 OCF 规范定义(即核心接口),也可以由 OCF 垂直域规范来定义(即垂直接口),还可以由制造商、终端用户或设备的开发人员来定义(即供应商定义的接口)。

资源的接口属性列出了该资源支持的所有接口。任何的资源至少有一个接口。默认接口应由 OCF 规范定义,并且从资源类型的定义中继承。在 OCF 规范定义中,与资源类型相关的默认接口,应该是在资源类型中定义的、可用枚举类型内首先被列出的支持接口。在一个 OCF 规范中指定的所有默认接口都应该是强制的。

除 OCF 规范定义的接口之外,所有的资源都应该支持基准接口("oic. if. baseline")。当请求选择使用一个接口时,在请求消息的资源 URI 中,该接口应该被指定为查询参数。如果没有指定查询参数,就会使用默认的接口。如果选择的接口不是资源所允许的接口,那么选择该接口就是错误的。

一个接口可以接收多个媒体类型，也可以用多个媒体类型来响应。被接收的媒体类型可以不同于响应的媒体类型。在传输协议中，媒体类型是使用合适的头参数来指定。需要注意的是该特征必须合理使用，并且允许在线上优化表示，每个接口都应该至少有一种媒体类型。

3.5.1　接口属性

一个资源所支持的接口应该使用接口公共属性（如表 3-6 所示）进行声明，格式为“if=<接口数组>”。一个接口属性值应该是一个分段的小写字符串，各段之间用“.”分隔开。当接口属性的第一段中使用“oic”时，该接口就被保留为 OCF 定义的接口。接口属性值也可以是对一个权威的引用，这样的引用可以用于找到一个接口的定义，一个资源类型应该支持一个或多个接口。

表 3-6　资源接口属性定义

属性	属性名	属性值类型	属性值规则	单位	访问模式	是否强制	描　述
接口	if	JSON	用点分隔开的字符串数组		读	是	声明一个资源所支持的接口的属性

3.5.2　接口方法

本节主要介绍接口的定义、基准接口、链路列表接口、批处理接口、执行器接口、传感器接口、只读接口、读写接口。

1. 概述

在 OCF 规范中，定义的接口如表 3-7 所示。

表 3-7　OCF 规范中定义的接口

接　口	名　称	可用的方法	描　述
基准接口	oic. if. baseline	检索、更新	基准接口定义了包括原属性在内的一个资源的所有属性的视图。该接口用于在一个资源的完全表示上进行操作
链路列表接口	oic. if. ll	检索	链路列表接口提供了一个集合（资源）中链接的视图。由于链接表示与其他资源的关系，该接口可能用于发现关于一个上下文的资源。这个发现是通过检索到其他资源的链接来完成的。例如，核心资源/oic/res 使用该接口，以允许发现托管在一个设备上的资源
批处理接口	oic. if. b	检索、更新	批处理接口用于同时与一组资源进行交互。这不需要客户端首先发现它正在操作的资源，服务器端会转发请求并聚合响应
执行器接口	oic. if. a	创建、检索、更新	执行接口用于读写一个驱动器资源的属性
传感器接口	oic. if. s	检索	传感器接口用于读一个传感器资源的属性
只读接口	oic. if. r	检索	只读接口暴露了可以被“读”的资源属性。该接口不提供方法来更新属性或资源，因此，只能用来读属性值
读写接口	oic. if. rw	检索、更新	读写接口只暴露了既可以被读又可以被写的属性，并提供了相关的方法来读写一个资源的属性

2. 基准接口

资源的所有属性表示，包括公共属性，对于基准接口是可见的。基准接口是为所有的资源类型定义的，即所有的资源都应该支持基准接口。

通过将"if=oic. if. baseline"添加到目标资源的 URI 的查询参数中，就选择了基准接口，例如 GET/oic/res? if=oic. if. baseline。

1）检索的使用

基准接口可以用来检索资源的所有属性。当一个客户端检索一个资源的所有属性时，就会使用基准接口。客户端将"? if=oic. if. baseline"加入到一个检索请求中，服务器端接收到请求时，会将此时所有属性的表示加到响应中。如果服务器端不能返回所有资源的表示，就会回复一个错误信息；也就是说，服务器端不会返回部分资源表示。

一个使用了基准接口的检索请求示例响应如下所示。

```
{
"rt": ["oic.r.temperature"],
"if": ["oic.if.a","oic.if.baseline"],
"temperature": 20,
"units": "C",
"range": [0,100]
}
```

2）更新的使用

使用基准接口，并且在更新请求中有属性及其期望值的列表时，一个资源的所有属性都可以被修改。

3. 链路列表接口

链路列表接口提供了一个集合(资源)中链路列表的一个视图。通过该接口可见的表示只有链路，该链路在属性值中定义。因此，该接口用于与一个集合中的链路列表进行操作或交互。通过使用该接口，可以检索到链路列表，接口定义和语义如下。

(1) 链路列表接口名称应该是"oic. if. ll"。

(2) 如果请求中有定义的话(通常在请求头中)，响应中的序列化应采用请求定义的预期格式。

(3) 在链路列表接口检索请求的响应中，被引用资源的 URI 应该作为一个 URI 引用被返回。

(4) 如果在一个资源中显示没有链路，则返回一个空列表。

(5) 由该接口视图定义的表示只包括链路属性值。

(6) 关于链路列表接口示例，即一个集合的请求表示，例如，检索在房间中链路的请求，链路可能引用灯、风扇和电插座等，如 GET oic://< devID >/a/room/1? if=oic. if. ll。

4. 批处理接口

批处理接口通过使用一个(相同的)请求与资源集合进行交互。批处理接口支持集合链路中的资源方法，并且可以使用一个资源表示来检索或更新被链接的资源属性。

批处理接口选择了一个集合链接的视图，将请求发送到该视图中所有的链路，并包含可能修改的链路参数。批处理接口定义如下。

(1) 批处理接口名称应该是"oic. if. b"。

(2) 具有批处理接口的资源会有很多链路，这些链路会有资源的引用，这些引用可能是 URI(对远程资源完全适用)或者是相对引用(用于本地资源)。

(3) 如果一个资源的链接并没有指明使用哪个接口(使用"bp"链接参数)，则该请求会被转发到被引用资源的默认接口。如果"bp"使用"q"关键字指明了一个查询，则该查询会被应用在 URI 的查询参数中(该 URI 是由引用形成的)，以选择在目标资源中的接口。

(4) 如果要针对资源链接中的每个对象修改原始请求以创建新的请求，则需要将原始请求的 URI

替换为链接中目标资源的 URI。新请求中的有效荷载可以直接复制原始请求中的有效荷载。

(5) 来自链接资源的所有响应都应集中到单个响应中,发送到服务器端。服务器端会根据时间窗口判断响应是否超时,如果时间窗口已经跟客户端进行了协商,则服务器端不会在时间窗口内超时;如果没有协商好的时间窗口,服务器端会根据情况选择合适的窗口。如果目标资源不能处理新的请求,就会返回空响应或错误响应。这些空的/错误的响应会包含在集中响应中,返回到原始客户端。

(6) 集合响应是对象的集合,每个对象都是单独的响应。集合中的每个响应包含至少两项:完全限定的 URI,表示为"href":< URI >;在响应中声明的表示,使用关键字"rep",即"rep":{<在单个响应中的表示>}。

(7) 通过向原始批处理接口请求处理的集合 URI 中添加一个过滤器,客户端可以限制请求转发的链接列表。

(8) 在特定链路请求表示,可能与目标资源上接口暴露的表示不匹配。在这种情况下,使用 PUT 方法的更新操作通常会失败,使用 POST 方法比较合适。在这种情况下,如果请求中的属性与暴露的资源视图属性相匹配,那么子集语义就会应用于目标资源中可以修改的属性,当然,属性可以修改的前提是该属性是可写的。

(9) 如果一个设备支持批处理接口,那么该设备应该既实现客户端角色,又实现服务器端角色。

一个批处理接口的例子如表 3-8 所示。

表 3-8 批处理接口的例子

资源	/a/room/1
	{
	"rt": ["acme.room"],
	"if": ["oic.if.baseline", "oic.if.b"],
	"color": "blue",
	"dimension": "15bx15wx10h",
	"links": [
	{"href": "/the/light/1",
	"rt": ["acme.light"],
	"if": ["oic.if.a", "oic.if.baseline"],
	"ins": 1},
	{"href": "/the/light/2",
	"rt": ["mycorp.light"],
	if = ["oic.if.a" , "oic.if.baseline"],
	"ins": 2},
	{"href": "/my/fan/1",
	"rt": ["hiscorp.fan"],
	if = ["oic.if.baseline", "oic.if.a"],
	"ins": 3 },
	{"href": "/his/fan/2",
	"rt": ["hiscorp.fan"],
	if = ["oic.if.baseline", "oic.if.a"],
	"ins": 4, "bp": {"q": "if = oic.if.a"}}
	]
	}
	/the/light/1
	{
	"rt": ["acme.light"],
	"if": ["oic.if.s", "oic.if.baseline"],

续表

资源	"state": 0,
	"colortemp": "2700K"
	}
	/the/light/2
	{
	"rt": ["mycorp.light"],
	"if": ["oic.if.a", "oic.if.baseline"],
	"state": 1,
	"color": "red"
	}
	/my/fan/1
	{
	"rt": ["hiscorp.fan"],
	"if": ["oic.if.a", "oic.if.baseline"],
	"state": 0,
	"speed": "10"
	}
	/his/fan/2
	{
	"rt": ["hiscorp.fan"],
	"if": ["oic.if.a", "oic.if.baseline"],
	"state": 0,
	"speed": "20"
	}

批处理的使用	Request: GET /a/room/1?if = oic.if.b
	在客户端中由设备将上面的请求处理为下面单个的请求。
	GET /the/light/1(注意：使用默认接口，即传感器接口)
	GET /the/light/2(注意：使用默认接口，即传感器接口)
	GET /my/fan/1(注意：使用默认接口，即基准接口)
	GET /his/fan/2?if = oic.if.a(注意：来自 bp 链接参数的接口，即执行器接口)
	响应：
	[
	{
	"href": "oic://<devID>/the/light/1",
	"rep": {"state": 0, "colortemp": "2700K"}
	},
	{
	"href": "oic://<devID>/the/light/2",
	"rep": {"state": 1, "color": "red" }
	},
	{
	"href": "oic://<devID>/my/fan/1",
	"rep": { "rt": ["hiscorp.fan"], "if": ["oic.if.a", "oic.if.baseline"], "state": 0,
	"speed": "10" }
	},
	{
	"href": "oic://<devID>/his/fan/2",
	"rep": { "state": 0, "speed": "20" }
	}
	]

续表

批处理的使用(有 POST 语义的 UPDATE)	UPDATE /a/room/1?if = oic.if.b
	{
	"state": 1
	}
	变为：
	UPDATE /the/light/1 { "state": 1 }
	UPDATE /my/fan/1 { "state": 1 }
	UPDATE /his/fan/2?if = oic.if.a { "state": 1 }
	该操作打开了房间内的所有灯(除了"/the/light/1"资源)和风扇,因为所有的资源都有"state"这一属性。"/the/light/1"默认接口是传感器,因此,POST 不支持传感器接口(设备主机"/a/room/1"不会发送该请求)

批处理的使用(有 POST 语义的 UPDATE)	UPDATE /a/room/1?if = oic.if.b
	{
	"state": 1,
	"color": "blue"
	}
	该操作打开了房间中所有的灯(除了"/the/light/1"资源之外)和风扇,也将"/the/light/2"的颜色设置为蓝色

表 3-9 更进一步展示了链路列表和批处理接口。

表 3-9　链路列表和批处理接口示例

示例	/myexample
	{
	"rt": ["oic.r.foo"],
	"if": ["oic.if.baseline", "oic.if.ll"],
	"links": [
	{"href": "/acme/switch",
	"di": "<deviceID1>",
	"rt": ["oic.r.switc.binary"],
	"if": ["oic.if.a"]},
	{"href": "oic://<deviceID1>/acme/fan",
	"rt": ["oic.r.fan"],
	"if": ["oic.if.a"] }
	]
	}

基准接口的使用	GET /myexample?if = oic.if.baseline
	{
	"rt": ["oic.r.foo"],
	"if": ["oic.if.baseline", "oic.if.ll"],
	"links": [
	{"href": "/acme/switch",
	"di": "<deviceID1>",
	"rt": ["oic.r.switc.binary"],
	"if": ["oic.if.a"]},
	
	{"href": "oic://<deviceID1>/acme/fan",
	"rt": "oic.r.fan", "if": "oic.if.a"}
	]
	}

续表

链路列表接口的使用	GET /myexample?if = oic.if.ll [{"href": "/acme/switch", "di": "< deviceID1 >", "rt": ["oic.r.switc.binary"], "if": ["oic.if.a"]}, {"href": "oic://< deviceID1 >/acme/fan", "rt": ["oic.r.fan"], "if": ["oic.if.a"]}]

5. 执行器接口

执行器接口是用于查看可以被驱动的资源接口,即改变由资源抽象出来的某个实体内的值或状态。

(1) 执行器接口的名称应该是"oic.if.a"。

(2) 执行器接口在资源表示中暴露所有强制属性,属性由可用的JSON定义;执行器接口会暴露可选属性,它由目标设备中实现的、可用的JSON Schema定义。

加热器资源如下所示,"prm"是参数属性的名称。

```
/a/act/heater
{
"rt": ["acme.gas"],
"if": ["oic.if.baseline", "oic.if.r", "oic.if.a"],
"prm": {"sensitivity": 5, "units": "C", "range": "0 .. 10"},
"settemp": 10,
"currenttemp" : 7
}
```

根据加热器资源,执行器接口说明如下。

(1) 检索一个驱动器的值。

```
Request: GET /a/act/heater?if = "oic.if.a"
Response:
{
"prm": {"sensitivity": 5, "units": "C", "range": "0 .. 10"},
"settemp": 10,
"currenttemp" : 7
}
```

(2) 驱动器的正确使用。

```
Request: POST /a/act/heater?if = "oic.if.a"
{
"settemp": 20
}
Response:
{
Ok
}
```

(3) 驱动器的不正确使用。

```
Request: POST /a/act/heater?if = "oic.if.a"
```

```
        {
          "if": "oic.if.s" ?对 baseline 接口可见
        }
Response:
        {
          Error
        }
```

(4) 使用该接口的检索请求，会返回符合可能存在的资源表示，这些资源表示可以针对任何查询和过滤器参数。

(5) 使用该接口的更新请求，会提供一个有效荷载或者是包体，其中包含目标资源可能或需要更新的属性。

(6) 如果一个资源使用了该接口，会返回使用一个媒体类型的表示，并使用 CBOR(Concise Binary Object Representation，简明二进制对象表示)编码，如 IETF RFC 7049 中所定义的。未来可能会定义其他的媒体类型，以修改在返回值中的细节。

6. 传感器接口

传感器接口用于资源检索被测量的、被感知的或者是特定能力的信息，主要包括：

(1) 传感器接口的名称是"oic.if.s"。

(2) 传感器接口会在资源表示中，暴露所有强制属性，由可应用的 JSON 定义；传感器接口也会暴露由目标设备实现的可选属性，也由 JSON 定义。

(3) 使用该接口的检索请求，会返回符合可能存在的资源表示，它可以相对于任何查询和过滤器参数。

(4) 如果一个资源使用了该接口，会返回使用一个媒体类型的表示，并使用 CBOR 编码，如 IETF RFC 7049 中所定义的，未来可能会定义其他的媒体类型，以修改在返回值中的细节。

传感器接口主要包括如下几方面。

(1) 检索一个传感器的值。

```
Request: GET /a/act/heater?if = "oic.if.s"
Response:
        {
          "currenttemp": 7
        }
```

(2) 传感器的不正确使用。

```
Request: PUT /a/act/heater?if = "oic.if.s"              //PUT 是不允许的
        {
          "settemp": 20                                 //这可以通过执行器接口实现
        }
Response:
        {
          Error
        }
```

(3) 传感器的不正确使用。

```
Request: POST /a/act/heater?if = "oic.if.s"             //POST 是不允许的
        {
```

```
        "currenttemp": 15                                //这可以通过执行器接口实现
    }
Response:
    {
        Error
    }
```

7. 只读接口

只读接口暴露可能会被“读”的属性，包括“只读”属性、“读写”属性等，但却不包括“只写”属性和“只设置”属性。能够使用的方法只有检索，客户端如果想尝试检索之外的方法，就会被拒绝，还会产生错误响应代码。

8. 读写接口

读写接口只会暴露可能会被“读”和“写”的属性。这表明，“只读”属性不会包含在读写接口的表示中。可应用的方法只有检索和更新。同样地，客户端如果想尝试其他的方法请求会被拒绝，会产生错误响应代码。

3.6 资源结构

在外部的可见性和资源属性的可操作快照以及在一个时间点上各自的值被称为资源表示。资源表示抓住了特定时间一个资源的状态。当资源进行交互时，资源表示在请求和响应中被交换。资源表示可以用于检索或更新一个资源的状态。资源表示不应该被数据连接协议和技术（例如，CoAP、UDP/IP或低功耗蓝牙）所操作。

在很多场景中，资源之间可能会有一个隐式或显式的结构。例如，一个结构可以是树、网格、扇出或扇入。框架提供了对这些结构、资源间的关系建模和映射方法。框架中资源结构的主要构件是集合。一个集合表示一个容器，该容器是可以扩展的，以便对复杂结构进行建模。

3.6.1 资源关系

资源关系表示为链接。链接包含并扩展了典型的Web链接概念，作为一种表示资源之间关系的方法。一个链接包含一系列参数，这些参数定义如下。

(1) 一个上下文URI。

(2) 一个目标URI。

(3) 从上下文URI到目标URI的关系。

(4) 提供关于目标URI、链接的关系或上下文元数据的元素。

除了目标URI是强制的外，链接中的其他项是可选的。链接中的附加项可能会根据不同上下文中链接的使用而被设置为强制的（例如，在集合中、发现中和桥接中等）。

链接的例子如下所示。

```
{"href":"/switch","rt":["oic.r.switch.binary"],"if":["oic.if.a",/room2"oic.if.baseline"], "p":{"bm":
3}, "rel":"item"}
```

只要有一个参数不同，这两个链接就是不同的。例如，以下两个链接是不同的，可以在同一个链路列表中出现。

```
{"href": "/switch", "rt": ["oic.r.switch.binary"], "if": ["oic.if.a","oic.if.baseline"], "p":{"bm":2},
"rel": "item"}
{"href": "/switch", "rt": ["oic.r.switch.binary"], "if": ["oic.if.a","oic.if.baseline"], "p":{"bm":2},
"rel": "activates"}
```

当需要特定的能力时，该规范可能要求很多参数和参数值。对于在"/oic/res"上的检索请求返回的所有链接，如果一个链接没有显式包括"rel"参数，则默认 rel＝hosts。hosts 相关值是由 IETF RFC 6690 定义，并在用于链接关系的 IANA（Internet Assigned Numbers Authority，互联网数字分配机构）进行注册，网址为 http://www.iana.org/assignments/link-relations/link-relations.xhtml。

一个链接中上下文 URI 和目标 URI 之间的关系使用"rel"JSON 元素进行声明，并且该元素的值指明了特定的关系。

链接的上下文 URI 应该隐式地成为包含链路资源（或者是一个集合）的 URI，除非该链路指明了一个"anchor"参数。"anchor"用于改变一个链路的上下文 URI——如果该参数被指明的话，与目标 URI 的关系是基于该参数 URI 的。该参数使用 OIC 1.1 链接的传输协议 URI（如"anchor"："coaps://[fe80::b1d6]:44444"）和 OCF 1.0 链接（如"anchor""ocf://dc70373c-1e8d-4fb3-962e-017eaa863989"）。

在集合的上下文中使用"anchor"的例子如下。一层楼有很多房间，房间里有灯可能会作为链接被定义在房间中，但是这些链接应该有"anchor"参数，该参数被设置为包含该灯房间的 URI（关系为包含关系）。这就允许一层楼上所有的灯同时被打开或关上，同时仍然有灯在定义时与包含它们的房间相关（灯也可以通过使用房间 URI 打开）。在链接中使用"anchor"的示例如下。

```
/a/floor {
  "links": [
    {
      "href": "/x/light1",
      "anchor": "/a/room1",
      //注意:/a/room1 具有与/x/light1 的关系,不是/a/floor
      "rel": "item"
    }
  ]
}
/a/room1 {
  "links": [
    {
    //注意:/a/room1 包含/x/light,由于/a/room1 是隐式的 URI
    "href": "/x/light1",
    "rel": "item"
    }
  ]
}
```

1. 参数

一个链接包含一系列参数，本节介绍"ins"（链接实例）参数、"p"（策略）参数、"type"（媒体类型）参数、"bp"（批处理接口）参数、"di"（设备 ID）参数和"eps"参数。

1）"ins"参数

"ins"参数指明了一个链接列表中一个特定的链接实例。"ins"参数常被用于在一个链接列表中修改或删除一个特定的链接。当一个链接被拥有链接列表的 OCF 设备（服务）实例化时，"ins"参数的值被设置，一旦被设置，只要该链接是其列表的成员，“ins”参数就不会被修改。

2）“p”参数

策略参数定义了正确获取一个目标URI所引用资源的各种规则，策略规则是由一系列键值对配置的，策略参数“p”定义如下。

“bm”键：对应于一个整型值，该整型值被解释为一个8位掩码。位掩码中的每一位对应于一个特定的策略规则，为“bm”指定了规则，如表3-10所示。

表3-10 “bm”键指定规则

比特位置	策略规则	注释
比特0	可发现	定义了链接是否会通过oic/res被包含在资源的发现消息中，如果链接会被包含在资源发现消息中，“p”会包括“bm”键，并且将可发现比特的值置为1。如果资源发现消息中不包括该链接，则“p”可以选择包含“bm”键并将可发现值置为0或者是省略“bm”键的值
比特1	可观察	定义了目标URI所引用的资源是否支持通知操作。如果资源支持通知操作，则“p”应该包括“bm”键，并且将可观察位的值置为1；如果该资源不支持通知操作，“p”可以选择包括“bm”键并将可观察位设置为0或者是省略“bm”键
比特2～7		保留用于以后使用，“bm”中的所有保留位的值都应该被设置为0

注意，如果“bm”中的所有比特值都置为0，则为提高效率，可以从“p”中将“bm”键完全省略。然而，如果有任何一比特值被置为1，则“p”就应该包含“bm”，所有的比特都应该被合适地定义。

“sec”和“port”仅在OIC 1.1有效载荷中使用。在OCF 1.0有效载荷中，不应使用“sec”和“port”，而是通过“eps”参数提供加密链接的信息。

“sec”键：对应于一个布尔值，该值定义了目标URI引用的资源是否是通过一种加密链接被获取到的。如果“sec”为真，则是通过加密链接访问资源，使用“port”规范（见下文）；如果“sec”为假，则资源是通过未加密的链接或通过加密链接来访问资源（如果这样的链接是使用另一个资源的“port”设置进行的，“sec”为真）。

“port”键：对应于一个整型值，该值用于指明目标URI所引用的资源可以通过加密链接被获取到的端口号。

如果该资源只能通过加密链接访问（例如DTLS），那么“p”中应该包括“sec”的值并设为真。“p”中应包含“port”键，并且将“port”的值设置为用于获取资源的端口号。

如果资源不是通过加密链接访问的，那么“p”应该包括“sec”键，它的值应该为假。或者“p”应该省略“sec”键；默认的“sec”值为假。“p”应该省略“port”键。

若该资源既不是通过加密链接也不是通过非加密链接访问的，则遵循本节定义的补充方案。通过端口键指定端口上的资源访问应通过加密链接（例如“coaps：//”）进行（请注意，在通过多播发现的单独端口上可能存在与资源的未加密链接）。

请注意，资源的访问由资源的接入控制列表控制，成功的加密链接不能确保所请求的操作成功。它是OCF安全访问控制的一部分。

例3-1：对于一个可以被发现但不可被观察，可以通过CoAPS端口33275进行认证访问的资源策略，参数示例如下。

```
"p": { "bm":1 }
```

例 3-2：如下展示一个自链接，例如，在自身内部的可以被发现和观察的"/oic/res"链接。

```
{
"href":"/oic/res",
"rel":"self",
"rt":["oic.wk.res"],
"if":["oic.if.ll","oic.if.baseline"],
"p":{"bm":3}
}
```

3）"type"参数

"type"参数可以用于指明一个特定的目标资源所支持的各种媒体类型。当"type"元素被省略时，"application/cbor"类型的默认值会被使用。一旦一个客户端对每个资源都发现了该信息，那么在请求或响应合适的头部域内就会选择其中可使用的一个。

4）"bp"参数

"bp"参数用于指定对目标 URI 的修改，因为批处理请求是通过该链路转发的。值中的"q"元素定义了要附加到"href"的查询字符串以构成目标 URI。"q"查询字符串可以包含在该上下文中有效的多个属性字符串。例如，给定一个集合如下。

```
/room2
{
"if": "oic.if.b",
"color": "blue",
"links":
[
  {"href": "/switch", "rt": ["oic.r.switch.binary"], "if": ["oic.if.a", "oic.if.baseline" ],"p":{"bm":
2}, "rel": "contains", "bp": { "q": "if = oic.if.baseline"}
  }
]
}
```

下面是到"/room2"的批处理请求序列 GET /room2? if＝oic. if. b，当批处理请求通过链接被传播到目标"/switch"时，该请求被传输到 GET /switch? if＝oic. if. baseline。

5）"di"参数

"di"参数指定了承载"href"参数中定义目标资源的设备 ID。设备 ID 可以用于限定在"href"中的相对引用或查找相对引用的端点信息。

6）"eps"参数

"eps"参数指出了目标资源的终端信息。"eps"用一个数组作为其值，每一个数组的"ep"和"pri"组合代表一个终端信息。"ep"是强制性的，"pri"是可选的。

多终端的"eps"示意如下。

```
"eps": [
  {"ep": "coap://[fe80::b1d6]:1111", "pri": 2},
  {"ep": "coaps://[fe80::b1d6]:1122"},
  {"ep": "coap + tcp://[2001:db8:a::123]:2222", "pri": 3}
]
```

当链接中存在"eps"时，可以使用它的终端信息来访问由"href"参数引用的目标资源。当"eps"存在时，最大时限信息决定了“eps”的存在时间。

2. 格式及链接列表

在JSON中，链接列表的格式是一个数组。资源中的链接列表将作为该资源的链路属性的值包括在该资源中。一个包含链接的资源是一个集合。具有链接列表的资源如下所示。

```
/Room1
{
    "rt": ["my.room"],
    "if": ["oic.if.ll", "oic.if.baseline" ],
    "color": "blue",
    "links":
    [
  {
    "href": "/oic/d",
    "rt": ["oic.d.light", "oic.wk.d"],
    "if": [ "oic.if.r", "oic.if.baseline" ],
    "p": {"bm": 1}
  },
  {
    "href": "/oic/p",
    "rt": ["oic.wk.p"],
    "if": [ "oic.if.r", "oic.if.baseline" ],
    "p": {"bm": 1}
  },
  {
    "href": "/switch",
    "rt": ["oic.r.switch.binary"],
    "if": [ "oic.if.a", "oic.if.baseline" ],
    "p": {"bm": 3},
    "mt": [ "application/cbor", "application/exi + xml" ]
  },
  {
    "href": "/brightness",
    "rt": ["oic.r.light.brightness"],
    "if": [ "oic.if.a", "oic.if.baseline" ],
    "p": {"bm": 3}
  }
]
}
```

3.6.2 集合

一个包括一或多个引用其他的资源被称作一个集合。这些引用可能是相关的，也可能只是一个列表；并且该集合使用单个句柄来引用整个集合的方法。任何资源都可以通过建立资源间的链接和指向变成集合。集合可以使用在层次、序号、组别等创建、定义和具体化中。

在一个集合的生命周期中，最少需要固定一个资源类型和一个接口。在集合创建之初就需要将资源类型和接口绑定至这个资源，并且这些初始的变量可以用类似改写资源变量的方式进行改写。绑定后加入的资源类型和接口，则应该在集合生命周期中完成。

集合需要定义共有链路属性，这个属性的值应为一个包含零条或多条链接的数组。链接中的目标

URI 应该指向另一个集合或资源，其中，被指向的集合或资源可以与链接的发起方在同一台设备上，这样的链接称为本地指向；被指向的资源或集合也可以挂载在另一台设备上，此时的链接称作远程访问。在链路数组中代表链接的 URI 内容应该是(隐含)那些包含链路属性的集合。隐含的 URI 内容可以在之后通过链接中的“anchor”属性进行改写，这里的“anchor”属性值是链接的基础。

一个资源可以被多个集合链接，因此，这样的链接和指向关系并不能保证唯一确定的“父子”关系。在一个确定的集合与其指向的资源间并没有预先定义好的关系，例如，可以使用集合来展现一个具体的关系，但这些关系并不是自动定义或隐含的。集合的生命周期和所指向的资源生命周期也是相互独立的。

如果集合的“drel”属性已经定义，那么所有没有指定联系的链接则会继承在这个资源配置指令中的默认值。关系的默认值定义了链接中集合和目标 URI 之间隐含的关系。

链路属性表示集合中的链接列表。链路属性有一个条目数组，每个条目都是 OCF 的链接，如图 3-2 所示。

```
---------------------------------------------------------------
/my/house                                                      } URI(资源)
---------------------------------------------------------------
{                                                              }
  "rt":["my.r.house"],                                         } 属性(资源)
  "color":"blue",                                              }
  "n":"myhouse",
"links":[ ------------------------------------
     {                                                         }
       "href":"/door",                                         } 参数(链接)
       "rt":["oic.r.door"],                                    }
       "if":["oic.if.b","oic.if.ll","oic.if.baseline"]
     }, ----------------------------------------------------
     {
     "href":"/door/lock",
     "rt":["oic.r.lock"],
     "if":["oic.if.b","oic.if.ll","oic.if.baseline"],
     "type":["application/cbor","application/exi+xml"]
  }
```

图 3-2 集合和链接示例

集合可以预先定义，也可以被优先定义，并且在这个集合的生命周期中都是静态(稳定)的全局变量。这样的集合可以用于模型(此处指一个案例)，例如，一个由多个设备或一些固定资源组成的应用可以实现固定功能。

一个集合若只能用于配置在这个 OCF 集合里的设备上，那么该集合称为这个设备的本地集合，这样的集合也可用作客户端上指向多个服务器端的简便形式。集中化的集合是指配置在一个 OCF 设备上，并能被其他 OCF 设备访问或更新的集合。宿主集合属于集中化集合，但由一个或多个代理服务器管理控制。

1. 集合的属性

集合中需要定义链路属性。除此之外，还需要通过资源类型定义多个其他的属性。一些强制的、推荐的集合公共属性如表 3-11 所示。当涉及的资源属性与作为资源定义的属性重复时，在集合中会重写这些属性。

表 3-11 OCF 集合的公共属性(作为公共属性的补充)

属 性	描 述	属性名称	属性值类型	是否强制
链路	集合中所有的链接	links	JSON、OCF 链接组成的数组	是
名称	便于人们理解的 OCF 名称	n	字符串	否
编号	OCF 集合的编号	id	UUID	否
资源类型	资源中允许链接使用的资源类型的列表。对于这个列表,增加使用链接列表或链接批处理接口链接的请求是有效的。如果这个属性没有定义或者是空字符串,则任何资源类型都是可用的	rts	CSV 格式:由逗号分隔的列表,列表中是 OCF 资源的类型名称	否
默认关系	具体规定用于集合中的 OCF 链接默认的关系,其中“rel”参数并没有被明确地定义。允许集合和 OCF 链接中没有定义“drel”属性和“rel”属性	rel	字符串	否

2. 默认资源类型

默认资源类型“oic. wk. co”在 OCF 集合中可用。这种资源类型只能用于集合中没有定义别的资源类型或在集合创建的时候没有声明资源类型的情况下。

默认资源类型支持公共属性和链路属性。在默认资源类型中,链路属性的值应该是一个简单的 OCF 链接构成的数组,不支持使用含有标记的链接。

默认资源类型需要支持基准接口和链路列表接口,默认的接口应该是链路列表接口。

3.7 第三方指定扩展

本节介绍第三方如何将设备类型、资源类型、第三方定义的属性添加到现有或第三方定义的资源类型、枚举值和属性中去。

第三方可以规定 OCF 设备中的附加(非 OCF)资源。第三方还可以在现有 OCF 定义的资源类型中规定其他属性。此外,第三方可以使用它定义的值,扩展 OCF 定义的枚举。

第三方定义的设备类型可能会暴露第三方和 OCF 定义的资源类型。第三方定义的设备类型必须暴露本规范中定义的所有 OCF 设备必须公开的资源。

被第三方定义的资源类型应包括本规范中定义的任何强制性属性以及任何垂直领域指定的强制属性。OCF 命名空间资源类型中被第三方定义的资源属性,不属于 OCF 规范中定义的常用属性,应该遵循表 3-12 中第三方定义的属性规则。

表 3-12 第三方定义的属性规则

第三方定义	资 源 元 素	供应商定义规则
设备类型	/oic/d 的资源类型属性值	x. <Domain_Name>. <resource identification>
资源类型	属性值	x. <Domain_Name>. <resource identification>
OCF 命名空间中的属性	资源属性值	x. <Domain_Name>. <property>
OCF 枚举规范中的值	枚举属性值	x. <Domain_Name>. <enum value>
OCF 属性规范中的参数值	参数键值	x. <Domain_Name>. <parameter keyword>

关于在此方案中使用的Domain_Name,其标签与它们在DNS或其他解析机制中的表现方式相反。第三方定义的设备类型和资源类型请遵循资源类型属性中定义的规则。第三方定义的资源类型应在IANA约束下的RESTful参数注册表中注册。举例如下。

```
x.com.samsung.galaxyphone.accelerator;
x.com.cisco.ciscorouterport;
x.com.hp.printerhead;
x.org.allseen.newinterfae.newproperty。
```

第4章 OCF资源的操作

CHAPTER 4

本章主要介绍 OCF 资源操作的方法。在 OCF 标准中,主要有五种操作,分别是创建、检索、更新、删除和通知。这是定义对 OCF 资源的几个操作,一般称为 CRUDN。OCF 客户端可以通过这些方法操作 OCF 服务器端上的资源。

4.1 概述

CRUDN 操作使用了在消息中携带的一系列参数,这些参数在表 4-1 中进行了定义。OCF 设备可以使用 CBOR 作为默认负载内容的编码方案,用于 CRUDN 操作和操作响应中包含的资源表示;同时,也可以协商出一个新的负载编码方案(例如 CoAP 消息)。本章的内容将会明确 CRUDN 的具体操作及相关参数的使用。这些操作的类型定义,在每个协议中都会映射到一个消息块中。

表 4-1 CRUDN 消息的参数

适用范围	名称	意　义	定　义
所有消息	fr	发送者	消息发送者的 URI
	to	接收者	消息接收者的 URI
	ri	请求标识符	唯一标识消息发送者和接收者间消息的标识符
	cn	内容	具体操作的信息
请求	op	操作	具体要求服务器端执行的操作
	obs	观察	观察请求的标志
回复	rs	响应代码	请求结果的标志,表示请求是否被接收以及操作的结果。CRUDN 操作的返回代码应遵守 IETF RFC 7252 规范
	obs	观察	观察响应的标志

4.2 创建

创建操作用来在服务器端请求新建一个资源。客户端初始化的创建操作包括三个步骤,如图 4-1 所示。

1. 创建请求

创建请求消息由 OCF 客户端发送到 OCF 服务器端,并由 OCF 服务器端创建新的资源。创建请求消息将包含以下参数。

(1) fr:OCF 客户端的唯一标识符。

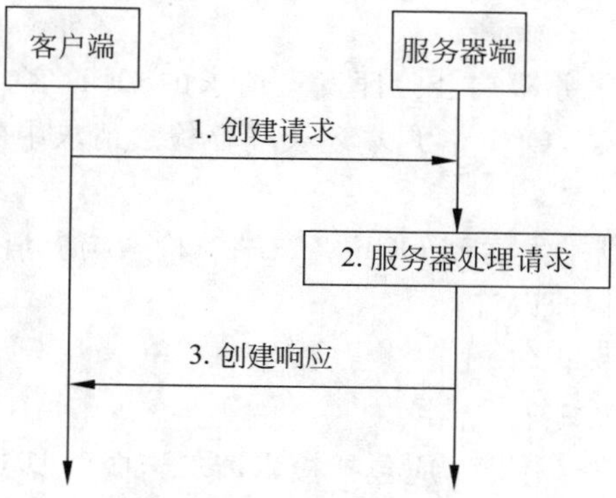

图 4-1　创建操作

（2）to：负责创建新目标资源的 URI。

（3）ri：创建请求的标识符。

（4）cn：有关服务器端将要创建的资源信息，包括将创建资源的 URI 和资源类型属性，可能包括将被创建资源的其他属性。

（5）op：创建。

2. OCF 服务器端处理请求

在收到创建请求之后，服务器端会验证发送请求的客户端是否具有创建所需资源的权限。如果有，服务器端不仅创建要求的资源，还会缓存创建请求中的 ri 参数，并在创建响应中使用。

3. 创建响应

创建响应消息由服务器端发送到客户端。创建响应消息将包含以下参数。

（1）fr：服务器端的唯一标识符。

（2）to：客户端的唯一标识符。

（3）ri：创建请求中包括的标识符。

（4）cn：有关服务器端已创建的资源信息，包括已创建资源的 URI 和已创建的资源表示。

（5）rs：创建操作的结果。

4.3　检索

检索操作用来请求现有的资源状态或表示。客户端初始化的检索操作包括三个步骤，如图 4-2 所示。

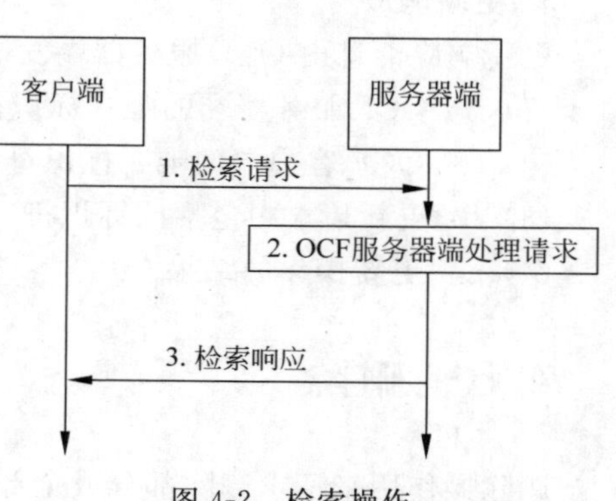

图 4-2　检索操作

1. 检索请求

检索请求消息由 OCF 客户端发送到 OCF 服务器端，以请求 OCF 服务器端上的 OCF 资源表示。检索请求消息将包含以下参数。

（1）fr：OCF 客户端的唯一标识符。

（2）to：OCF 客户端指向资源的 URI。

（3）ri：检索请求的标识符。

（4）op：检索。

2. OCF 服务器端处理请求

在收到检索请求之后，OCF 服务器端会验证发送请求的 OCF 客户端是否具有获取所需资源的权限，以及资源的有关属性是否可读。OCF 客户端还会缓存检索请求中的 ri 参数，并在检索回复中使用。

3. 检索响应

检索响应消息由 OCF 服务器端发送到 OCF 客户端。检索响应消息将包含以下参数。

（1）fr：OCF 服务器端的唯一标识符。

（2）to：OCF 客户端的唯一标识符。

（3）ri：检索请求中包括的标识符。

（4）cn：OCF 客户端请求的资源信息，应包括检索请求指向资源 URI。

（5）rs：检索操作的结果。

4.4 更新

更新操作用来请求替换部分或全部 OCF 资源的信息。OCF 客户端初始化的更新操作包括三个步骤，如图 4-3 所示。

1. 更新请求

更新请求消息由 OCF 客户端发送到 OCF 服务器端，以请求更新 OCF 服务器端上的 OCF 资源信息。更新请求消息将包含以下参数。

（1）fr：OCF 客户端的唯一标识符。

（2）to：OCF 客户端指向需要更新信息的资源 URI。

（3）ri：更新请求的标识符。

（4）op：更新。

（5）cn：信息，包括目标资源上需要更新的资源属性。

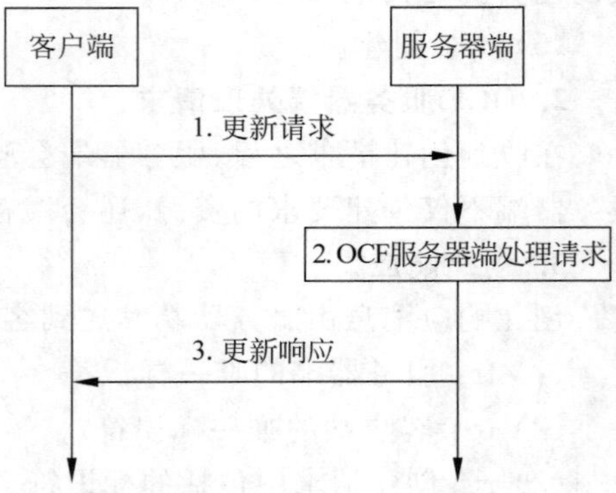

图 4-3　更新操作

2. OCF 服务器端处理请求

在收到更新请求之后，OCF 服务器端会验证发送请求的 OCF 客户端是否具有更新有关资源的权限。如果有，OCF 客户端就会根据更新请求消息中 cn 参数的值来更新目标资源的信息。OCF 客户端还会缓存更新请求中的 ri 参数，并在更新响应中使用。

3. 更新响应

更新响应消息由 OCF 服务器端发送到 OCF 客户端，更新响应消息将包含以下参数。

（1）fr：OCF 服务器端的唯一标识符。

（2）to：OCF 客户端的唯一标识符。

（3）ri：更新请求中包括的标识符。

（4）rs：更新操作的结果。

4.5 删除

删除操作用来请求删除部分或全部 OCF 资源的信息。OCF 客户端初始化的删除操作包括三个步骤，如图 4-4 所示。

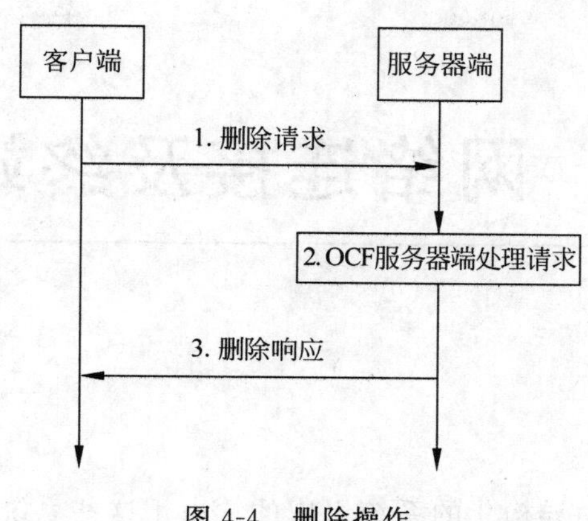

图 4-4 删除操作

1. 删除请求

删除请求消息由 OCF 客户端发送到 OCF 服务器端，以删除 OCF 服务器端上的 OCF 资源。删除请求消息将包含以下参数。

(1) fr：OCF 客户端的唯一标识符。

(2) to：OCF 客户端指向需要删除的资源 URI。

(3) ri：删除请求的标识符。

(4) op：删除。

2. OCF 服务器端处理请求

在收到删除请求之后，OCF 服务器端会验证发送请求的客户端是否具有删除有关资源的权限，以及相关资源是否存在。如果验证通过，OCF 客户端就会删除请求资源以及所有相关的信息。OCF 客户端还会缓存删除请求中的 ri 参数，并在删除响应中使用。

3. 删除响应

删除响应消息由 OCF 服务器端发送到 OCF 客户端。删除响应消息将包含以下参数。

(1) fr：OCF 服务器端的唯一标识符。

(2) to：OCF 客户端的唯一标识符。

(3) ri：删除请求中包括的标识符。

(4) rs：删除操作的结果。

4.6 通知

通知操作用来请求状态改变的异步通知，使用通知响应信息。通知消息由 OCF 服务器端发送到 OCF 客户端，以通过 OCF 客户端上的 URL 告知 OCF 客户端有状态改变。通知响应消息将包含以下参数。

(1) fr：OCF 服务器端的唯一标识符。

(2) to：需要通知消息 OCF 目标资源的 URI。

(3) ri：通知请求中包括的标识符。

(4) op：通知。

(5) cn：更新后的资源状态。

第 5 章 网络连接及终端发现

CHAPTER 5

OCF 所处的物联网环境是由异构化的系统组成的。由于这些系统通常被定制成处理专用需求的系统，所以它们都是由非常多样的产品和服务组成的。这些产品的范围很广，既涉及有限的、只能依靠电池运行的设备，也涉及用户可以从市场上购买到的日常使用的科技设备。现阶段缺少并亟待创立一个全球化的标准，以使得致力于研究 OCF 的不同项目组可以在一个通用网络标准下进行精简操作。

IETF 发现了市场的变化并意识到了 IPv4 已经不能满足使用需求。不只是新的科技领域需要新技术的支持，管理更多样的设备、日益复杂的多种子网、更高的安全和隐私要求也需要一系列新技术标准的出现。认识到物理层/数据链路层的存在需求后，IETF 建立了专门的工作组来精简、提炼各种现有的网络层技术。根据这些市场的现实情况，这个规范也意味着可以充分利用现有的无线网（如蓝牙、WiFi 或 802.15.4），并集中研究网络层和由 IETF 所产生的相关协议。

5.1 网络连接架构

IPv4 中心网络已经发展到支持复杂的拓扑结构，其部署主要由单一的互联网服务提供商作为单一的网络提供。而常出现于家居住宅的更复杂的网络拓扑，大多是通过收购更多的家庭网络设备实现的，这依赖于技术的支持，如私有网络地址转换等。这些技术在搭建和设置时需要专业人员的帮助，并应避免在家庭网络中使用，因为它们经常导致路由结构、命名和发现等服务的故障。

多段生态系统的 OCF 地址，不仅会引发新设备和有关路由器的激增，同时也会增加那些引入额外边缘路由器的新服务。所有这些新的要求都需要先进的系统架构，以解决复杂的网络拓扑，如图 5-1 所示，深色的部分表示非 OCF 部分。

图 5-1 中所示的设备承担以下几个角色之一。

(1) IETF RFC 6434 CE 路由器（用户端边缘路由器）中定义的 IPv6 节点，IPv6 路由器。

(2) IPv6 主机：在 IETF RFC 7084 中具体定义。

(3) 6LN(6LoWPAN 节点)、6LR(6LoWPAN 路由器)、6LBR(6LoWPAN 边界路由器)，在 IETF RFC 6775 中定义。

(4) IPv6 转换器，用以在 IPv6 网络、非 IPv6 网络间翻译和路由相关的设备，图 5-1 中的网关就是一个转换器的实例。

(5) 约束节点：由于受约束的环境（有限的处理能力、存储器、非易失性存储介质和传输容量）需要 IP 网络层下的特别适配层，并需要专门的路由协议的节点。例如，在低功率下传输的设备、IEEE 802.14.5、ITU G9959、低功耗蓝牙和 NFC 等。

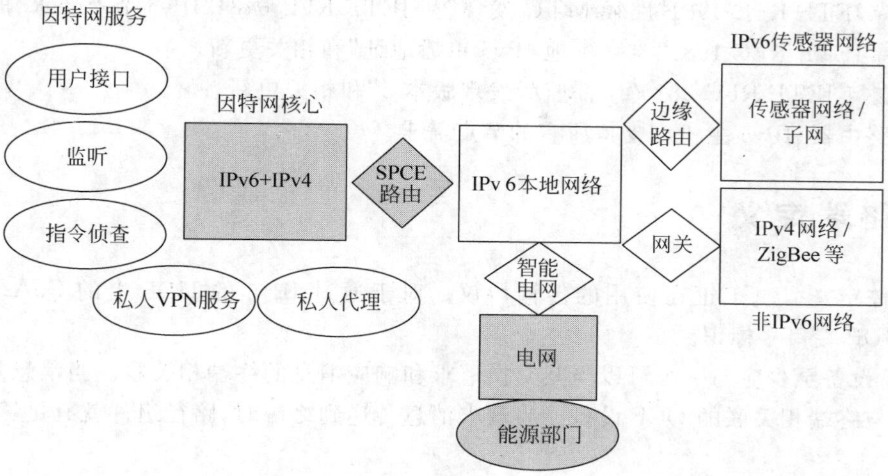

图 5-1 高层网络和连接架构

5.2 IPv6 网络层需求

预测表明，数百亿新的物联网终端及相关服务将在未来几年内联机。这些端点功能范围将从使用电池供电的具有有限的计算、存储和带宽的节点跨越到拥有更丰富的资源，通过以太网和 WiFi 链路工作的器件。

大约 30 年前部署的互联网 IPv4 已经成熟，并支持多种应用，如 Web 浏览、电子邮件、语音、视频和关键系统的监测和控制。但是，IPv4 的能力濒临用尽，并不仅仅只是可用地址空间已被消耗的程度。

IETF 开发 IPv4 的继任者 IPv6。OCF 建议在网络层使用 IPv6。其原因如下。

(1) 更大的地址空间，大大减少网络接入转换的需要。

(2) 更灵活的地址结构，每个结构可以使用多个地址和类型，如本地链路、ULA、GUA 和各种范围的组播地址等；更好地支持多归属网络，拥有更好的重新编号能力等。

(3) 更强大的自动配置功能，如 DHCPv6、SLAAC 和路由器发现等。在技术约束节点上实现 IP 连接的操作也是基于 IPv6 的。

(4) 所有主流的消费者操作系统，如 iOS、Android、Windows 和 Linux 都已经支持 IPv6。全球各地的主要服务提供商也都已经部署 IPv6。

为了保证网络层服务从节点到节点的互操作性，在所有节点上强制统一公共网络层协议是至关重要的。该协议应使网络能够成为安全的、可管理的、可扩展的网络，并包括约束节点和自组网状节点。OCF 建议使用 IPv6 作为公共的网络层协议，以保证所有 OCF 设备间的互操作性。本章将关注 IPv6 主机、约束主机和路由器的互操作需求。

IPv6 节点应支持 IPv6。若一个节点支持 IPv6，则应该遵守以下在本地网络中通信的要求。

(1) 应支持 IETF RFC 2460"IPv6 规范"和类似 IETF RFC 6434"IPv6 节点要求"的相关更新。

(2) 应支持 IETF RFC 4291"IPv6 寻址体系结构"和类似 IETF RFC 6434"IPv6 节点要求"的相关更新。

(3) 应支持 IETF RFC 4861"IPv6 邻近发现"和 IETF RFC 6434 "IPv6 节点要求"。

(4) 应支持 IETF RFC 1981“路径 MTU 发现”和 IETF RFC 6434“IPv6 节点要求”的相关更新。

(5) 应支持 IETF RFC 1981“唯一本地 IPv6 单播地址”和相关更新。

(6) 应当支持 IETF RFC 3810“组播监听发现版本 2”和相关更新。

(7) IPv6 路由器、IPv6 主机应支持所有的节点需求。

5.3 终端定义

终端的具体定义取决于正在使用的传输协议。对于通过 IPv6 的 UDP 上的 CoAP 示例，终端由 IPv6 地址和 UDP 端口号标识。

每个 OCF 设备至少应与一个可以与其交换请求和响应消息的终端相关联。当消息发送到终端时，它将被传递到与终端相关联的 OCF 设备。当请求消息传递到终端时，路径组件就有足够的能力找到目标资源。

OCF 设备能与多个终端相关联。例如，一个 OCF 设备可以拥有几个 IP 地址或者端口号，它也可以同时支持 HTTP 协议和 CoAP 协议。

另外，当有一种方法能去清楚地用 URI 指定目标资源时，一个终端也可以被多个 OCF 设备共享。例如，当一个 CoAP 服务器端对托管于自身的资源使用了唯一不同路径，那么它就可以被多个 OCF 设备共享。然而，这对于 OCF 1.0 和 OCF 1.1 是不可能的，因为一些预定义 URI（如“oic/d”）对于某些资源是强制性的。

终端由终端信息来表示。其中，终端信息是由“ep”和“pri”两个键值对组成。

1. “ep”

“ep”表示传输协议和终端定位器，指定如下内容。

(1) 传输协议（例如 CoAP + UDP + IPv6）的组合，可以与 RESTful 操作（即 CRUDN）交换请求和响应消息。传输协议套件应由 IANA 注册方案表示，还允许供应商或 OCF 定义的方案（如“org. ocf. foo”或“com. samsung. bar”）。

(2) 终端定位器，通过该地址（如 IPv6 地址＋端口号）可以将消息发送到终端，然后将相关联的 OCF 设备发送到该地址。“CoAP”“CoAPS”“CoAP＋TCP”“CoAPS＋TCP”“HTTP”和“HTTPS”的终端定位器应指定为“IP 地址＋端口号”。不应使用临时地址，因为终端定位器是为了接收传入的会话，而临时地址用于启动传出会话。此外，它包含在“/oic/res”中可能会导致隐私问题。

(3) “ep”应具有一个如方案组件所说明的传输协议的 URI。例如，"ep":"coap://[fe80::b1d6]:1111"。

各传输协议中的“ep”值如表 5-1 所示。

表 5-1 各传输协议中的“ep”值

传输协议	方案	终端定位器	“ep”值示例
CoAP＋UDP＋IP	CoAP	IP 地址＋端口号	coap://[fe80::b1d6]:1111
CoAPS＋UDP＋IP	CoAPS	IP 地址＋端口号	coaps://[fe80::b1d6]:1122
CoAP＋TCP＋IP	CoAP＋TCP	IP 地址＋端口号	coap＋tcp://[2001:db8:a::123]:2222
CoAPS＋TCP＋IP	CoAPS＋TCP	IP 地址＋端口号	coaps＋tcp://[2001:db8:a::123]:2233
HTTP＋TCP＋IP	HTTP	IP 地址＋端口号	http://[2001:db8:a::123]:1111
HTTPS＋TCP＋IP	HTTPS	IP 地址＋端口号	https://[2001:db8:a::123]:1122

2. "pri"

当有多个终端的时候,"pri"用于指出它们之间的优先级。"pri"应当由一个正整数来表示(例如,"pri":1),而且值越小,优先级越高。默认的"pri"值是1,例如,当"pri"没有表示出来时,它应当等于1。

3. "eps"参数中的终端信息

为了传输终端信息,在第3章中定义了一个新的链接参数"eps"。"eps"中以项目数作为其值,每个项目以"ep"和"pri"两个键值对来表示终端信息,其中,"ep"是必需的,"pri"是可选的。具有"eps"的链接如下。

```
{
"anchor": "ocf://light_device_id",
"href": "/myLightSwitch",
"rt": ["oic.r.switch.binary"],
"if": ["oic.if.a", "oic.if.baseline"],
"p": {"bm": 3},
"eps": [{"ep": "coap://[fe80::b1d6]:1111", "pri": 2}, {"ep":
"coaps://[fe80::b1d6]:1122"}]
}
```

其中,"anchor"代表OCF主机设备,"href"代表目标资源,"eps"代表目标资源的两个终端。如果一个目标资源要求一个安全连接(如"coaps"),在OCF 1.0的有效载荷中"eps"参数应该被用于指出必要的信息(如端口号),因为"sec"和"port"只能在OCF 1.1的有效载荷中使用。

5.4 终端发现

终端发现被定义为一个客户端向一个OCF设备或者资源请求终端信息的一个过程。

1. 隐式发现

如果设备是CoAP消息的源(如"/oic/res"响应),则可以通过组合源IP地址和端口号形成设备的终端定位器。根据CoAP方案和默认的"pri"值,可以构建设备的终端信息。

换句话说,具有CoAP的"/oic/res"响应消息可以隐含携带响应设备的终端信息,反过来又可以使用相同的CoAP传输协议访问所有在主机上托管的资源。

2. 使用"/oic/res"响应进行显式发现

终端信息可以使用"/oic/res"中链接的"eps"参数明确指出。"/oic/res"响应可以隐式表示由响应设备托管目标资源的终端信息。但是,"/oic/res"可能会暴露属于另一个设备的目标资源。当链接目标资源的端点不能被隐式推断时,应包含"eps"参数提供客户端可以访问目标资源的显式终端信息。

这种方法适用于资源目录或桥接设备的"/oic/res",该设备通常携带另一台设备所承载的资源链接。下面是链接中"eps"参数的"/oic/res"响应。

```
[
{
    "anchor": "ocf://e61c3e6b-9c54-4b81-8ce5-f9039c1d04d9",
    "href": "/oic/res",
    "rel": "self",
    "rt": ["oic.wk.res"],
    "if": ["oic.if.ll", "oic.if.baseline"],
```

```
        "p": {"bm": 3},
        "eps": [{"ep": "coap://[2001:db8:a::b1d4]:55555"},
                {"ep": "coaps://[2001:db8:a::b1d4]:11111"}]
    },
    {
        "anchor": "ocf://e61c3e6b-9c54-4b81-8ce5-f9039c1d04d9",
        "href": "/oic/d",
        "rt": ["oic.wk.d", "oic.d.bridge"],
        "if": ["oic.if.r", "oic.if.baseline"],
        "p": {"bm": 3},
        "eps": [{"ep": "coap://[2001:db8:a::b1d4]:55555"},
                {"ep": "coaps://[2001:db8:a::b1d4]:11111"}]
    },
    {
        "anchor": "ocf://e61c3e6b-9c54-4b81-8ce5-f9039c1d04d9",
        "href": "/oic/p",
        "rt": ["oic.wk.p"],
        "if": ["oic.if.r", "oic.if.baseline"],
        "p": {"bm": 3},
        "eps": [{"ep": "coaps://[2001:db8:a::b1d4]:11111"}]
    },
    {
        "anchor": "ocf://e61c3e6b-9c54-4b81-8ce5-f9039c1d04d9",
        "href": "/mySecureMode",
        "rt": ["oic.r.securemode"],
        "if": ["oic.if.rw", "oic.if.baseline"],
        "p": {"bm": 3},
        "eps": [{"ep": "coaps://[2001:db8:a::b1d4]:11111"}]
    },
    {
        "anchor": "ocf://e61c3e6b-9c54-4b81-8ce5-f9039c1d04d9",
        "href": "/oic/sec/doxm",
        "rt": ["oic.r.doxm"],
        "if": ["oic.if.baseline"],
        "p": {"bm": 1},
        "eps": [{"ep": "coap://[2001:db8:a::b1d4]:55555"},
                {"ep": "coaps://[2001:db8:a::b1d4]:11111"}]
    },
    {
        "anchor": "ocf://e61c3e6b-9c54-4b81-8ce5-f9039c1d04d9",
        "href": "/oic/sec/pstat",
        "rt": ["oic.r.pstat"],
        "if": ["oic.if.baseline"],
        "p": {"bm": 1},
        "eps": [{"ep": "coaps://[2001:db8:a::b1d4]:11111"}]
    },
    {
        "anchor": "ocf://e61c3e6b-9c54-4b81-8ce5-f9039c1d04d9",
        "href": "/oic/sec/cred",
        "rt": ["oic.r.cred"],
        "if": ["oic.if.baseline"],
        "p": {"bm": 1},
        "eps": [{"ep": "coaps://[2001:db8:a::b1d4]:11111"}]
    },
```

```
{
    "anchor": "ocf://e61c3e6b-9c54-4b81-8ce5-f9039c1d04d9",
    "href": "/oic/sec/acl2",
    "rt": ["oic.r.acl2"],
    "if": ["oic.if.baseline"],
    "p": {"bm": 1},
    "eps": [{"ep": "coaps://[2001:db8:a::b1d4]:11111"}]
},
{
    "anchor": "ocf://e61c3e6b-9c54-4b81-8ce5-f9039c1d04d9",
    "href": "/myIntrospection",
    "rt": ["oic.wk.introspection"],
    "if": ["oic.if.r", "oic.if.baseline"],
    "p": {"bm": 3},
    "eps": [{"ep": "coaps://[2001:db8:a::b1d4]:11111"}]
},
{
    "anchor": "ocf://dc70373c-1e8d-4fb3-962e-017eaa863989",
    "href": "/oic/res",
    "rt": ["oic.wk.res"],
    "if": ["oic.if.ll", "oic.if.baseline"],
    "p": {"bm": 3},
    "eps": [{"ep": "coap://[2001:db8:a::b1d4]:66666"},
           {"ep": "coaps://[2001:db8:a::b1d4]:22222"}]
},
{
    "anchor": "ocf://dc70373c-1e8d-4fb3-962e-017eaa863989",
    "href": "/oic/d",
    "rt": ["oic.wk.d", "oic.d.light", "oic.d.virtual"],
    "if": ["oic.if.r", "oic.if.baseline"],
    "p": {"bm": 3},
    "eps": [{"ep": "coap://[2001:db8:a::b1d4]:66666"},
           {"ep": "coaps://[2001:db8:a::b1d4]:22222"}]
},
{
    "anchor": "ocf://dc70373c-1e8d-4fb3-962e-017eaa863989",
    "href": "/oic/p",
    "rt": ["oic.wk.p"],
    "if": ["oic.if.r", "oic.if.baseline"],
    "p": {"bm": 3},
    "eps": [{"ep": "coaps://[2001:db8:a::b1d4]:22222"}]
},
{
    "anchor": "ocf://dc70373c-1e8d-4fb3-962e-017eaa863989",
    "href": "/myLight",
    "rt": ["oic.r.switch.binary"],
    "if": ["oic.if.a", "oic.if.baseline"],
    "p": {"bm": 3},
    "eps": [{"ep": "coaps://[2001:db8:a::b1d4]:22222"}]
},
{
    "anchor": "ocf://dc70373c-1e8d-4fb3-962e-017eaa863989",
    "href": "/oic/sec/doxm",
    "rt": ["oic.r.doxm"],
```

```
        "if": ["oic.if.baseline"],
        "p": {"bm": 1},
        "eps": [{"ep": "coap://[2001:db8:a::b1d4]:66666"},
        {"ep": "coaps://[2001:db8:a::b1d4]:22222"}]
    },
    {
        "anchor": "ocf://dc70373c-1e8d-4fb3-962e-017eaa863989",
        "href": "/oic/sec/pstat",
        "rt": ["oic.r.pstat"],
        "if": ["oic.if.baseline"],
        "p": {"bm": 1},
        "eps": [{"ep": "coaps://[2001:db8:a::b1d4]:22222"}]
    },
    {
        "anchor": "ocf://dc70373c-1e8d-4fb3-962e-017eaa863989",
        "href": "/oic/sec/cred",
        "rt": ["oic.r.cred"],
        "if": ["oic.if.baseline"],
        "p": {"bm": 1},
        "eps": [{"ep": "coaps://[2001:db8:a::b1d4]:22222"}]
    },
    {
        "anchor": "ocf://dc70373c-1e8d-4fb3-962e-017eaa863989",
        "href": "/oic/sec/acl2",
        "rt": ["oic.r.acl2"],
        "if": ["oic.if.baseline"],
        "p": {"bm": 1},
        "eps": [{"ep": "coaps://[2001:db8:a::b1d4]:22222"}]
    },
    {
        "anchor": "ocf://dc70373c-1e8d-4fb3-962e-017eaa863989",
        "href": "/myLightIntrospection",
        "rt": ["oic.wk.introspection"],
        "if": ["oic.if.r", "oic.if.baseline"],
        "p": {"bm": 3},
        "eps": [{"ep": "coaps://[2001:db8:a::b1d4]:22222"}]
    },
    {
        "anchor": "ocf://88b7c7f0-4b51-4e0a-9faa-cfb439fd7f49",
        "href": "/oic/res",
        "rt": ["oic.wk.res"],
        "if": ["oic.if.ll", "oic.if.baseline"],
        "p": {"bm": 3},
        "eps": [{"ep": "coap://[2001:db8:a::b1d4]:77777"},
              {"ep": "coaps://[2001:db8:a::b1d4]:33333"}]
    },
    {
        "anchor": "ocf://88b7c7f0-4b51-4e0a-9faa-cfb439fd7f49",
        "href": "/oic/d",
        "rt": ["oic.wk.d", "oic.d.fan", "oic.d.virtual"],
        "if": ["oic.if.r", "oic.if.baseline"],
        "p": {"bm": 3},
        "eps": [{"ep": "coap://[2001:db8:a::b1d4]:77777"},
              {"ep": "coaps://[2001:db8:a::b1d4]:33333"}]
```

```
    },
    {
        "anchor": "ocf://88b7c7f0-4b51-4e0a-9faa-cfb439fd7f49",
        "href": "/oic/p",
        "rt": ["oic.wk.p"],
        "if": ["oic.if.r", "oic.if.baseline"],
        "p": {"bm": 3},
        "eps": [{"ep": "coaps://[2001:db8:a::b1d4]:33333"}]
    },
    {
        "anchor": "ocf://88b7c7f0-4b51-4e0a-9faa-cfb439fd7f49",
        "href": "/myFan",
        "rt": ["oic.r.switch.binary"],
        "if": ["oic.if.a", "oic.if.baseline"],
        "p": {"bm": 3},
        "eps": [{"ep": "coaps://[2001:db8:a::b1d4]:33333"}]
    },
    {
        "anchor": "ocf://88b7c7f0-4b51-4e0a-9faa-cfb439fd7f49",
        "href": "/oic/sec/doxm",
        "rt": ["oic.r.doxm"],
        "if": ["oic.if.baseline"],
        "p": {"bm": 1},
        "eps": [{"ep": "coap://[2001:db8:a::b1d4]:77777"},
               {"ep": "coaps://[2001:db8:a::b1d4]:33333"}]
    },
    {
        "anchor": "ocf://88b7c7f0-4b51-4e0a-9faa-cfb439fd7f49",
        "href": "/oic/sec/pstat",
        "rt": ["oic.r.pstat"],
        "if": ["oic.if.baseline"],
        "p": {"bm": 1},
        "eps": [{"ep": "coaps://[2001:db8:a::b1d4]:33333"}]
    },
    {
        "anchor": "ocf://88b7c7f0-4b51-4e0a-9faa-cfb439fd7f49",
        "href": "/oic/sec/cred",
        "rt": ["oic.r.cred"],
        "if": ["oic.if.baseline"],
        "p": {"bm": 1},
        "eps": [{"ep": "coaps://[2001:db8:a::b1d4]:33333"}]
    },
    {
        "anchor": "ocf://88b7c7f0-4b51-4e0a-9faa-cfb439fd7f49",
        "href": "/oic/sec/acl2",
        "rt": ["oic.r.acl2"],
        "if": ["oic.if.baseline"],
        "p": {"bm": 1},
        "eps": [{"ep": "coaps://[2001:db8:a::b1d4]:33333"}]
    },
    {
        "anchor": "ocf://88b7c7f0-4b51-4e0a-9faa-cfb439fd7f49",
        "href": "/myFanIntrospection",
        "rt": ["oic.wk.introspection"],
```

```
    "if": ["oic.if.r", "oic.if.baseline"],
    "p": {"bm": 3},
    "eps": [{"ep": "coaps://[2001:db8:a::b1d4]:33333"}]
  }
  ]
```


5.5 基于 CoAP 的终端发现

本节主要是一些基于 CoAP 终端发现的总结描述。

(1) 所有正在广播和发布的设备都应加入"所有 CoAP 节点"多播分组，如 IPv6 中的 FF0X: FD 或监听端口 5683。

(2) 需要发现资源的 OCF 客户端应首先加入"所有 CoAP 节点"多播分组。

(3) OCF 客户端应发送一个发现请求(GET 请求)给多播分组"所有 CoAP 节点"和端口 5683，请求中的 URI 应为/oic/res。

(4) 若 OCF 客户端正处于发现指定资源类型的过程中，则它应使用带有键"rt"的问询机制，"rt"的值应为需要发现的目标。

(5) 如果问询请求中不带有"rt"键，则所有的 OCF 设备都应回复这个请求。

(6) 处理多播请求的注意事项应与在 IETF RFC 7252 规范和 IETF RFC 6690 规范中定义的一样。收到请求的 OCF 设备应该使用 CBOR 作为负载(内容)编码方式进行回复。OCF 设备应使用 CBOR 作为额外的多播发现负载(内容)编码方式。OCF 设备也应该使用 CBOR 作为负载(内容)编码方式回复一个已收到支持 CBOR 的多播发现消息。在之后的版本中，可以被其他方式使用(如 JSON、XML/EXI 等)。

第6章 OCF的功能交互

CHAPTER 6

功能交互是指客户端和服务器端之间相互信息交换的过程。功能交互使用 CRUDN 消息,包括发现、通知和设备管理。这些功能需要支持表 6-1 中定义的核心资源。

表 6-1　核心资源列表

预定义的 URI	资源类型标题	相关的功能交互	要求
/oic/res	默认	发现	强制
/oic/p	平台	发现	强制
/oic/d	设备	发现	强制
/oic/rts	资源类型	发现	条件
/oic/ifs	接口	发现	条件
/oic/con	配置	设备管理	条件
/oic/mon	监测	设备管理	条件
/oic/mnt	维护	设备管理	条件

6.1　服务开通

在 OCF 框架中服务开通包括两个不同的过程：On-Boarding 和配置。On-Boarding 是向加入 OCF 网络的设备传递所需信息的过程。当 On-Boarding 过程完成时,设备具有必要的信息并且能够加入 OCF 网络(图 6-1 中的状态＃1)。配置是向设备提供所需信息以访问 OCF 服务的过程。在配置过程结束时,设备具有所有必要的信息,并且能够访问 OCF 服务(图 6-1 中的状态＃2)。

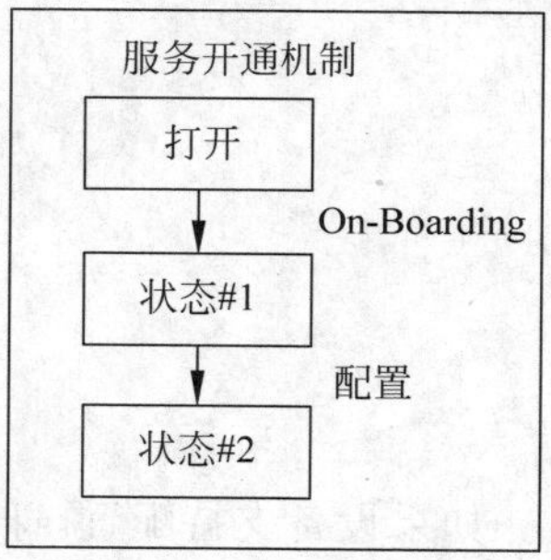

服务开通机制

打开

On-Boarding

状态#1

配置

状态#2

图 6-1　服务开通状态改变

1. ＃1 On-Boarding

OCF 框架适用于具有不同能力的多种类型设备,包括丰富的用户接口设备,这种设备可以从用户处接收输入,如智能电话；还包括没有接收用户输入装置的设备,如传感器。另外,设备可以支持不同的通信和连接技术,如蓝牙和 WiFi 等,不同的通信和连接技术提供特定的登录机制。

由于设备能力的差异和多样性,OCF 1.1 规范不要求特定的 On-Boarding 过程,而是在完成 On-Boarding 过程时指定设备的状态。

作为 On-Boarding 过程的一部分,设备会获取详细信息和所需的参数值(如 WiFi 的 SSID 以及认证证书),以便能够连接到网络,从而在 On-Boarding 过程结束时成功建立到网络的连接。OCF 规范的后续版本可以指定跨不同通信和连接技术的 On-Barding 的通用过程。

2. ♯2 配置

一旦设备成功连接到 OCF 网络，它需要额外的配置信息来访问 OCF 或订阅 OCF 服务。所需的信息可以包括地理位置、时区和安全要求等。该信息可以预先加载在设备上，或者可以从另一设备(如配置源)上的配置服务获取。关于配置服务资源的信息，如配置源的 URI，被预先配置在设备上。

配置信息也在核心资源"/oic/con"中。当完成 On-Boarding 过程并且设备连接到网络，如果没有预加载配置信息，则将启动配置过程，通过获取或推送的交互，并用当前配置的状态信息填充其指定的配置资源，通过这一过程设备获取相关配置信息。指定的配置资源保持最新的配置状态，配置的更新是通过指定资源来完成的。

如果配置信息未加载，则设备将从配置源检索它们。在设备的生存期间，客户端可以检索或更新设备的配置状态。某些配置信息是只读的，可以由配置源修改，具体取决于"/oic/con"资源中属性的访问模式。

图 6-2 描述了设备从配置源(其可以位于远程设备或本地)检索其配置信息而触发的交互。这些相互作用在完成 On-Boarding 过程后立即发生；设备可以在生存期内的任何时间检索其配置(见图 6-3)。图 6-4 表述了当一个设备的配置信息被一个客户端(如配置源)更新了的交互。

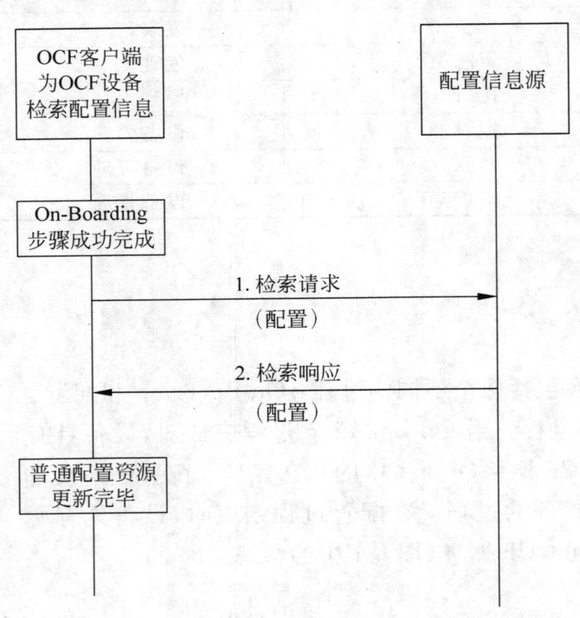

图 6-2　设备从配置源检索其配置信息而触发的交互

如果设备支持配置，即配置信息可以动态更新，则应支持核心资源"/oic/con"作为指定的配置资源，如表 6-2 所示。表 6-3 定义了"oic. wk. con"资源类型。

设备或平台可能最初是在引导中设置或提供信息进行配置。此外，根据变化的条件或上下文环境，设备和平台可以由引导后的外部代理进一步配置。核心资源"/oic/con"暴露可用于配置更改属性。

一个配置是通过设置与该配置中相关的全体属性来确定的。设置新配置的结果由该集合中特定属性的值确定。通过"/oic/con"设置新配置的启动进程，可能导致在其他资源中产生副作用。

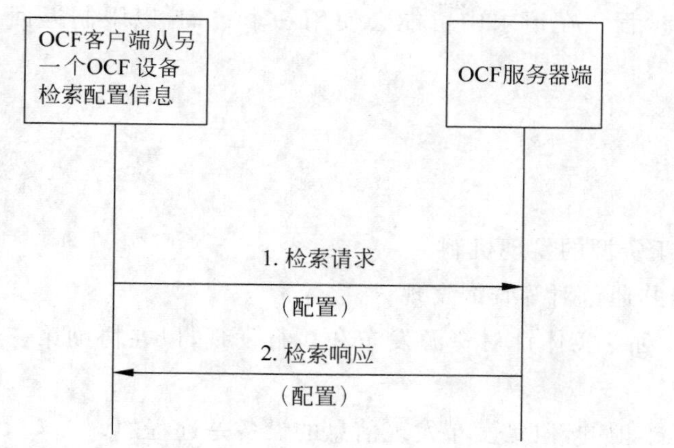

图 6-3 检索一个设备的配置状态的交互

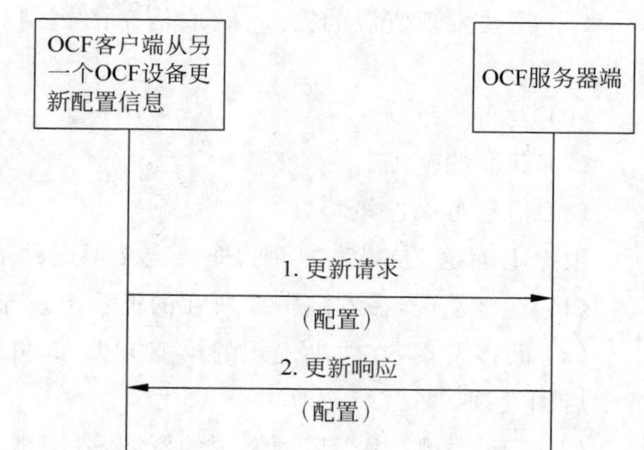

图 6-4 一个设备配置的更新

表 6-2 配置资源

预定义的 URI	资源类型标题	资源类型 ID（"rt"值）	接口	描述	相关的功能交互
/oic/con	配置	oic. wk. con	oic. if. rw	资源通过特定设备的可配置信息被暴露出来；由"/oic/con"暴露的资源属性在表 6-3 中列出	配置

表 6-3 "oic. wk. con"资源类型定义

属性	属性名	属性值类型	属性值规则	单位	访问模式	是否强制	描述
设备名	n	字符串			读写	是	终端用户可配置的人性化名称（例如，Bob 的恒温器）
位置	loc	JSON			读写	否	在可用时提供位置信息
位置名	locn	字符串			读写	否	人性化的位置名称，例如客厅
货币	c	字符串			读写	否	指明用于交易的货币
地区	r	字符串			读写	否	自由格式文本指示设备在地理上所处的当前区域，自由格式文本不应以引号开头

6.2 资源发现

作为发现的一部分，客户端可以找到关于其他 OCF 对等体的适当信息。该信息可以是资源实例、资源类型，或者 OCF 对等体期望另一个 OCF 对等体发现的资源模型信息。至少，基于资源的发现会使用以下内容。

（1）应定义一个启用发现的资源，该资源的表示应包含可以被发现的信息。

（2）能够发现的资源应是指定的和事先已知的，或在引导区（如在垂直规范中指定），或可以被发现的（例如使用其他方法）。

（3）包含发现资源的设备应该被标识。

（4）需要公布发现资源信息的机制、过程以及启用发现资源。

(5) 需要发现资源的接入、获取信息的机制和过程。在请求中可能会使用一个查询,以限制返回信息。

(6) 发布的范围。

(7) 获取的范围。

(8) 信息可见性策略。

根据上面定义的基本原则,框架定义了三种基于资源的发现机制。

(1) 直接发现,在本地资源所在的设备上发布,并通过对等查询发现。

(2) 间接发现,在协助发现的第三方发布,并且对等实体针对资源发布和执行发现,以在协助第三方上启用发现。

(3) 广播发现,用于启用发现的资源,发起者是本地的,但是发布发现信息的设备是远程的。

6.2.1 直接发现

一个设备应该支持直接发现,在直接发现中主要完成以下功能。

(1) 提供信息的设备应该托管用于启用发现的资源。

(2) 设备使用本地资源发布可发现的信息,以启用发现(即本地范围)。

(3) 对此设备发现信息感兴趣的客户端应直接向资源发出检索请求,该请求可以作为单播或多播。请求可以是通用的,或者通过在请求中使用适当的查询来限定或限制。

(4) 接收请求的服务器端设备应当将发现信息的响应直接发送回请求的客户端设备。

(5) 请求中包含的信息由一些策略来决定,这些策略是为在响应设备上本地发现资源而设置的。

6.2.2 间接发现/基于资源目录

资源的间接发现也称为基于资源目录的发现,在间接发现中,有关要发现资源的信息托管在非资源所在的服务器端上。在间接发现中主要完成以下功能。

(1) 被发现的资源既不在发起发现客户端的设备上,也不在提供或发布要发现信息的设备上。该设备可以使用相同的资源来为多个代理提供发现,也可以为多个代理提供发现信息。

(2) 被发现的设备或具有发现信息的设备会使用另一设备上发现的资源来发布该信息。共享信息的策略(包括生命周期/有效性)由发布设备指定。发布设备可以根据需要修改这些策略。

(3) 执行发现的客户端可以向托管发现信息的设备发送单播发现请求,或者发送由设备进行监视和响应的多播请求。在这两种情况下,托管发现信息的设备都代表发布设备。

(4) 发现策略可以由托管发现信息的设备设置,也可以由正在发布发现信息的一方来设置。发现响应中返回的发现信息应遵守请求时生效的策略。

6.2.3 广播发现

在广播发现中主要完成以下功能。

(1) 发现资源托管在启动发现请求(客户端)的设备本地。用于启用的发现资源可以作为核心资源或作为引导的一部分被发现。

(2) 请求可以是实现相关的查找,或者是对启用发现资源的本地检索请求。

(3) 要被发现信息的设备应该向发现资源发布适当的信息。

(4) 发布设备负责发布信息。发布设备可以通过发送发布请求来更新资源上的信息,以根据其需要实现发现过程。被发现信息的策略,包括生命周期,都是由发布设备决定的。

6.2.4 资源信息发布过程

使用资源发布信息以实现发现的机制,既可以在本地实现,也可以远程实现。发布过程如图 6-5 所示。具有要发布的发现信息设备,如果在本地托管,则更新发现资源,或者向托管发现资源的设备发出更新请求。承载了用于发现资源的设备会添加/更新资源,以使用所提供的信息来进行发现,然后使用更新响应已请求公布资源的设备。

6.2.5 资源发现信息

发现过程如图 6-6 所示,初始化是以对资源检索请求完成发现的。该请求可以被发送到单个设备(如在单播)或多个设备(如在多播)中。用于进行单播或多播的具体机制,由数据连接层的支持来确定。对请求的响应具有被发现资源的策略信息。策略可以确定共享哪些信息、何时向代理请求以及向哪个请求代理。发现的信息可以是资源、类型、配置、其他标准或用户方面,这取决于对资源请求和请求形式。请求者可以选择使用 URI 查询中的查询参数,来缩小在请求中返回的信息。

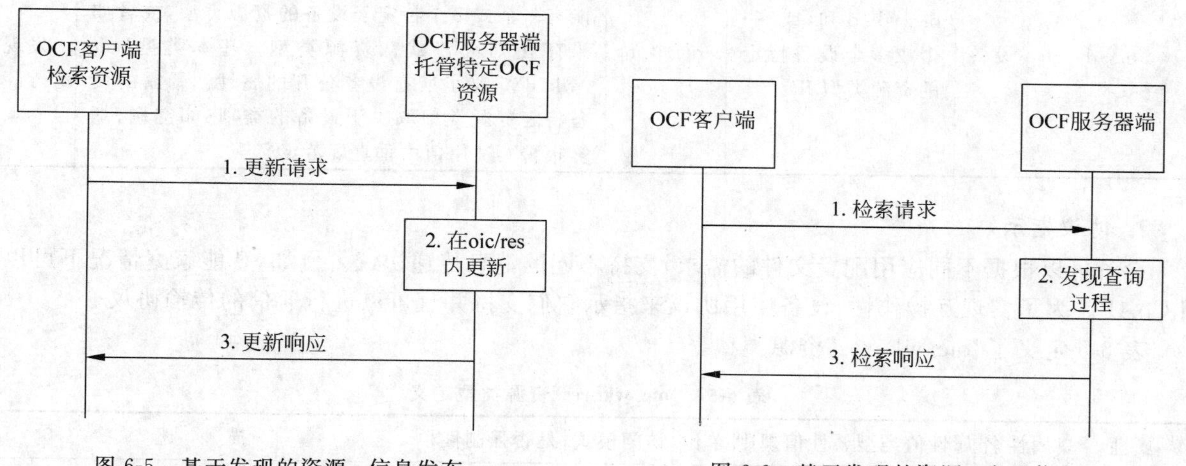

图 6-5 基于发现的资源:信息发布

图 6-6 基于发现的资源:发现信息

一些核心资源可能在所有的设备上都会实现,以支持发现过程。用来支持发现过程的核心资源有:

(1) /oic/res,用于资源的发现。

(2) /oic/p,用于平台的发现。

(3) /oic/d,用于设备信息的发现。

这些强制发现核心资源的详细信息如表 6-4 所示。

1. 平台资源

OCF 认为可以在单个平台上承载多于一个的设备实例。客户需要一种方法来发现和访问平台上的信息。核心资源"/oic/p"暴露平台特定的属性。同一平台上的所有设备实例应具有相同的公开属性值(即设备可以选择在"/oic/p"中公开可选属性,但当暴露时,该属性值应与平台上其他设备的该属性值相同)。

2. 设备资源

设备资源应具有预定义的 URI。资源"/oic/d"暴露了与表 6-4 中定义设备相关的属性。暴露的属性由特定的设备实例确定,并由该设备上的"/oic/d"资源类型定义。因为"/oic/d"的所有资源类型都不是已知的,所以,"/oic/d"的资源类型应通过核心资源"/oic/res"的发现来确定。设备资源"/oic/d"应该有一个默认资源类型,其有助于引导与此设备的交互(默认类型如表 6-4 所示)。

表 6-4 强制发现的核心资源

预定义的 URI	标题	资源类型 ID("rt")	接口	描　述	功能
/oic/res	默认	oic. wk. res	oic. if. ll	资源通过该接口发现相应的服务器端,并对可用资源进行检查。"/oic/res"将暴露设备上可发现的资源。当服务器端接收到以"/oic/res"(如 GET /oic/res)为目标的检索请求时,它将用自身所有可发现资源的链接列表进行响应。"/oic/d"和"/oic/p"是可发现的资源,因此,它们的链接包含在"/oic/res"响应中	发现
/oic/p	平台	oic. wk. p	oic. if. r	通过平台特定信息发现资源	发现
/oic/d	设备	oic. wk. d 和(或)一个或多个设备特定的资源类型 ID	oic. if. r	可通过"/oic/res"发现资源,暴露了设备实例特有的属性。"/oic/d"除了默认资源类型之外还可能有一个或多个特定于设备的资源类型,或者如果存在则覆盖默认资源类型。基本类型"oic. wk. d"定义了所有设备公开的属性。暴露的设备特定资源类型取决于设备的类别(如空调、烟雾报警),适用值由垂直规范定义	发现

3. 协议指示

设备可以根据不同应用配置文件的需求,支持不同的消息传递协议。例如,智能家庭情况下可以使用 CoAP。为了实现互操作性,设备使用协议来指示它们支持并且可以进行通信的传输协议。

表 6-5 定义了"oic. wk. res"资源类型。

表 6-5 "oic. wk. res"资源类型定义

属性	属性名	属性值类型	属性值规则	单位	访问模式	是否强制性	描　述
名称	n	字符串			读	否	由供应商定义的人性化名称
设备标识符	di	UUID			读	是	由设备的"/oic/d"资源指示的设备标识符。"/oic/res"中可能有多个"di"实例,但每个"di"都有唯一的值。"di"值的唯一性意味着设备的资源应当是单个"di"的组合
链接	links	数组			读	是	链接数组描述了 URI,支持的资源类型、接口以及访问策略
消息协议	mpro	SSV			读	否	消息传递协议的空间分隔值(SSV)的字符串是作为一个 SI 数被支持的。例如,"1 和 3"表示设备支持 CoAP 和 HTTP 作为消息协议

如果一个设备要指示其消息协议能力，可以在对“/oic/res”的请求响应中添加属性“mpro”。设备应支持基于 CoAP 作为基准发现机制。在发现响应中看到此属性的客户端，可以选择任何支持的消息传递协议，以便与服务器端进行通信获取更多消息。例如，如果支持多个协议的设备指示它在发现响应中支持“mpro”属性的值“1,3”，不能假定存在隐含的排序或优先级。但是，垂直服务规范可以选择指定隐含的顺序或优先级。如果响应中没有“mpro”属性，则客户端将使用垂直规范中指定的默认消息传递协议进行进一步通信。表 6-6 提供了协议方案的 OCF 注册表。

表 6-6　协议方案的 OCF 注册表

SI 数	协　议	SI 数	协　议
1	CoAP	4	HTTPS
2	CoAPS	5	CoAP+TCP
3	HTTP	6	CoAPS+TCP

注意：特定协议使用的端点发现不在讨论范围内。除发现之外，客户端用于在不同消息协议中形成请求的机制不在讨论范围内。

以下规则适用于“/oic/d”的使用：垂直规范可以选择暴露其设备类别(如冰箱)，通过将与设备类别相对应的资源类型添加到与“/oic/d”相关联的资源类型列表来实现。

例如，“/oic/d”的 rt 变成“["oic. wk. d","oic. d. < thing >"];”。其中，"oic. d. < thing >"在另一个规范中定义，如智能家居垂直规范。

这意味着“/oic/d”公开的属性默认为表 6-7 中的强制属性。垂直规范可以选择扩展资源类型“oic. wk. d”定义的属性列表。在这种情况下，垂直规范将分配新的设备类特定资源类型 ID。表 6-7 中定义的强制属性应始终存在。

注意：根据现有的核心规范定义，资源类型 ID 可以是资源类型 ID 的列表；在这种情况下，“/oic/d”的默认资源类型 ID 是列出的第一个资源类型。所以，一个垂直规范可以先列出“oic. d. thing”。这意味着 GET /oic/d 会返回“oic. d. thing”的属性，并且 GET /oic/d? rt=< some rt >返回查询中列出的“rt”属性。在表 6-7 中，“oic. wk. d”资源类型定义了“/oic/d”的基本资源类型。

表 6-7　“oic. wk. d”资源类型定义

属性	属性名	属性值类型	属性值规则	单位	访问模式	是否强制	描　述
设备名称	n	字符串			读	否	由供应商定义的人性化名称
规范版本	icv	字符串			读	是	该设备实现的核心规范版本，语法是 core.< major >.< minor >.< sub-version >，其中，< major >、< minor >和< sub-version >分别是主版本号、次版本号和子版本号，该版本的字符串值应该是"core. 1. 1. 0"
设备 ID	di	UUID			读	是	设备的唯一标识符。此值应按照设备 ID 在 OCF 安全中进行定义

续表

属性	属性名	属性值类型	属性值规则	单位	访问模式	是否强制	描　述
数据模型版本	dmv	CSV			读	是	实现该设备数据模型的资源规范版本；如果是针对垂直特定资源规范来实现的，那么就是实现该设备模型的规范版本的垂直规范。语法是以逗号分隔的列表< res >. < major >. < minor >. < sub-version >或< vertical >. < major >. < minor >. < sub-version >。< res >是字符串"res"，< vertical >是在垂直特定资源规范中定义的垂直名称。< major >、< minor >和< sub-version >分别是规范的主版本号、次版本号和子版本号，该版本的字符串值应该是"core. 1. 1. 0"

“/oic/d”资源的附加资源类型由垂直规范定义。表 6-8 定义了“oic. wk. p”资源类型。

表 6-8　“oic. wk. p”资源类型定义

属　性	属性名	属性值类型	访问模式	是否强制	描　述
平台 ID	pi	字符串	读	是	物理平台的 UIUID 是根据 IETF RFC 4122 得到的，推荐使用 RFC 中特定的随机生成方案来创建 UUID
制造商名称	mnmn	字符串	读	是	制造商的名称
制造商细节链接	mnml	URI	读	否	对制造商的引用，表示为 URI
模型号	mnmo	字符串	读	否	制造商指定的模型号
制造日期	mndt	日期	读	否	设备的制造日期
平台版本	mnpv	字符串	读	否	平台的版本，字符串(由制造商定义)
OS 版本	mnos	字符串	读	否	平台 OS 版本，字符串(由制造商定义)
硬件版本	mnhw	字符串	读	否	平台硬件版本
固件版本	mnfv	字符串	读	否	设备固件版本
支持链接	mnsl	URI	读	否	指向制造商的支持信息的 URI
系统时间	st	日期时间	读	否	设备的参考时间
供应商 ID	vid	字符串	读	否	供应商为平台定义的字符串，字符串没有格式，由供应商决定填充的文本

4. 复合设备

物理设备可以被建模为单个设备或者其他设备的组合。例如，冰箱可以被建模为组合设备，其定义的一部分可能包括一个恒温设备，因为它自身是由可调温度计装置组成的。

将服务器端建模为组合设备的方法有很多。一种示例方法是使代表复合设备的平台产生具有多个设备的实例。每个设备实例表示组合中的不同设备之一。设备的每个实例可以自身具有或承载其他资源的多个实例。

不管它是如何组成的，一个具体实现只应为每个逻辑服务器端公开带有资源类型选项的“/oic/d”单个实例。

因此，上述冰箱实例如果被建模为单个服务器端；“/oic/res”将使用适合于冰箱的资源类型名称来暴露“/oic/d”。附属恒温器和温度计设备会通过一个实现指定相关的 URI，被简单地暴露为具有设备适当资源类型的资源实例。例如,/MyHost/MyRefrigerator/Thermostat 和/MyHost/MyRefrigerator/Thermostat/Thermometer。

5. 附加的发现资源

附加的发现资源可以被实现,支持附加发现的核心资源有:"/oic/rts",用于资源类型的发现;"/oic/ifs",用于接口的发现。这些可选核心资源的详细信息如表 6-9 所示。

表 6-9 可选的发现核心资源

预定义的 URI	标题	资源类型 ID	接口	描述	功能
/oic/rts	资源类型	oic. wk. rts	oic. if. r	标识设备上支持的资源类型所使用的资源。资源类型可以在构建时预加载或预配置,或者可以在运行时下载。此资源是只读的,这意味着资源类型不能推送到设备(这不排除设备从资源类型注册表/目录下载新类型,然后将其暴露为支持)。通过"/oic/rts"发现的资源类型用于指定被创建的资源类型。此资源被建模为一个简单的资源	发现
/oic/ifs	接口	oic. wk. ifs	oic. if. r	通过其识别支持的接口资源。这是作为资源发现的一部分,返回每个资源的接口信息。该资源宣布所支持的接口可以用在该服务器端创建资源	发现

表 6-10 定义了"oic. wk. rts"资源类型。

表 6-10 "oic. wk. rts"所支持的资源类型定义

属性	属性名	属性值类型	属性值规则	单位	访问模式	是否强制	描述
类型列表	tl	BSV	0 到多种资源类型 ID		读	是	设备支持的资源类型列表

表 6-11 定义了"oic. wk. ifs"资源类型。

表 6-11 "oic. wk. ifs"资源类型定义

属性	属性名	属性值类型	属性值规则	单位	访问模式	是否强制	描述
接口列表	il	BSV	0 到很多资源接口		读	是	设备支持的资源接口列表

6.2.6 使用"/oic/res"的资源发现

使用"/oic/res"的资源发现是默认的发现机制,所有的设备都会支持,具体实现如下。

(1) 每个设备更新其本地"/oic/res"与那些可发现的资源。每次在设备上实例化新资源时,如果该资源可由远程设备发现,则与设备本地的"/oic/res"资源(作为实例化资源)一起发布。

(2) 如果一个设备想要在一个或多个远程设备上发现资源或资源类型,则该设备就会向远程设备上的"/oic/res"发出检索请求。只要探测到一个特定主机,则该请求就可以被多播(默认)或单播发送。还可以选择在请求的查询部分使用适当子句来限制检索请求。查询可以基于资源类型、接口或属性进行选择。

(3) 查询应用于资源的表示。"/oic/res"是唯一表示中有"rt"的资源。所以"/oic/res"是可以用于传输协议层多播发现的唯一资源。

(4) 接收检索请求的设备以资源列表、每个资源的类型和支持的接口进行响应。此外,还可以发送关于资源活动上的策略信息,支持的策略包括可观察性和可发现性。

(5) 基于"/oic/res"请求返回的资源,接收设备可以进行更深层的发现。

"/oic/res"发现时返回的信息至少应该包括以下几点。

(1) 资源的 URI(相对或完全限定 URL):每个资源的类型。如果资源启用多种类型,则可以返回多种资源类型。要访问多种类型的资源,应在请求中指定目标的特定资源类型。

(2) 该资源支持的接口。可以返回多个接口。要访问特定接口,应在请求中指定接口。如果未指定接口,则假定为默认接口。

(3) 针对该资源定义的策略。这些策略可以是与安全相关的访问模式、交互类型等。除了请求/响应类型的交互,规范允许资源被"观察"。

前文已经描述了使用"/oic/res"发现的 JSON 模式。关于在 CoAP 传输上使用"/oic/res"多播发现的详细信息,参见第 5 章的终端发现。

使用"/oic/res"执行发现功能后,客户端可以通过使用"/oic/p""/oic/rts"等执行发现功能,发现有关服务器的更多详细信息。如果客户端已经知道服务器端,它可以使用其他资源发现,而不会通过"/oic/res"。

6.2.7 基于资源目录的发现

直接发现是当前用来在网络中查找资源的机制。当需要时,在特定节点处直接查询资源,或者向所有节点发送组播分组。每个被查询的节点直接用其可发现的资源对发现设备做出响应。本地可用的资源在同一设备上注册。

在某些情况下,间接发现也是需要的。间接发现是指第三方设备(发现设备和被发现设备除外)协助发现过程。第三方设备仅提供代表另一个设备的资源信息,但不会在该设备上承载资源。图 6-7 是通过资源目录间接发现资源的过程。其中,设备 B 作为设备 A 和设备 D 的资源目录;设备 A 和设备 D 不回复多播发展。

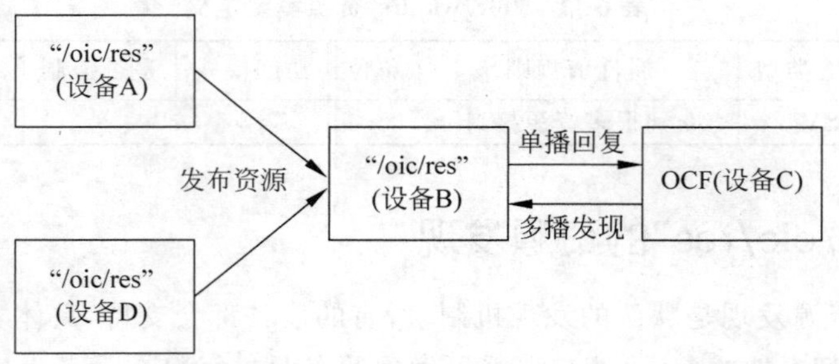

图 6-7 通过资源目录间接发现资源的过程

间接发现对于资源受限设备是有用的,例如,设备需要休眠进行功率管理而不能处理每个发现请求,或者当设备可能不在相同的网络上而需要对发现进行优化。一旦使用间接发现功能,发现了资源,则可以直接向承载该资源的设备发送请求,完成对资源的访问。

资源目录(RD)是辅助间接发现的设备。可以在其"/oic/res"资源中查询 RD,以查找承载在其他设备上的资源。这些设备可以是睡眠节点或不能响应发现请求的任何设备。设备可以将它们承载的所有或部分资源列表发布到一个 RD。然后,RD 代表发布设备响应对资源发现的查询(例如当设备进入睡眠时)。对于一般资源的发现,RD在响应对"/oic/res"的请求方面与其他服务器一样。

任何服务或作为RD的设备应公开一个众所周知的资源"/oic/rd"。想要发现RD的设备将使用该资源和资源发现机制,发现RD并获取RD的参数。通过该资源发现的信息,应该用于选择适当的RD,进一步用于资源发布。加权信息应包括以下标准:电源(AC、电池供电或安全/可靠),连接(无线、有线),CPU,内存和负载统计(处理发布和从设备查询)。此外,RD将返回0~100的偏差因子。作为可选,RD还可以返回上下文,它是字符串的值,上下文的语义在本书中不讨论,但是它将用于建立对应用、部署、使用有意义的领域、区域或类似的范围。

通过使用这些标准或偏置因子,设备应选择一个RD(每个上下文)以公布其资源。上下文是来自另一个用于资源发现、一组客户端或者任何一组的情况。上下文通常在部署时根据应用程序需求确定。例如,上下文可以是多播组,作为多个多播组的成员设备,可能必须在每个多播组(即每个上下文)中找到,并选择RD以发布其信息。设备可以决定在其生存期内选择其他RD,但是设备在任何情况下,都不会将其资源信息发布到具有上下文的多个RD。与移动电话设备相比,诸如电视、网络路由器、台式机之类的设备将具有更高的权重或偏置因子。

6.2.7.1 资源目录发现

OCF网络中的RD应支持RD发现、提供相关设施允许设备向RD发布其资源信息、更新资源信息,以及从RD删除资源信息的功能。如图6-8所示,希望广播其资源的设备首先发现资源目录,然后发布所需的资源信息。一旦一组资源被发布到RD,则当RD在相同的多播域上时,发布设备将不响应针对这些发布资源的多播资源发现查询。在这种情况下,只有RD根据发布给它的资源,响应多播资源发现请求。

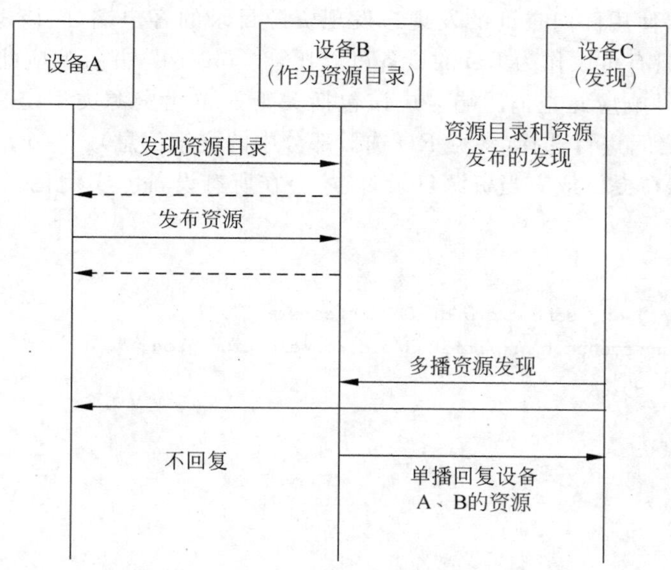

图6-8 RD发现和RD支持的资源支持查询

OCF网络允许多个设备充当RD。有多个RD支持的原因是使网络可扩展,处理网络故障和集中的设备故障瓶颈。并不排除这样一种情况:其中用例或部署环境可能需要将环境中的单个设备部署为唯一资源目录(例如网关模型)。除此之外,在平台上可能有多个设备充当RD。

RD的发现可能导致来自多个RD的响应。发现设备应选择一个RD。该选择可以基于来自RD的响应中提供的权重参数。

RD对应用是不可知的，即应用不应该知道资源目录是否被查询以获得资源信息。所有检索处理对于应用是不透明的。执行资源发现的客户端使用RD，就像它可以使用任何其他服务器端进行发现一样。当它仅需要在RD上广播资源时，可以向RD发送单播请求，或者当它不需要或具有RD的显示知识时，可以进行多播查询。资源目录的部署方式如图6-9所示。

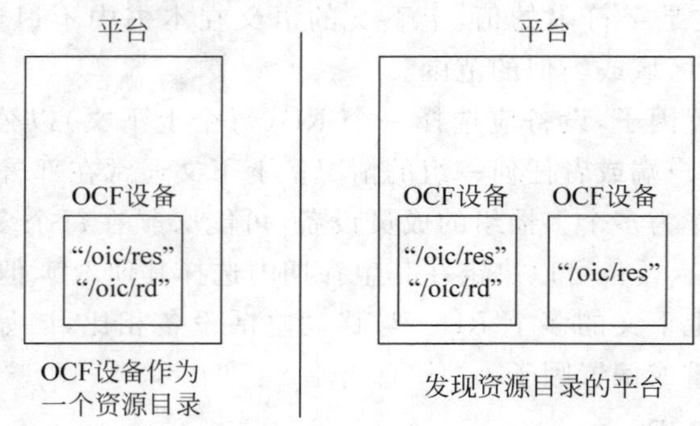

图6-9 资源目录的部署方式

当然，也可以用以下方式发现资源目录。

(1) 预配置：希望发布资源信息的设备可以预先配置特定资源目录的信息(例如IP地址、端口和传输等)。预配置可以在On-Boarding时完成，或者可以使用其他方法在设备上更新，预配置可以由制造商或由用户/设备管理器来完成。

(2) 面向查询：想要使用面向查询的发现来发现资源目录的客户端，应该发出针对"/oic/rd"资源的多播资源发现请求。只有可以作为RD的设备可以托管"/oic/rd"，并且响应此查询，所有这些设备都应该可以托管"/oic/rd"并响应此查询。响应应该包括关于RD(由资源类型定义)和权重参数的信息，以允许发现设备在RD之间进行选择(参见RD选择部分中的详细信息)。"/oic/rd"资源在某些条件下是强制性核心资源，只有在提供或充当资源目录时，才会在所有设备上实例化。对一个"oic/rd"查询响应中的相关信息，如下所示。

```
{
  "$schema": "http://json-schema.org/draft-04/schema#",
  "id":"http://openinterconnect.org/schemas/oic.rd.selection.json#",
  "title" : "RD Selection",
  "definitions": {
  "oic.rd.attributes": {
    "type": "object",
    "properties": {
      "n": {
        "type": "string",
        "description": "A human friendly name for the Resource Directory",
        "format": "UTF8"
      },
      "di": {
        "type": "string",
        "description": "A unique identifier for the Resource Directory",
```

```
            "format": "uuid"
        },
        "sel": {
            "description": "Selection criteria that a device wanting to publish to any RD can use to choose this Resource Directory over others that are discovered",
            "oneOf": [
            {
                "type": "object",
                "properties": {
                    "pwr": {
                    "type": "string",
                    "enum": [ "ac", "batt", "safe" ],
                    "description": "A hint about how the RD is powered. If AC then this is stronger than battery powered. If source is reliable (safe) then appropriate mechanism for managing power failure exists"
                },
                "conn": {
                    "type": "string",
                    "enum": [ "wrd", "wrls" ],
                    "description": "A hint about the networking connectivity of the RD. * wrd * if wired connected and * wrls * if wireless connected."
                    },
                    "cpu": {
                        "type": "integer",
                        "description": "Memory available at request in MHz units"
                    },
                    "memory": {
                        "type": "integer",
                        "description": "A processing capacity of the CPU specified in MB units"
                    },
                    "load": {
                        "type": "array",
                        "items": {
                            "type": "number",
                        },
                        "minitems": 3,
                        "maxitems": 3,
                        "description": "Current load capacity of the RD. Expressed as a load factor 3 - tuple (upto two decimal points each). Load factor is based on request processed in a 10 minute window, in the last hour and hourly average over the RD current lifetime"
                    }
                }
            },
            {
                "type": "integer",
                "minimum": 0,
                "maximum": 100,
                "description": "A bias factor calculated by the Resource directory - the value is in the range of 0 to 100 - 0 implies that RD is not to be selected. Client chooses RD with highest bias factor or randomly
```

```
between RDs that have same bias factor"
          }
        ]
      }
    },
    "required": [ "sel" ]
  }
  },
  "type": "object",
  "$ref": "#/definitions/oic.rd.attributes"
  }
```

(3) 广播/存在：RD 可以向设备广告其存在。它是存在和广播包的组合。已经发布到 RD 的设备可以将其用作 RD 的存在或心跳消息。如果 RD 广播未以规定的间隔到达，则发布设备开始搜索网络中的其他 RD，因为该信号表示 RD 不在线。这个消息的其他用法是将其作为广播，对设备寻求 RD 并发布其资源。来自广播的细节可以直接用于查询 RD，以获得加权细节，而不是在网络中发送组播分组。按照预期，这是以规则间隔发送的，并且不包括用于保持分组大小的权重信息。

RD 的一个重要优点是使服务在不支持站点范围的组播，但支持站点范围的路由网络中可被发现。这样网络的典型例子就是家庭网络。为了在这样的网络上启用 RD 功能，需要站点发现机制来发现 RD 服务(IP 地址和端口号)。支持混合代理的家庭网络允许基于"dns-sd/ mDNS"站点范围内的发现。为了使自身在链路本地范围之外可被发现，具有可路由 IP 地址的 RD 将实现 mDNS 响应者需求，其在 IETF RFC 6762 中定义。RD 将响应 PTR 类型、服务名称等于"_rd._sub._oic._udp.local"的 mDNS 查询。响应应该包括所有可路由的 IP 地址。具有可路由 IP 地址的设备应通过一个服务名称为"_rd._sub._oic._udp.local"的 DNS-SD PTR 查找(IETF RFC 6763 中定义)，发现所有可用的 RD 实例。响应应包括所有可路由地址/端口，可通过它们使 RD 服务被访问。

当设备发现多个 RD 时，它将基于选择标准决定使用哪个 RD。设备将在给定时间仅使用多播域内的一个 RD，或向多播域内的一个 RD 发布信息。这是为了最小化资源发现阶段处理重复信息的负担。

有两种方法可以选择 RD：一种方法是基于加权或偏差因子(RD 生成)；另一种方法是基于服务器端(生成的客户端/设备)提供客户端的粒度参数确定。设备可以使用其中一种或两种方法来选择 RD。

偏差因子：由服务器端生成的正数，范围为 0～100，其中 0 是最低的，100 是最高的。如果两个 RD 具有相同的偏差因子，则设备可以基于辅助标准或者随机选择。无论哪种方式，每次只能选择和使用一个 RD。在 OCF1.1 规范中并没有明确定义 RD 偏差因子的具体方法。该数可以在 RD 的 On-Boarding 或后续配置时预配置，也可以由 RD 实现的公式来确定，OCF 将在未来版本中为此计算提供标准公式。

偏差因子应由 RD 计算，将表 6-12 中每个参数确定的值(贡献)加到一起，并除以参数的数量。当 RD 具有足够能力时，例如，服务提供商网关，RD 可以广播大于计算值的偏差因子。

参数：可以在发现响应中选择返回表 6-12 中定义的参数，如直接电源、网络连接、负载条件、CPU 功率和内存等。发现设备可以使用细节，基于客户端定义的 RD 参数策略和标准，做出粒度选择决策。例如，工业部署中的设备可以将电力连接所占比重降低，而在家庭环境中，设备电力权重应该高一些。

表 6-12 选择参数

参数	值(贡献)	描述
功率	安全(100) AC(70) 电池(40)	安全意味着电源可靠,并且备有电池预防停电等。实现时可能会根据 RD 设备运行的电池类型降低电池数量。如果电池节电是重要的,那么这个数字应该降低
移动性	固定(100) 移动(50)	实现可以基于 RD 设备的移动性来进一步对移动性数量进行分级;较低的数字对应于高度移动性,较大数量对应于有限移动性;移动性数字不应该大于 80
网络乘积	类型: 有线(10) 无线(4) 带宽: 高(10) 低(5) 有损耗(3) 接口数量	网络乘积=[Σ(类型 * 每个网络接口的带宽)]/接口的数量,归一化到 100
内存因子	可用的/总的	存储器是用于存储资源信息的易失性或非易失性存储器;内存因子=[可用的]/[总的];归一化到 100(即以百分数表示)
请求加载因素	1min 5min 15min	RD 当前的请求负载,与 UNIX 负载因子类似(使用可观察、待决定和处理请求,而不是可运行进程);表示为负载因子 3 元组(每个最多两个小数)。因子基于在 1min(L1)、5min(L5)和 15min(L15)窗口中处理的请求见 http://www.teamquest.com/import/pdfs/whitepaper/ldavg1.pdf;因子=100-(L1 * 3+L5 * 7+L15 * 10)/3

发布设备使用广播和查询机制来查找 RD。基于选择过程将如何工作,有如下 4 种情况。

(1) 当新设备启动时,单个或多个 RD 已经存在于网络中。

(2) 网络中不存在 RD,并且新设备启动。

(3) 另一个设备启动,具有 RD 的能力。

(4) 两个 RD 同时启动,例如电源发生故障后。

第 1 种情况下,RD 已经存在了,设备监听广播分组或者查询多播地址上的 RD。如果接收到单个或多个响应,则发布设备使用来自查询响应的权重信息来选择要发布资源的 RD。

第 2 种情况下,设备将监听广播。一旦接收到 RD 广播分组,就可以接收权重信息或检索它,如果偏置因子满足其标准,则可以在 RD 上注册其资源。

第 3 种情况下,如果发布设备正在向现有 RD 发布并且发现新的 RD,相对偏置因子有利于新的 RD,则发布设备可以选择移动到新的 RD。如果决定选择新的 RD,则发布设备将从先前的 RD 中删除其资源信息,然后将该信息发布到新的 RD(在过渡期间,一旦发送了删除请求,发布设备将响应资源发现请求)。

第 4 种情况下,每个 RD 启动并且广播它们的存在。基于权重标准的发布设备,选择适当的 RD 用于发布其资源信息。

6.2.7.2 资源发布

在 RD 的选择过程之后,设备可以选择以下机制:将其资源信息推送到所选的 RD 或请求 RD 通过对其"/oic/res"执行单播发现请求，获取资源信息。

发布设备可以决定在资源目录上发布所有资源还是少量资源。发布设备只能发布通过其他方式发布到自己"/oic/res"的资源。对于不向RD发布的资源，发布设备可以响应发现请求（在其"/oic/res"资源上）。尽管如此，强烈建议在使用RD时，将发布者的所有可发现资源发布到RD。

1. 发布：推送资源信息

使用"oic.wk.rdpub"资源类型、"oic.if.baseline"接口，进行CRUDN操作将资源信息发布到"/oic/rd"。

一旦发布设备向RD发布了资源，它可以不响应针对自己"/oic/res"相同资源的多播发现查询，特别是当与RD处于相同的多播域时。在发布资源之后，RD负责回复对已发布资源的查询。

如果发布设备处于睡眠模式，并且RD已经代表发布设备进行了答复，则发现设备将尝试访问所提供的URI上的资源。

可能资源目录和发布设备都响应了来自发现设备的多播查询，这将创建重复的分组，但是可以用于不可靠网络的一种选择。我们并不推荐这种方式，不过对于工业场景，这确实是一种可能性。无论哪种方式，发现客户端应始终准备在响应多播发现请求时，处理重复的信息。发布及推送资源信息的过程如下。

```
{
  "$schema": "http://json-schema.org/draft-04/schema#",
  "id": "http://openinterconnect.org/schemas/oic.rd.publish.json#",
  "title": "RD Publish & Update",
  "definitions": {
    "oic.rd.publish": {
      "type": "object",
      "description": "Publishes resources as Links into the resource directory",
      "properties": {
        "n": {
          "type": "string",
          "description": "Readonly, Human friendly name of the publishing Device"
        },
        "di": {
          "type": "string",
          "description": "ReadOnly, Unique identifier (UUID) for device that is publishing",
          "format": "uuid"
        },
        "$ref": "oic.collection.json#/definitions/oic.collection.setoflinks/properties",
        "lt": {
          "type": "integer",
          "description": "Time to indicate a RD, how long to keep this published item. After this time (in seconds) elapses, the RD invalidates the links. To keep link alive the publishing device updates the ttl using the update schema"
        }
      },
      "dependencies": {
        "links": [ "lt" ]
      }
    }
  },
  "type": "object",
```

```
    "$ref": "#/definitions/oic.rd.publish",
    "required": [ "di" ]
  }
```

2. 更新资源信息

服务器端将保存发布资源信息，直到生存时间字段中指定的时间。如果设备寻求 RD 来保持资源并代表其回复查询，则设备可以发送更新，可用于在 RD 上发布的所有资源，或者根据已发布的每个资源进行更新。

使用与初始发布相同的资源类型和接口完成更新，但只在有效载荷中提供要更新的信息。

3. 删除资源信息

资源目录中保持的资源信息可以由发布设备随时删除。它可以是整个设备信息，也可以是特定资源。该资源应仅在设备满足特定要求时允许，因为它可能会产生潜在的安全问题。

当要删除来自设备的所有资源信息时，使用设备 ID 作为删除请求查询中的标签进行删除。在特定资源的情况下，删除请求将在查询中包括实例“ins”标签以及设备 ID。

在 RD 提取资源信息的情况下，不能选择性删除单个资源的信息。发布设备可以请求删除，但是仅针对 RD 已经从该设备提取的所有资源信息。在这种情况下，删除请求在查询中具有设备“id”的标记。

4. 将资源信息进行 RD 转移

当发布设备识别出更适合的 RD 时，它可能会决定向该 RD 发布。由于设备每次只能发布一个 RD，因此，客户端应确保先前发布的信息在发布到新选择的 RD 之前，从当前使用的 RD 中删除，可以通过允许生存时间到期或显式地删除资源信息，完成资源的删除。多个 RD 之间不应相互传送资源信息，客户端有责任选择 RD 并管理已发布的资源。

6.2.7.3 资源发现

基于发现过程的查询与没有 RD 的情况保持相同。可以通过发送多播或单播请求来查询“/oic/res”资源，进一步发现资源。在多播发现请求的情况下，RD 将对承载资源的设备做出响应。客户端应准备处理来自多个 RD、具有相同信息的重复资源信息，或者来自一个 RD 的重复资源信息，并且主机设备(公布资源信息)都应该响应请求。与使用 RD 发现的资源交互相同，通过查询承载资源设备的“/oic/res”资源也使用相同的机制和方法(例如，连接到资源和对资源执行 CRUDN 操作)。

资源目录应该支持 DTLS。设备和资源目录之间的通信应该尽可能地基于 DTLS。设备和资源目录应该尽可能使用 PSK、证书和原始公钥进行身份验证。

设备在注册之后应该使用 UUID 与资源目录进行通信，即检索、更新和删除应当使用 UUID，因为 IP 地址可以改变资源所在的位置并且会混淆资源目录。

访问控制列表应该用于发布和查找目的，因为它们可能不同。查找应检查请求来源、网络或资源级别。

请求通配符查找时，资源目录应该能够进行路由检查，以避免 DDoS 攻击的情况下返回。由于通过 UDP 不可能检查路由，因此，通配符查找应仅支持 DTLS、TCP 或 TLS。

6.3 通知

服务器端应支持通知操作，以使客户端能够以异步方式请求并获取一个或多个资源的状态通知。那么，如何实现将更新传递给请求者？在 OCF 规范中是通过观察机制实现的。

在观察机制中，客户端利用检索操作，在资源状态改变的情况下，要求服务器端进行更新。观察机制如图 6-10 所示。注意，观察机制只能用于具有可观察属性的资源。

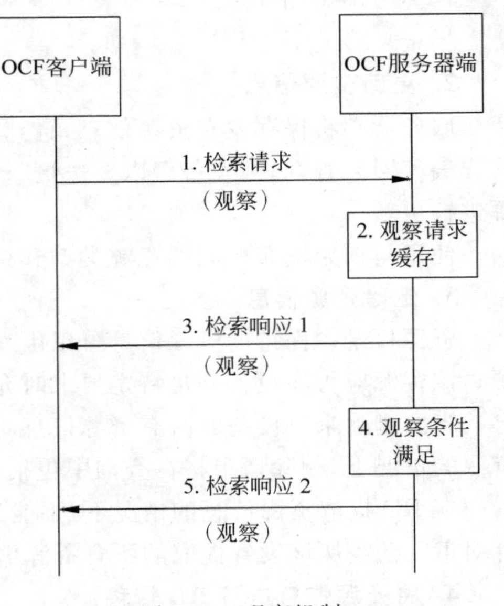

图 6-10 观察机制

1. 有观察指示的检索请求

如果状态发生改变，则客户端向服务器端发送检索请求消息，以请求对服务器端上的资源更新。检索请求消息会携带以下参数。

(1) fr：客户端的唯一标识符。

(2) to：客户端请求观察的资源。

(3) ri：检索请求的标识符。

(4) op：检索。

(5) obs：观察请求的指示。

2. 服务器端的处理

在接收到检索请求之后，服务器端会验证客户端是否具有所请求的操作权限，并且属性是可读和可观察的。如果验证成功，则服务器端缓存与观察请求相关的信息。服务器端还会缓存来自检索请求"ri"的参数值，以便在状态改变的情况下用于初始响应和未来响应。

3. 有观察指示的检索响应

服务器端会传输一个检索响应消息，以响应来自客户端的检索请求消息。如果验证成功，则响应包括一个观察指示。如果没有验证成功，则响应中会省略观察指示，向请求客户端通知不允许通知注册。检索响应消息应包括以下参数。

(1) fr：服务器端的唯一标识符。

(2) to：客户端的唯一标识符。

(3) ri：检索请求中包含的标识符。

(4) cn：客户端请求的信息资源表示。

(5) rs：检索操作的结果。

(6) obs：响应对一个观察请求做出的指示。

4. 服务器端的资源监测

服务器端应监视来自客户端观察请求中标识的资源状态。当观察到的资源状态发生改变时，服务器端发送具有观察指示的另一个检索响应。该机制不允许客户端指定触发通知的任何边界或限制，该决定完全留给服务器端处理。

5. 其他有观察指示的检索响应

在观察到由客户端指示的资源状态发生改变之后，服务器端将会传送更新检索的响应消息。

6. 取消观察

客户端可以通过向其观察的服务器端上资源发送没有观察指示字段的检索请求，显式地取消观察。对于某些协议映射，客户端还能够通过停止检索响应来取消观察。

6.4 设备管理

设备管理包括诊断和维护。本版规范中指定的设备管理功能旨在解决基本设备管理功能。预期在未来版本的规范中会增加新的设备管理功能。框架中的诊断和维护功能旨在供管理员使用,以解决现场操作时遇到的设备问题。如果设备支持诊断和维护,则应支持核心资源"/oic/mnt",如表 6-13 所示。

表 6-13 可选的诊断和维护设备管理核心资源

预定义的 URI	资源类型标题	资源类型 ID ("rt"值)	接口	描 述	相关的功能交互
/oic/mnt	维护	oic. wk. mnt	oic. if. rw	设备维护所用的资源,可用于诊断目的	设备管理

表 6-14 定义了"oic. wk. mnt"资源类型。Factory_Reset、Reboot 和 StartStatCollection 三个属性中至少应该实现一个。

表 6-14 "oic. wk. mnt"资源类型定义

属 性	属性名	属性值类型	属性值标准	单位	访问模式	是否强制	描 述
出厂重设	fr	布尔型			读写	是	当写该属性时,"0"表示无任何指示(初始默认值[①]);"1"表示开始工厂设定;在工厂重设后,该值会恢复到初始值,工厂重设的参数和数据将会被清理。当读取该属性时,值"1"表示待恢复的出厂复位,否则在出厂复位后该值为"0"
重启	rb	布尔型			读写	是	当写该属性时,"0"表示无任何指示(初始默认值);"1"表示开始重启,重启后该值将会被恢复到初始值状态,且重启将会在 60s 内结束
开始状态收集	ssc	布尔型			读写	是	"0"表示无任何统计资料记录;"1"表示开始进行统计;在数据采集和无任何终端采集设备之间的触发器的值应该被设定为"0"或者"1"

注:① 默认值表示设备重新启动或恢复出厂设置后此属性的值。

6.5 场景

场景是某些自动化操作的机制。一个场景是存储资源的集合,是一组定义的资源属性值的静态实体。场景提供了一种机制来存储由多个独立 OCF 服务器端托管的多个资源设置,一旦成立,可以由多个客户端来恢复设置。场景可以分组和重复使用,一组场景也是一个场景。简而言之,场景是用户捆绑设置。

1. 简介

场景是通过资源来描述的。场景资源由 OCF 的服务器端托管,最高级别资源在"/oic/res"上。这意味着,OCF 客户确定场景功能是通过一个检索"/oic/res"或资源发现 OCF 服务器端上的承载。场景的设置由 OCF 客户端交互驱动,包括创建新的场景,这是场景的一部分 OCF 服务器端资源属性的映射。

场景功能是由多个资源创建并具有如图 6-11 所示的结构。场景列表和场景集合是重载的资源集合。场景集合包含场景的列表，这个列表包含零个或多个场景集合。场景成员资源包含场景和控制资源之间的映射。

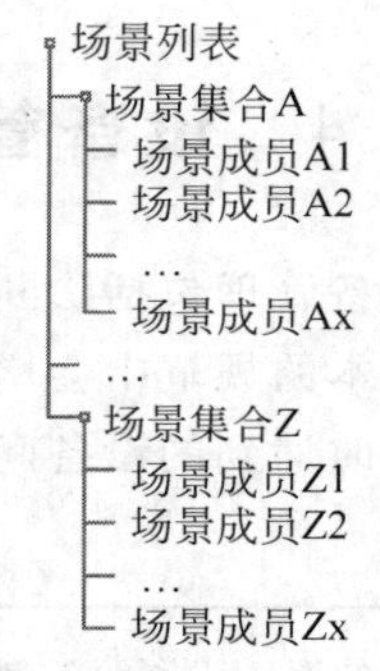

图 6-11　通用场景资源架构

2. 场景建立

想要创建一个场景，OCF 客户端需要验证一个 OCF 的服务器端支持场景的特征；场景成员不需要在同一 OCF 服务器端支持该特征。客户端通过检查"/oic/res"包含场景列表资源进行确认，如图 6-12 所示，将场景值与场景资源进行映射。用于场景集合的每一个资源映射，被称为场景成员资源描述。该场景资源包含链接资源，并在场景值属性中列出，由链接表示 OCF 实际资源属性值之间的映射。

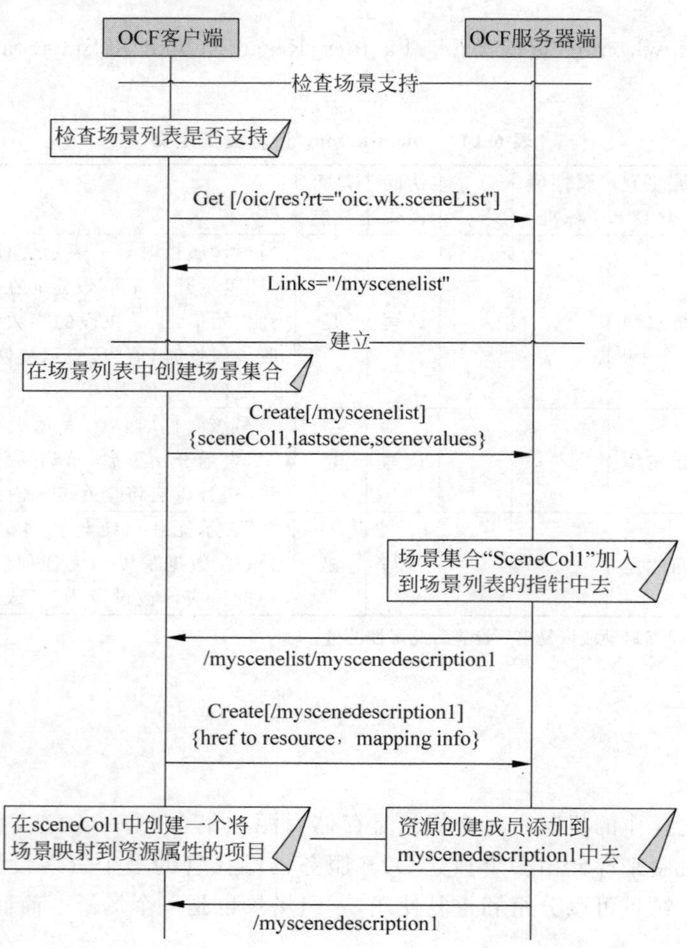

图 6-12　检查场景支持和特定场景支持的交互

3. 场景互动

具有 OCF 能力的客户端可以与场景互动。允许的场景值和最后应用场景值可以从托管现场 OCF 的服务器端进行检索。现场值应通过最后一次场景属性设置，现场值通过一个有效载荷的更新操作来改变。这些步骤在图 6-13 中说明，被描述的最后一次场景值并不意味着是场景资源当前状态的映射值。这是由于该设置现场的值并不建模为系统的实际状态。这意味着另一个 OCF 客户端可以改变场景的值,只是资源中部分无须反馈场景状态的值被改变。

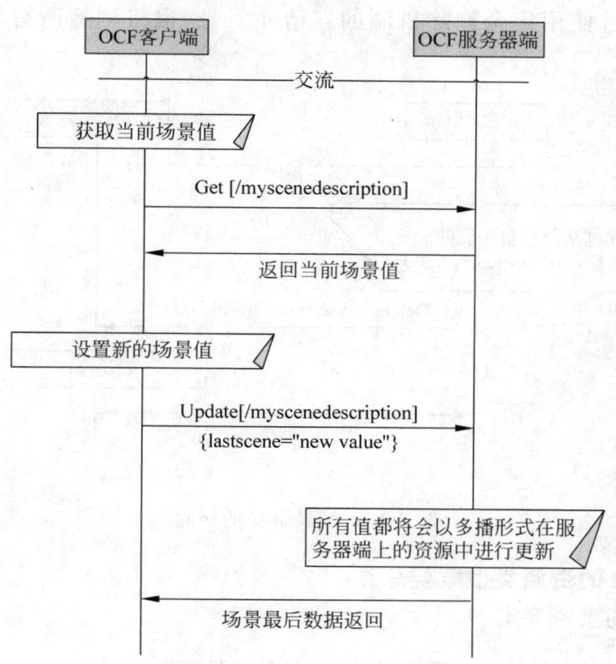

图 6-13 在特定场景下的客户交互

如前所述,一个场景可引用存在于一个或多个 OCF 服务器端的一个或多个资源,每一个场景发生改变时重新评估。这个评估要么嵌入 OCF 服务器端的一部分现场,要么单独通过一个检索操作,由具有识别场景的 OCF 客户端触发,观察该机制引用的资源。在评估过程中的新场景值映射将被应用到 OCF 服务器端。此行为如图 6-14 所示。

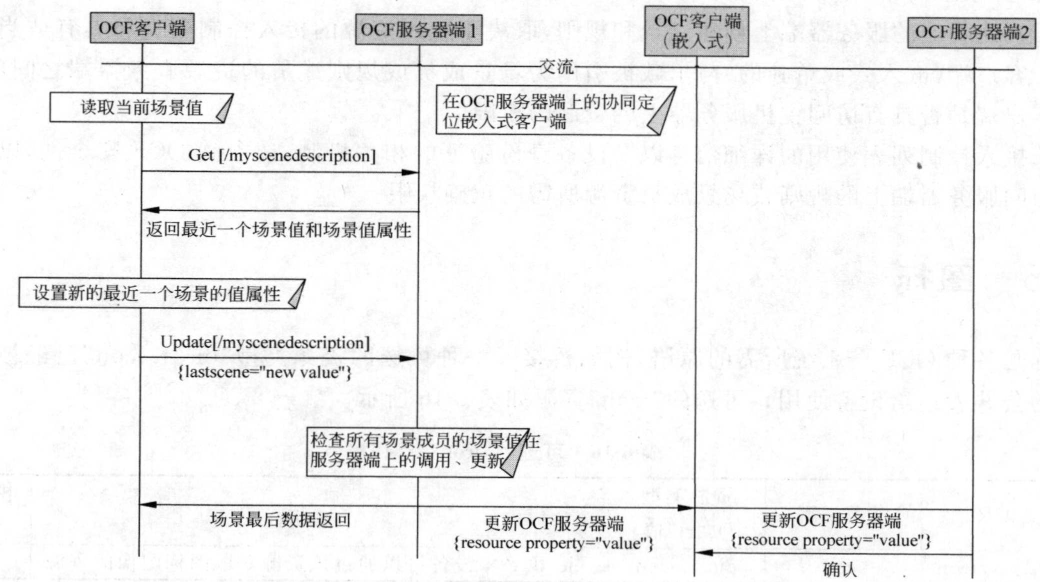

图 6-14 由于一个场景的改变可能发生的交互

4. 一个场景的删除

当不再需要场景集合时，最终用户可以从 OCF 服务器端删除场景集合。由于场景集合是 OCF 集合的特殊化，那么它是通过使用集合删除机制的。请注意，这也将删除所有场景资源，包括场景成员。删除过程如图 6-15 所示。

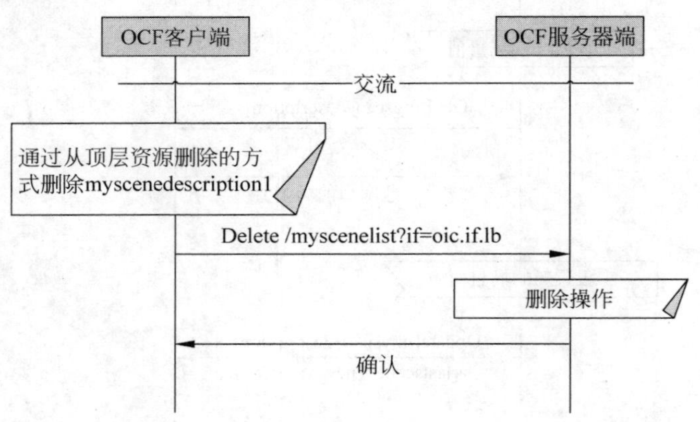

图 6-15 场景资源的删除

5. 对于场景功能定义的资源类型概要总结

表 6-15 汇总了部分场景资源类型。

表 6-15 部分场景资源类型

人性化名称	资源类型(rt)	简 单 描 述
场景列表	oic. wk. sceneList	包含场景集合的顶级集合
场景集合	oic. wk. sceneCollection	零个或多个场景的描述
场景成员	oic. wk. sceneMember	场景集合中每个特定资源部分的映射描述

6. 安全考虑

在具有此功能的服务器端上创建场景和规则，取决于应用资源的接入控制列表和具有适当权限的客户端。客户端(嵌入式或单独的)与承载被引用为场景成员或规则成员的资源服务器端之间的交互，取决于客户端是否具有访问主机服务器上的资源的权限。

有关接入控制列表使用的详细信息以及设备身份验证的相关机制，请参阅 OCF 安全性，以确保客户端有访问服务器端上的规则或场景成员资源所需的正确权限。

6.6 图标

图标是各种 OCF 子系统所需的原语，例如桥接。一种可选的资源类型"oic. r. icon"已经被作为图标资源的公共表达给设备使用。可送的"icon"资源如表 6-16 所示。

表 6-16 可选的"icon"资源

预定义 URI	资源类型	资源类型ID("rt"值)	接口	描 述	相关功能交互
/example/oic/icon	icon	oic. r. icon	oic. if. r	设备可以通过该资源获取图标图像的资源	图标

表 6-17 详细定义了"oic. r. icon"资源类型。

表 6-17 "oic. r. icon"资源类型定义

属性	属性名	属性值类型	属性值标准	单位	访问模式	是否强制	描述
Mime 类型	mimetype	字符串			读	是	指定图标的格式(媒体类型)。它应该是 IANA 媒体类型中指定的模板字符串
宽度	width	整型	>=1		读	是	图标的宽度(以像素为单位)大于或等于 1
高度	height	整型	>=1		读	是	图标的高度(以像素为单位)大于或等于 1
图标	media	URI			读	是	用于定位图标图像的 URI

6.7 内省

内省是宣布在设备商托管资源的功能机制。内省设备数据的预期用途是启用动态客户端。例如，可以使用内省设备数据的客户端动态生成用户界面，或者动态地将托管资源翻译创建到另一个生态系统。内省的其他用途是可以使用该信息来生成客户端代码。内省设备数据旨在扩大线上已有的数据。这意味着需要使用现有的机制来全面了解在设备中实现的内容。例如，内省设备数据不表示关于观察的信息，因为已经使用"/oic/res"中链接上的"p"属性表示了该信息。

内省设备数据推荐作为静态数据表示。这意味着数据在设备正常运行时间内不会改变。然而，当数据不是静态时，内省资源将指示为可观察的，并且"oic. wk. introspection"资源的 URL 属性值应更改为指示内省设备数据已更改。

内省设备数据描述组成的资源，所包含资源如表 6-18 所示。内省设备数据被描述为 JSON 格式的 Swagger2. 0 文件。Swagger2. 0 文件将按照以下定义包含资源描述。

(1) 内省设备数据中的资源 URL 不应具有端点描述，例如，它不应该是一个完整的 URL，而只是来自端点的相对路径，相对路径应与"/oic/res"的表达相同。

(2) "/oic/res"资源不应该被列在内省设备数据中。

(3) 资源"/oic/d""/oic/p"和安全资源允许出现在内省设备数据中，但不是必需的。当供应商定义或可选资源应用了，那么应该包括"/oic/d""/oic/p""/oic/res"和安全资源。

(4) 所有其他资源必须包含在内省设备数据中。

(5) 每个资源将会包含：所有被应用的方法，所支持的每种方法中应用的查询参数；支持接口"if"作为枚举值；方法中的请求和响应载体方案，方案数据应该表示为 Swagger 方案。

(6) 应当执行 Swagger2. 0 方案：它是完全解决方案，在 Swagger2. 0 文件外应该不存在任何引用；方案应该列出所支持的接口；方案应该列出属性是可选的还是必需的；方案应该指出属性是只读的还是可读写的；通过属性标记，默认的"rt"属性应当被用于指出其支持的资源类型。允许使用 oneOf 和 anyOf 结构作为 Swagger2. 0 方案的一部分。

动态资源(例如，可以根据客户端请求创建的资源)的 URL 定义中应该包含 URL 标识符(例如使用{}语法)。具有{}的 URL 标识资源可以用于创建整个资源组。实际路径中可以包含与该资源相链接的集合点路径。具有标识符的URL举例如下。

```
/SceneListResURI/{SceneCollectionResURI}/{SceneMemberResURI}:
```

当允许在一个集合内创建不同的资源类型时，那么不同的创建方法应该将所有可能被创建的资源类型都进行定义。oneOf的结构允许使用可选资源定义模式。oneOf结构允许所有模式进行集成，现有的子模式可以用于表示可创建资源的定义，如下所示。

```
{
"oneOf":[
  {<< subschema 1 definition>>},
  {<< sub schema 2 definition>>}
]
}
```

使用内省设备数据的客户端，应检查设备支持的内省设备数据版本。每个文件中都标明Swagger版本，标签为“swagger”。Swagger2.0版本的标签示例是"swagger":"2.0"。该规范的较新版本可能会引用Swagger较新版本，例如Swagger3.0。

一个服务器端将支持一个内省资源，资源类型为“oic.wk.introspection”，如表6-18所定义。资源类型为“oic.wk.introspection”的资源应包含在资源“/oic/res”内。

表6-18 内省资源

预定义URI	资源类型	资源类型ID(“rt”值)	接口	描 述	相关功能交互
无	introspection	oic.wk.introspection	oic.if.r	告知内省文件的URL的资源	内省

表6-19定义了“oic.wk.introspection”资源类型。

表6-19 “oic.wk.introspection”资源类型定义

属 性	属性名	属性值类型	属性值规则	访问模式	是否强制	描 述
URL信息	urlinfo	数组		读	是	对象数组
URL	url	字符串	url	读	是	托管负载的URL
协议	protocol	字符串	enum	读	是	从URL检索内省设备数据的协议
内容类型	content-type	字符串	enum	读	否	URL的内容类型
版本	version	整型	enum	读	否	内省协议的版本。当前值为1

内省设备数据检索应该按照以下步骤进行，如图6-16所示。

(1) 检查内省资源是否支持并检索资源的URL。

(2) 检索内省资源的内容。

(3) 从内省资源中指定的URL上下载内省设备数据。

(4) 客户端使用内省设备数据。

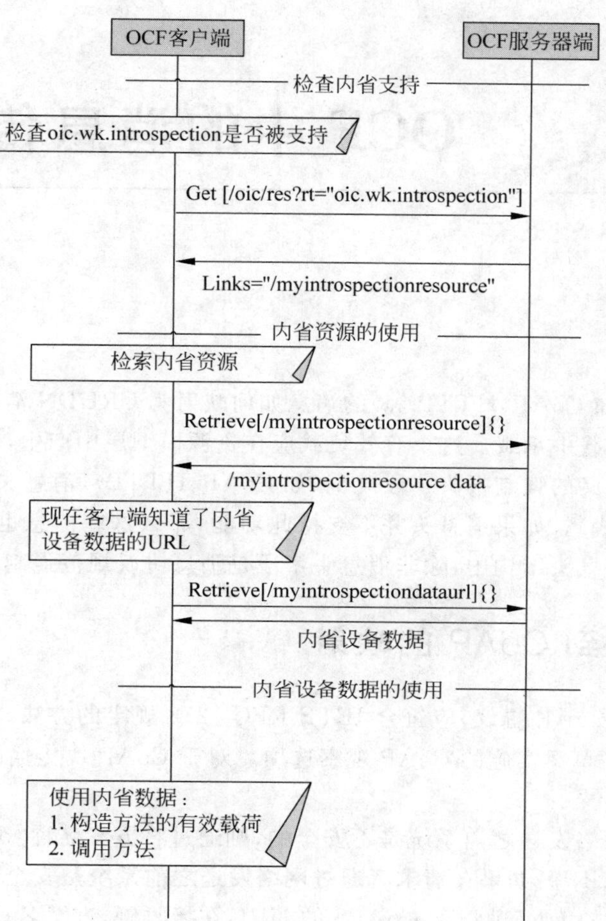

图 6-16 检查内省支持以及下载内省设备数据的交互

第 7 章

CHAPTER 7

OCF 中的消息传递

本章规定每种协议(如 CoAP、HTTP 等)的消息如何映射为 CRUDN 消息的传递操作,所有的资源模型属性信息应在消息负载中承载。这个有效负载应在资源模型层中产生,并在数据连接层进行封装。除了在消息收发协议中定义的强制报头字段规范(如 CoAP、HTTP),消息头仅用于描述该消息的有效载荷(如消息有效负载格式)。如果消息头并不支持此功能,那么这些信息也应在消息负载中携带。资源模型信息不应包含在消息头结构中,除非消息头字段是消息协议规范强制的。

7.1 CRUDN 到 CoAP 的映射

一个 OCF 设备实施 CoAP 协议,应符合 IETF RFC 7252 规定的方法。同时,一个 OCF 设备实施 CoAP 协议,应符合 IETF 草案实施的 CoAP 观察选项。对于 CoAP 块传输的支持,系统规定了最大有效载荷 MTU。

如果在请求者通过网络发送之前系统是不安全的,则通过将方案名称"oic"替换为"coap"来将 OCF 的 URI 映射到 CoAP 的 URI。如果在请求者通过网络发送之前系统是安全的,则通过将方案名称"oic"替换为"coaps"来将 OCF 的 URI 映射到 CoAPS 的 URI,在接收侧,方案名称被替换为"oic"。

7.1.1 具有请求和响应的 CoAP 方法

每个请求都具有实现该请求的 CoAP 方法。主要方法及其含义如表 7-1 所示,其中,提供如何映射传统的 GET/PUT/POST/DELETE 方法到 OCF 规范中的 CREATE/RETRIEVE/ UPDATE/ DELETE 操作方法。使用这些操作方法,相关的描述提供一般的操作行为,但资源的接口可以修改这些通用语义(默认接口体现了通用语义)及其通用的行为。

表 7-1 CoAP 主要方法及其含义

用于 CRUDN 的方法	(强制的) 请求数据	(强制的) 响应数据
用于检索的 GET	方法代码：GET(0.01) 请求 URI：要检索资源的现有 URI	响应代码：成功(2.××)或失败(4.××) 负载：目标资源的资源表示(成功时)
用于创建的 POST	方法代码：POST(0.02) 请求 URI：负责创建资源的现有 URI 负载：要被创建资源的资源表示	响应代码：成功(2.××)或失败(4.××) 负载：新创建资源的 URI(成功时)
用于创建的 PUT	方法代码：PUT(0.03) 请求 URI：要被创建资源的新 URI 负载：要被创建资源的资源表示	响应代码：成功(2.××)或失败(4.××)

续表

用于 CRUDN 的方法	（强制的）请求数据	（强制的）响应数据
用于更新的 POST	方法代码：POST(0.02) 请求 URI：将被部分更新的资源的现有 URI 负载：要被部分更新资源的(部分)资源表示	响应代码：成功(2.××)或失败(4.××)
用于更新的 PUT	方法代码：PUT(0.03) 请求 URI：要被完全替换资源的现有 URI 负载：要被更新资源的资源表示	响应代码：成功(2.××)或失败(4.××)
用于删除的 DELETE	方法代码：DELETE(0.04) 请求 URI：要被删除资源的现有 URI	响应代码：成功(2.××)或失败(4.××)

1. 通过 POST/PUT 进行 OCF 的创建

本节介绍如何使用 POST 或者 PUT 方法创建 OCF 的相关资源。

1）通过 POST 进行 OCF 的创建

POST 仅在请求 URI 有效的情况下使用，也就是说，它是正在处理请求服务器端现有资源的 URI。如果不存在这样的资源，则服务器端将以 4.××的错误代码进行响应。通过 POST 进行 OCF 的创建将使用的现有请求 URI，它负责创建服务器端的资源。创建的资源 URI 由服务器端确定，并在响应中提供给客户端。

客户端应在请求有效载荷中包括新资源的表示。有效载荷中新资源的表示应具有创建有效资源实例的所有必要属性，即所创建的资源应能够正确响应具有强制接口的有效请求（例如，GET with? if =oic. if. baseline)。

一旦接收到 POST 请求，服务器端会执行以下操作之一：使用新的 URI 创建新的资源，使用新创建资源的 URI 和成功响应代码(2.××)进行响应；错误响应代码(4.××)。POST 是不安全的，是无法预期或保证幂等行为时支持的方法。

2）通过 PUT 来创建

PUT 被用于创建一个新的资源或完全替代一个已存在资源的完整表示。在 PUT 请求的负载中，资源的表示应该是完整的表示。通过 PUT 来创建应该使用一个新的请求 URI 来标识要创建的新资源。

有效载荷中的新资源表示，应具有要创建有效资源实例的所有必要属性，即所创建的资源应能够使用强制接口对有效请求进行正确响应(例如，GET with? if =oic. if. baseline)。

一旦接收到 PUT 请求，服务器端会执行以下操作之一：创建具有 PUT 请求中提供请求 URI 的新资源，并发回具有成功响应代码(2.××)的响应；错误响应代码(4.××)。PUT 是一种不安全的方法，但它是幂等的，因此，当重复 PUT 请求时，每次的结果是相同的。

2. 使用 GET 进行 OCF 的检索

GET 可以用于检索操作，GET 方法检索由请求 URI 标识的目标资源表示。一旦接收到 GET 请求，服务器端会执行以下操作之一：使用成功响应代码(2.××)发回具有目标资源表示的响应；错误响应代码(4.××)或忽略它(例如不适用的多播 GET)。GET 是一种安全的方法，是幂等的。

3. 使用 POST 进行 OCF 的更新

POST 仅在请求 URI 有效的情况下使用，也就是说，它是正在处理请求服务器端现有资源的 URI。如果不存在这样的资源，则服务器端将以 4.××的错误响应代码进行响应。客户端应该使用 POST 用于现有资源的部分更新,此时有效载荷包含了目标资源的一部分属性。

一旦接收到POST请求,服务器端会执行以下操作之一:根据使用的接口将请求应用于由请求URI标识的资源(即忽略不存在属性的POST),并发回具有成功响应代码(2.××)的响应;错误响应代码(4.××)。注意,如果有效载荷中的表示与使用所应用接口的POST目标资源不兼容(即由于有效载荷中的只读属性,"覆盖"语义不能被执行),则会返回错误响应代码4.××。POST是一种不安全的方法,是无法预期或保证幂等行为时支持的方法。

4. 使用DELETE进行OCF的删除

DELETE用于OCF的删除操作。DELETE方法请求删除由请求URI标识的资源。一旦接收到删除请求,服务器端会执行以下操作之一:删除目标资源并发回具有成功响应代码(2.××)的响应;错误响应代码(4.××)。DELETE是一种不安全的方法,但它是幂等的(除非URI被重新用于新实例)。

7.1.2 内容类型

该OCF设备框架任务支持CBOR,如果支持多于一种编码类型(CBOR或者JSON),由一个实现支持,它允许有效载荷体协商。在这种情况下,在IETF RFC 7252中定义的接收选项,应被用来表示OCF客户端请求的内容格式。支持的内容格式如表7-2所示。

表7-2 OCF支持的内容格式

媒体类型	ID	媒体类型	ID
application/cbor	60	application/vnd.ocf+cbor	10000

客户端应在有效载荷的消息中包括内容格式选项。服务器端应包含一个内容格式选项,用于所有成功(2.××)响应与有效载荷主体。根据IETF RFC 7252,服务器端应包含与有效载荷主体的所有错误(4.××或5.××)响应的内容格式选项,除非它们包括诊断有效载荷;诊断有效载荷的错误响应不包括内容格式选项。内容格式选项应使用表7-2中的ID列数值。OCF可以直接授权特定的内容格式选项。

客户端还应在每个请求消息中包含接收选项。接收选项应指示响应消息在表7-3中定义的所需内容格式。服务器端应返回所需的内容格式(如果可用),如果不能返回所需的内容格式,则服务器端将使用适当的错误消息进行响应。

服务器端、客户端的请求和相应消息应该包含OCF内容格式版本。客户端应在请求消息中包含OCF接收内容格式版本。OCF内容格式版本和OCF接收内容格式版本在CoAP头中指定为选项号码,如表7-3所示。

表7-3 OCF内容格式版本和OCF接收内容格式版本的选项号码

CoAP选项号码	名称	格式	长度/字节
2049	OCF接收内容格式版本	uint	2
2053	OCF内容格式版本	uint	2

OCF内容格式版本和接收内容格式版本的值是2字节长度的无符号整型数,用来定义主要版本、次要版本以及副版本。主要版本和次要版本占据5位,副版本占据6位,如表7-4所示。表7-5解释了几个实例。

表 7-4　OCF 接收内容格式版本和 OCF 内容格式版本表示

主要版本						次要版本					副版本					
位	15	14	13	12	11	10	9	8	7	6	5	4	3	2	1	

表 7-5　OCF 内容格式版本和 OCF 接收内容格式版本表示的实例

OCF 版本	二进制表示	整型值
1.0.0	0000100000000000	2048
1.1.0	0000100001000000	2112

在 OCF 1.0 版本中的 OCF 接收内容格式版本和 OCF 内容格式版本应该是 1.0.0(如 0B0000100000000000)。

为了实现本规范中不同版本设备之间的兼容性,设备应遵循如图 7-1 所示的策略。所有设备都应该支持现在的、以往所有的内容格式选项和版本。客户端应该发送带有所有版本内容格式的发现请求消息,直到其发现所有在网络中的服务器端。

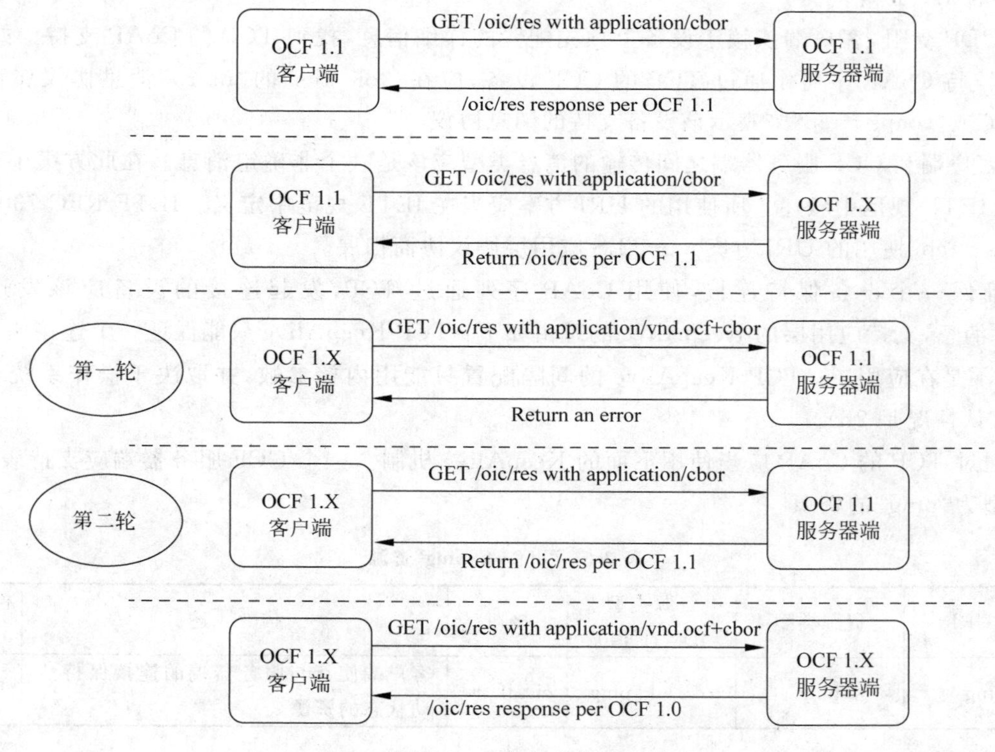

图 7-1　内容格式策略

7.1.3　CoAP 响应代码及块传输

CRUDN 操作响应代码到 CoAP 响应代码的映射与 IETF RFC 7252 中定义的响应代码相同。

基本 CoAP 消息的小型有效载荷是典型的轻量级,受限于物联网设备的正常工作。但是有些情景应用程序需要传输大量的有效载荷。如 IETF 定义 CoAP 逐块传输,OCF 所有服务器端使用处理任何定义 CRUDN 操作的结果，其生成内容的有效载荷将超过 CoAP 数据包的大小。

同样，如果 IETF 定义 CoAP 逐块传输应由 OCF 所有客户端支持。利用块传输有效载荷以及有效载荷超过一个 CoAP 数据块大小的传输，两者均可以接收。正在使用该机制传输的单个实例发送所有块，都应具有相同的可靠性设定(即可确定或可证实)。

由 IETF 描述的 OCF 的客户端可同时支持块 1(描述性)和块 2(对照)选项。一个 OCF 服务器端可以同时支持块 1(对照)和块 2(描述性)的选项。

7.2 CoAP 序列通过 TCP

在 TCP 已经可用的环境中，可以利用 CoAP 提供的可靠性。此外，在某些环境中 UDP 流量被阻挡，因此，部署可能使用 TCP 协议。例如，考虑云应用程序充当客户端 OCF，OCF 服务器端位于用户的家中。它已经支持 CoAP 作为消息传递协议 OCF 的服务器端(例如，智能家居垂直应用)，可以很容易地支持 CoAP 序列通过 TCP，而不是添加另一种消息传递协议。一个 OCF 设备通过 TCP 实现序列化 CoAP 应符合 IETF 规范。

如果 UDP 受阻，客户端依赖于设备上预先配置的详细信息，找到 TCP 的 CoAP 支持。如果 UDP 没有受阻，支持 CoAP 序列化通过 TCP 的 OCF 设备，应在“/oic/res”的“mpro”消息协议属性中填充“coap+tcp”或“coaps+tcp”来表示该设备支持的消息协议。

OCF 客户端和 OCF 服务器端之间传输的消息类型应该是一个非确定消息。在此方案中使用的协议栈，应在 IETF 规范中叙述。所使用的 URI 方案应当在 IETF 规范中定义。IETF RFC 7301 描述了对于“coaps+tcp”使用的 URI 方案——TLS 应用层协议协商扩展。

为了确保一个设备保持连接，使用 CoAP 序列通过 TCP，发起连接的设备应该发送应用层 KeepAlive 消息，支持应用层的 KeepAlive 的理由如下：TCP KeepAlive 只能保证一个连接始终存在于网络层，而不是在应用层；TCP KeepAlive 的间隔配置只能用内核参数，并取决于操作系统(例如，在 Linux 中默认情况是 2h)。

支持通过 TCP 的 CoAP 应当使用下面的 KeepAlive 机制。一个 OCF 服务器端应支持表 7-6 定义的资源类型为“ping”的资源。

表 7-6 “oic. wk. ping”资源

预定义 URI	资源类型	资源类型 ID(“rt”值)	接口	描　述	相关的功能交互
/oic/ping	ping	oic. wk. ping	oic. if. rw	客户端使其与服务器端的连接保持活动状态的资源	KeepAlive

表 7-7 定义了“oic. wk. ping”资源类型。

表 7-7 “oic. wk. ping”资源类型定义

属　性	属性名	属性值类型	属性值规则	单位	访问模式	是否强制	描　述
间隔	in	整型	min		读写	是	连接保持活动和未关闭的时间间隔

下列步骤详细描述了客户端和服务器端的 KeepAlive 机制。

(1) 客户端希望保持与 OCF 服务器端的连接状态,应在 OCF 的服务器端更新其连接时间间隔上发送一个 POST 请求到"/oic/ping"的资源。本时间间隔应开始为 2min,并以 2 的倍数增加,最多用时 64min,然后停留在 64min。

(2) 一个 OCF 服务器端接收到请求应在 1min 内做出回应。

(3) 如果 OCF 客户端未能在 1min 内收到答复,应当终止连接。

(4) 在指定的时间间隔内,如果 OCF 服务器端没有收到 POST 请求到"ping"的资源,OCF 服务器端将终止连接。

7.3 CBOR 中的负载编码

OCF 实现将根据 IETF RFC 7049 执行从 JSON 定义的模式到 CBOR 的转换以及从 CBOR 到 JSON 的转换,除非在规范中另有规定。

定义为 JSON 整数的属性应在 CBOR 中编码为整数(CBOR 主要类型 0 和 1)。定义为 JSON 数字的属性应编码为整数、单精度浮点数或双精度浮点数(CBOR 主要类型 7,子类型 26 和 27);选择是依赖于实现的。不应使用单精度浮点数(CBOR 主要类型 7,子类型 25)。整数的闭合范围为$[-2^{53},2^{53}]$。定义为 JSON 数字的属性应尽可能编码为整数;如果不能,在精度损失不影响服务质量的情况下,定义为 JSON 数字的属性应该使用单精度,否则属性将使用双精度。

在接收到 CBOR 有效载荷时,实现应该能够在任何位置解释 CBOR 整数值。如果定义为 JSON 整数的属性接收到的编码不是整数,那么实现可以使用适用于底层传输的最终响应(例如 CoAP 为 4.00,HTTP 为 400)拒绝此编码,因此需要对整数情况进行优化。如果一个属性被定义为一个 JSON 数值,则实现应该接收整数、单精度浮点数和双精度浮点数。

第 8 章 OCF 的应用实例

CHAPTER 8

本章通过描述参与实体间的一系列操作，来说明 OCF 的场景实例，包括 OCF 交互场景与部署模型、其他资源模型与 OCF 的映射。

8.1 OCF 操作例程

本节描述一个参与实体操作的场景。在例子中，“Light”是一个 OCF 服务器端，“智能手机”是一个 OCF 客户端。在环境中，“Garage”也是 OCF 服务器端。所有例子都遵循以下示例资源类型定义：“oic. example. light”，如表 8-1 所示。

表 8-1 “oic. example. light”资源类型定义

属　性	属性名	属性值类型	属性值规则	单位	访问模式	是否强制	描　述
名称	n	字符串			读写	否	
开关	of	布尔型			读写	是	on/off 控制：0＝off，1＝on
调光器	dm	整型	0～255		读写	是	该资源的取值范围为[0,255]

“oic. example. garagedoor”资源类型定义，如表 8-2 所示。

表 8-2 “oic. example. garagedoor”资源类型定义

属　性	属性名	属性值类型	属性值规则	单位	访问模式	是否强制	描　述
名称	n	字符串			读写	否	
开关	oc	布尔型			读写	是	open/close 控制：0＝open，1＝close

下面例子中用的“/oic/mon”（“rt＝oc. wk. mon”）和“/oic/mnt”（“rt＝oc. wk. mnt”）在资源模型中定义。在家用智能手机中打开单个灯，如图 8-1 所示，这个序列强调了 OCF 智能手机对 OCF 灯资源的发现和控制。

发现请求可以发送给 CoAP 多播地址 224. 0. 1. 187 或者可以直接发送给管理灯资源设备的 IP 地址。

(1) 智能手机发送一个 GET 请求给“/oic/res”资源来发现目标终端控制的所有资源。

(2) 终端(灯泡)用资源 URI、资源种类和该终端支持的接口作为应答(其中一个资源是“/light”，“rt＝oic. example. light”)。

(3) 智能手机发送一个 GET 请求给"/light"从而了解其现在的状态。终端用光资源的表示来应答({n=bedlight; of=0})。

(4) 智能手机通过发送一个 POST 请求给"/light"资源来改变光资源的"of"特性({of=1})。

(5) 一旦该请求成功执行,终端用改变的资源表示来回应;否则,返回错误代码。错误代码的细节在消息中定义。

图 8-2 强调了检测和维护的设备管理功能。前提条件是管理设备具有不同的安全权限,因此,可以在该 OIC 执行设备管理操作。

(1) 管理设备发送一个 GET 请求给"/oic/res"资源去发现目标终端的所有资源(在这个例子中是灯泡)。

(2) 终端(灯泡)用资源 URI、资源类型和终端支持的接口应答(其中一个资源是"/oic/mon","rt=oc. wk. mon"且另一个资源是"/oic/mnt","rt=oc. wk. mnt")。

(3) 在看见终端数据之后,管理设备通过发送 POST 请求给"/oic/mnt"资源({fr=1})来改变维修资源的"fr"特性,触发了终端(灯泡)的恢复出厂设置。

(4) 一旦请求被成功执行,终端用改变了的资源表示来回应;否则,将返回错误代码。错误代码的细节在消息中定义。

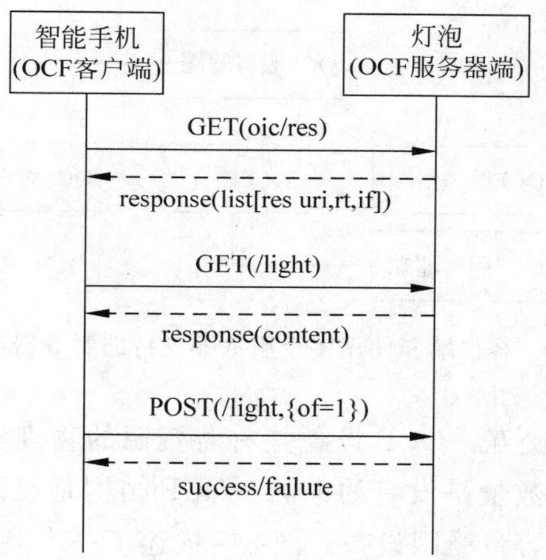

图 8-1 在家用智能手机打开单个灯

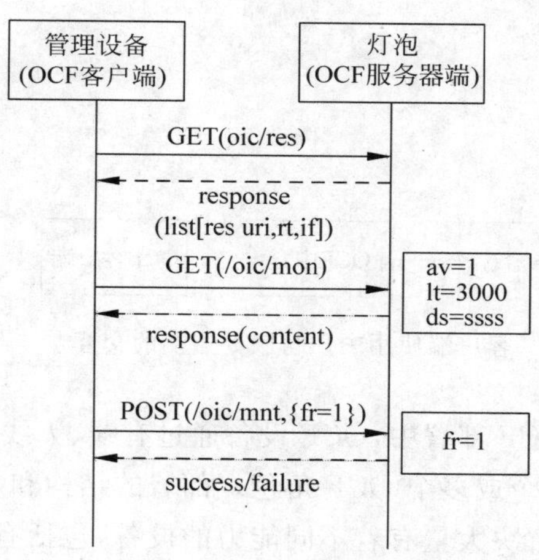

图 8-2 设备管理(检测和维护)

8.2 OCF 交互场景与部署模型

一个 OCF 客户端连接一个或多个 OCF 服务器端,以便于访问这些 OCF 服务器端提供的资源。图 8-3 表示 OCF 角色之间可能交互的场景。

在这个场景中,OCF 客户端和 OCF 服务器端直接交流,不用其他任何 OCF 设备的参与。例如,一个控制执行器的智能手机直接使用这个场景。

OCF 客户端和使用另一个 OCF 服务器端的 OCF 服务器端的交互,如图 8-4 所示。在这个场景中,需要另一个 OCF 服务器端提供所需要的支持,OCF 客户端直接访问一个特定 OCF 服务器端的资源。这个场景应用中,当一个智能手机首先访问了一个发现服务器端,找到特定应用程序的寻址信息,然后直接访问应用程序进行控制。

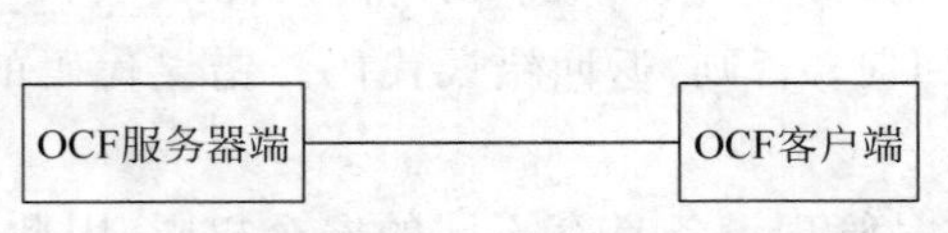

图 8-3 服务器端和客户端之间的直接交互

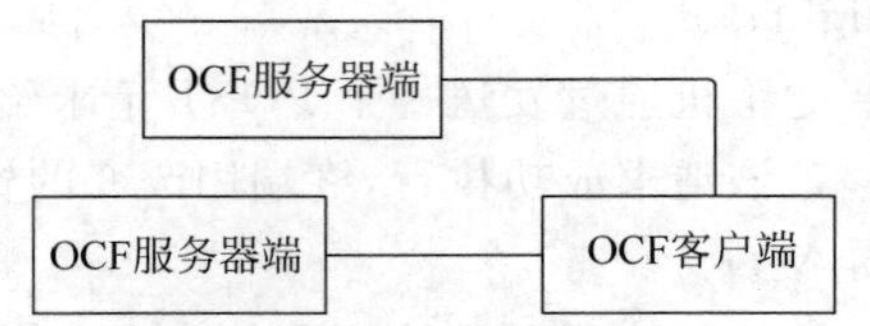

图 8-4 客户端和使用另一个服务器端间的交互

OCF 客户端和使用 OCF 中介，与 OCF 服务器端进行交互，如图 8-5 所示。在这个场景中，一个 OCF 中介连接了 OCF 客户端和 OCF 设备之间的交互，通过代理，控制智能家居电器，智能手机使用这个场景。

使用多个 OCF 服务器端和中介，OCF 客户端和 OCF 服务器端之间的交互如图 8-6 所示。在这个场景中，OCF 服务器端和 OCF 中介角色是用来促进 OCF 客户端与一个特定 OCF 服务器端的交互。在实际场景中，智能手机首先访问资源目录服务器端去找到特定家用电器的地址，然后利用代理给该家用电器传递命令消息。该智能手机可以利用资源目录中定义的机制(例如，默认位置、多播地址或者 DHCP)去发现资源目录信息。

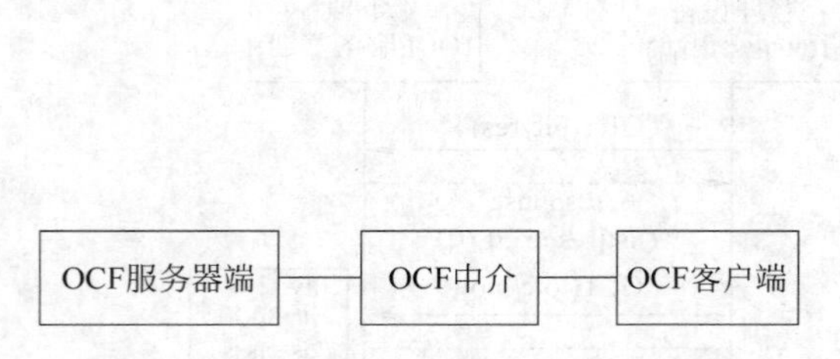

图 8-5 客户端使用中介与服务器端的交互

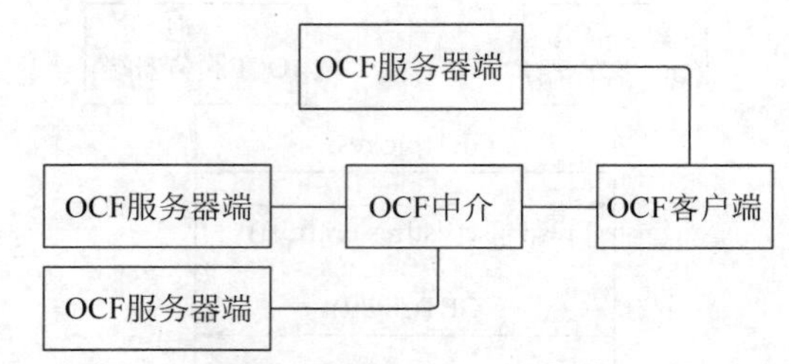

图 8-6 客户端和使用多个服务器支持的服务器端的交互

在 OCF 部署中，OCF 设备通过有线、无线来部署和交互。OCF 设备是有着资源的物理实体，并且承担着一个或多个 OCF 角色。部署的结构和 OCF 设备数量是没有约束的。OCF 结构是灵活的，可扩展的，能解决大量有着不同能力的设备，包括有限内存和容量受限设备。图 8-7 描述了一个典型的 OCF 部署和一套 OCF 设备，可分为以下几类。

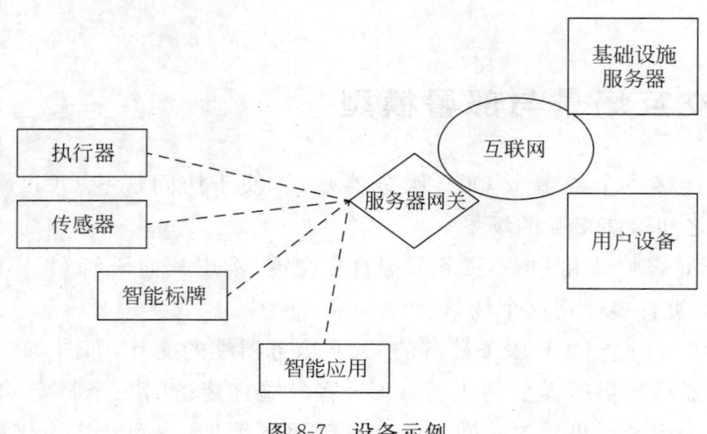

图 8-7 设备示例

物体：能够与物理环境接口的网络设备，是主要被控制和监测的设备，如智能家电、传感器以及执行器。物体大部分都承担OCF服务器端的角色，也承担OCF客户端的角色，例如，在机器对机器的通信中。

用户设备：被用户使用的设备，允许用户区访问资源和服务，如智能手机、笔记本电脑和其他穿戴式设备。用户设备主要承担OCF客户端的角色，但也承担OCF服务器端和OCF中介的角色。

服务器网关：承担OCF中介的网络设备，如家庭网关。

基础设施服务器：驻留在云基础设备的数据中心，通过提供网络服务，如AAA、NAT遍历，来促进OCF设备之间的交互。

8.3 其他资源模型与OCF映射

本节主要包括多资源模型、支持多资源模型的OCF方法、资源模型指示以及配置文件实例。

8.3.1 多资源模型

RESTful交互是依赖于资源模型定义的。因此，为了互操作性，OCF设备需要对资源模型有共同的理解。不同的组织(包括OCF、IPSO和oneM2M联盟)定义了多种资源模型，并且将其用在工业中。这限制了各自生态系统之间的互操作性。不同资源模型的差别如下。

(1) 资源结构：资源被定义为拥有属性(如oneM2M)，或被定义为一个原子实体并且不可以被分解为属性(如IPSO)。例如，智能灯可以表示为有开关属性的资源或者一个包括开关资源的集合。前者的开关属性没有自己的URI，并且只可以通过这个资源来间接访问。而后者本身成为一个资源，开关资源被分配了自己的URI并且可以被直接操作。

(2) 资源名称和种类：资源可以允许自由地命名，且用不同的资源类型属性来表明它的特性(如oneM2M)。但是，有的组织就预先固定了资源的名称，用名字本身去表明其特性(如IPSO)。例如，智能灯可以在oneM2M中被命名为“LivingRoomLight_1”，但是IPSO联盟是具有固定的物体名称和物体ID的，例如“IPSO Light Control(3311)”。结果是，URI的数据路径，oneM2M使用自由定义，而IPSO使用预先定义。

(3) 资源分层：一些允许资源按分层来管理，以便于一个资源本身可以包括其他资源，以一种父子关系(如oneM2M)；其他的联盟以扁平结构来托管资源，只能通过应用其链接来访问其他资源。

除了以上不同之外，不同的组织用不同的语法，并且定义不同的特性(例如资源接口)，这将抑制物联网的互操作性。

8.3.2 支持多资源模型的OCF方法

为了拓展物联网生态系统，OCF架构采用包容的方法来促进现有的资源模型之间的互通。具体来说，OCF架构定义了OCF资源模型，同时也提供了一种机制，可以将该资源模型映射到其他模型。通过包容现有的资源模型，OCF包容现有的生态系统，且在向所有生态系统的资源模型过渡。下面的OCF特性实现了对其他资源模型的支持。

(1) OCF资源模型是多个模型的超集：OCF资源模型被定义为现有资源模型的超集。换句话说，任何现存的资源模型可以被映射为OCF资源模型概念的子集。

(2) OCF框架允许资源模型协商：OCF客户端和服务器端交换各自支持的资源模型信息。基于交换的信息,OCF 客户端和服务器端选择一个资源模型去执行 RESTful 交互或执行转化。下面对资源模型协商进行高层描述。

8.3.3 资源模型指示

OCF 客户端和服务器端交换各自支持的资源模型信息。基于交换的信息,OCF 客户端和服务器端选择一个资源模型去执行 RESTful 交互或者执行转化,交换是发现和协商的一部分。基于交换,OCF 客户端和服务器端遵循一个过程来保证它们之间的互操作性。它们会选择一个通用的资源模型或者执行资源模型之间的转化。

资源模型模式交换：OCF 客户端和服务器会在初始化 RESTful 交互时共享资源模型信息。交换关于他们支持的资源模型信息,作为会话建立过程的一部分。另外,每个请求或者回应信息会携带正在用哪个资源模型的指示。例如,CoAP 定义了"内容格式选项"来指示"表示格式",如"application/json"。可以拓展"内容格式选项"来指示资源模型,如"application/ipso-json"。

附属程序：在 OCF 客户端和服务器端交换资源模型信息之后,执行合适的程序去保证他们之间的互操作性。最简单的方法就是选择一个 OCF 客户端和服务器端都支持的资源模型,OCF 客户端和服务器端可以通过第三方交互。

除了资源密集型的转化,基于一个配置文件的方法可以用于容纳多个配置文件的 OCF 实现中,也可以用于多个生态系统中。

资源模型配置文件：框架定义资源模型配置文件,实施者或用户选择活动配置文件。所选择的配置文件将设备限制为如何定义、实例化和交互资源的规则。这将允许与已标识生态系统(如 IPSO、oneM2M 等)设备之间的互操作。尽管这使得设备能够参与并成为任何给定生态系统的一部分,但是该方案不允许在运行时通用互操作性。尽管该方法可能适合于资源受限设备,但是非受限资源设备期望支持多个配置文件。

8.3.4 配置文件示例

IPSO 定义了具有特定资源的智能对象,并且这些智能对象获取该资源数据类型确定的值。智能对象规范定义了这样的对象类别,每个资源表示是被建模智能对象的特性。

虽然术语可能不同,但在 OCF 中有等同的概念来表示这些术语。本节提供了等效的 OCF 术语,然后以 OCF 术语描述 IPSO 智能对象。以 IPSO 对象 Light Control 作为参考示例。

IPSO 智能对象的定义,等同于资源类型的定义,也就是定义了要建模的实体相关特性。具体的 IPSO 资源相当于 IPSO 资源属性,具有定义的数据类型、可接收值的枚举、单位、一般描述和接入模式(基于接口)。

IPSO 智能对象定义、开发等资源类型的一般方法是忽略对象 ID,并用包含 IPSO 对象的 OCF"."(点)分隔的名称替换对象 URN(Uniform Resource Name,统一资源名称)。或者,对象 URN 可以作为资源类型 ID 使用[只要 URN 不包含任何"."(点)],使用与资源类型 ID 相同的对象 URN 在与 IPSO 交互时是兼容的。基于对象 URN 的命名对于 OCF 到 OCF 的互操作性没有任何影响,因此优选 OCF 格式,对于 OCF 到 OCF 的互操作性,只需要数据模型的一致性。

有两种模型可用于将 IPSO 对象转换为 OCF。

(1) IPSO 智能对象表示资源的地方。在这种情况下,IP 智能对象被视为资源,其类型与智能对象的描述匹配。此外,IPSO 定义中的每个资源都表示为资源类型中的一个属性(IPSO 资源 ID 被替换为表示该属性的字符串)。当在资源模型中表达 IPSO 数据模型时,这是优选的方法。

(2) 将 IPSO 智能对象建模为集合。将每个 IPSO 资源建模为资源,其具有与 IPSO 资源定义匹配的资源类型。然后,每个资源实例都绑定到表示此 IPSO 智能对象的集合。

下面是一个示例,显示如何将 IPSO Light Control 对象建模为资源。

资源类型为"Light Control",此对象用于控制光源,例如 LED 或其他光源。它允许开关灯,并且其调光器设置控制为 0～100%。可选的颜色设置,允许使用字符串来指示所需的颜色。表 8-3 和表 8-4 分别定义了资源类型及其属性。

表 8-3 "Light Control"资源类型定义

资源类型	资源类型 ID	多实例	描述
Light Control	oic.light.control 或者 urn:oma:lwm2m:ext:3311	是	带有开/关、可选调光和能量监视器的光控制对象

表 8-4 "Light Control"资源属性定义

属性	属性名	属性值类型	属性值规则	单位	访问模式	是否强制	描述
开/关	on-off	布尔型			读写	是	on/off 控制: 0=off 1=on
调光器	dim	整型		%	读写	否	比例控制,0～100 的整数值作为百分比
颜色	color	字符串	0-100	属性定义	读写	否	字符串表示颜色空间的一些值
单位	units	字符串			读	否	测量单位定义,例如,"Cel"表示温度(摄氏度)
计时	ontime	整型		s	读写	否	灯开着的时间。将值设置为 0 可以将计数器重置
累积有功功率	cumap	符点型		W·h	读	否	自上次累积能量复位或设备启动以来的累积有功功率
功率因数	powfact	浮点型			读	否	负载的功率因数

第 9 章 RAML 定义核心资源类型

CHAPTER 9

OCF 规范中的资源类型定义如表 9-1 所示，包含了所定义的核心资源列表。

表 9-1 按字母顺序的核心资源列表

人性化名称	资源类型(rt)	注 释
Collections	oic. wk. col	集合
Device Configuration	oic. wk. con	设备配置
Platform Configuration	oic. wk. con. p	平台配置
Device	oic. wk. d	设备
Discoverable Resources Baseline Interface	oic. wk. res	可发现资源基准接口
Discoverable Resources Link List Interface	oic. wk. res	可发现资源链表接口
Introspection	oic. wk. introspection	内省
Maintenance	oic. wk. mnt	维护
Resource Directory	oic. wk. rd	资源目录
Platform	oic. wk. p	平台
Ping	oic. wk. ping	Ping
Icon	oic. r. icon	图标
Scenes(Top Level)	oic. wk. sceneList	场景(顶层)
Scenes Collections	oic. wk. sceneCollection	场景集合
Scenes Member	oic. r. switch. binary	场景成员

9.1 OCF 集合

OCF 集合资源类型包含属性和链接。“oic. if. baseline”接口暴露了集合资源本身的链接和属性表示。URI 示例为“/CollectionBaselineInterfaceURI”，资源类型(rt)为“oic. wk. col”。

1. RAML 定义

```
# %RAML 0.8
  title: Collections
  version: 1.0
  traits:
    - interface-ll :
      queryParameters:
        if:
          enum: ["oic.if.ll"]
```

```
    - interface - b :
        queryParameters:
          if:
            enum: ["oic.if.b"]
    - interface - baseline :
        queryParameters:
          if:
            enum: ["oic.if.baseline"]
    - interface - all :
        queryParameters:
          if:
            enum: ["oic.if.ll", "oic.if.baseline", "oic.if.b"]
/CollectionBaselineInterfaceURI:
  description: |
  OCF Collection Resource Type contains properties and links.
  The oic.if.baseline interface exposes a representation of
  the links and the properties of the collection resource itself
  is : ['interface - baseline']
  get:
    description: |
      Retrieve on Baseline Interface
    responses :
      200:
        body:
          application/json:
            schema: |
              {
                "$schema": "http://json - schema.org/draft - 04/schema#",
                "description" : "Copyright (c) 2016 Open Connectivity Foundation, Inc. All rights
reserved.",
                "id": "http://www.openconnectivity.org/ocf - apis/core/schemas/oic.collection -
schema.json#",
                "title": "Collection",
                "definitions": {
                    "oic.collection.setoflinks": {
                        "description": "A set (array) of simple or individual OIC Links. In addition
to properties required for an OIC Link, the identifier for that link in this set is also required",
                        "type": "array",
                        "items": {
                            "$ref": "oic.oic - link - schema.json#/definitions/oic.oic - link"
                        }
                    },
                    "oic.collection.alllinks": {
                        "description": "All forms of links in a collection",
                        "oneOf": [
                            {
                                "$ref": "#/definitions/oic.collection.setoflinks"
                            }
                        ]
                    },
                    "oic.collection": {
                        "type": "object",
                        "description": "A collection is a set (array) of tagged - link or set (array)
of simple links along with additional properties to describe the collection itself",
```

```
                    "properties": {
      "id": {
                        "anyOf": [
                                    {
                                        "type": "integer",
                                        "description": "A number that is unique to that collection;
like an ordinal number that is not repeated"
                                    },
                                    {
                                        "type": "string",
                                        "description": "A unique string that could be a hash or
similarly unique"
                                    },
                                    {
                                        " $ ref": "oic.types - schema.json # /definitions/uuid",
                                        "description": "A unique string that could be a UUIDv4"
                                    }
                                ],
                                "description": "ID for the collection. Can be an value that is unique
to the use context or a UUIDv4"
                            },
                            "di": {
                                " $ ref": "oic.types - schema.json # /definitions/uuid",
                                "description": "The device ID which is an UUIDv4 string; used for
backward compatibility with Spec A definition of /oic/res"
                            },
                            "rts": {
                                " $ ref":"oic.core - schema.json # /definitions/oic.core/properties/rt",
                                "description": "Defines the list of allowable resource types (forTarget
and anchors) in links included in the collection; new links being created can only be from this list"      },
                            "drel": {
                                "type": "string",
                                "description": " When specified this is the default relationship to
use when an OIC Link does not specify an explicit relationship with * rel * parameter"
                            },
                            "links": {
                                " $ ref": " # /definitions/oic.collection.alllinks"
                            }
                        }
                    }
                },
                "type": "object",
                "allOf": [
                    {" $ ref": "oic.core - schema.json # /definitions/oic.core"},
                    {" $ ref": " # /definitions/oic.collection"}
                ]
            }
        example: |
            {
              "rt": ["oic.wk.col"],
              "id": "unique_example_id",
              "rts": [ "oic.r.switch.binary", "oic.r.airflow" ],
              "links": [
                {
```

```
                    "href": "switch",
                    "rt": ["oic.r.switch.binary"],
                    "if": ["oic.if.a", "oic.if.baseline"],
                    "eps": [
                        {"ep": "coap://[fe80::b1d6]:1111", "pri": 2},
                        {"ep": "coaps://[fe80::b1d6]:1122"},
                        {"ep": "coap+tcp://[2001:db8:a::123]:2222", "pri": 3}
                    ]
                },
                {
                    "href": "airFlow",
                    "rt": ["oic.r.airflow"],
                    "if": ["oic.if.a", "oic.if.baseline"],
                    "eps": [
                        {"ep": "coap://[fe80::b1d6]:1111", "pri": 2},
                        {"ep": "coaps://[fe80::b1d6]:1122"},
                        {"ep": "coap+tcp://[2001:db8:a::123]:2222", "pri": 3}
                    ]
                }
            ]
        }
    post:
      description: |
        Update on Baseline Interface
      body:
        application/json:
          schema: |
            {
              "$schema": "http://json-schema.org/draft-04/schema#",
              "description" : "Copyright (c) 2016 Open Connectivity Foundation, Inc. All rights
reserved.",
              "id": "http://www.openconnectivity.org/ocf-apis/core/schemas/oic.collection-schema.
json#",
              "title": "Collection",
              "definitions": {
                "oic.collection.setoflinks": {
                    "description": "A set (array) of simple or individual OIC Links. In addition to
properties required for an OIC Link, the identifier for that link in this set is also required",
                    "type": "array",
                    "items": {
                        "$ref": "oic.oic-link-schema.json#/definitions/oic.oic-link"
                    }
                },
                "oic.collection.alllinks": {
                    "description": "All forms of links in a collection",
                    "oneOf": [
                        {
                            "$ref": "#/definitions/oic.collection.setoflinks"
                        }
                    ]
                },
                "oic.collection": {
                    "type": "object",
                    "description": "A collection is a set (array) of tagged-link or set (array) of
```

```
simple links along with additional properties to describe the collection itself",
                        "properties": {
                            "id": {
                                "anyOf": [
                                    {
                                        "type": "integer",
                                        "description": "A number that is unique to that collection; like
an ordinal number that is not repeated"
                                    },
                                    {
                                        "type": "string",
                                        "description": "A unique string that could be a hash or similarly
unique"
                                    },
                                    {
                                        "$ref": "oic.types-schema.json#/definitions/uuid",
                                        "description": "A unique string that could be a UUIDv4"
                                    }
                                ],
                                "description": "ID for the collection. Can be an value that is unique to
the use context or a UUIDv4"
                            },
                            "di": {
                                "$ref": "oic.types-schema.json#/definitions/uuid",
                                "description": "The device ID which is an UUIDv4 string; used for backward
compatibility with Spec A definition of /oic/res"
                            },
                            "rts": {
                                "$ref": "oic.core-schema.json#/definitions/oic.core/properties/rt",
                                "description": "Defines the list of allowable resource types (for Target
and anchors) in links included in the collection; new links being created can only be from this list"      },
                            "drel": {
                                "type": "string",
                                "description": "When specified this is the default relationship to use
when an OIC Link does not specify an explicit relationship with *rel* parameter"
                            },
                            "links": {
                                "$ref": "#/definitions/oic.collection.alllinks"
                            }
                        }
                    }
                },
                "type": "object",
                "allOf": [
                    {"$ref": "oic.core-schema.json#/definitions/oic.core"},
                    {"$ref": "#/definitions/oic.collection"}
                ]
            }
    responses:
      200:
        body:
          application/json:
            schema: |
              {
```

```
"$schema": "http://json-schema.org/draft-04/schema#",
"description" : "Copyright (c) 2016 Open Connectivity Foundation, Inc. All rights reserved.",
"id": "http://www.openconnectivity.org/ocf-apis/core/schemas/oic.collection-schema.json#",
"title": "Collection",
"definitions": {
    "oic.collection.setoflinks": {
        "description": "A set (array) of simple or individual OIC Links. In addition to properties required for an OIC Link, the identifier for that link in this set is also required",
        "type": "array",
        "items": {
            "$ref": "oic.oic-link-schema.json#/definitions/oic.oic-link"
        }
    },
    "oic.collection.alllinks": {
        "description": "All forms of links in a collection",
        "oneOf": [
            {
                "$ref": "#/definitions/oic.collection.setoflinks"
            }
        ]
    },
    "oic.collection": {
        "type": "object",
        "description": "A collection is a set(array) of tagged-link or set(array) of simple links along with additional properties to describe the collection itself",
        "properties": {
            "id": {
                "anyOf": [
                    {
                        "type": "integer",
                        "description": "A number that is unique to that collection; like an ordinal number that is not repeated"
                    },
                    {
                        "type": "string",
                        "description": " A unique string that could be a hash or similarly unique"
                    },
                    {
                        "$ref": "oic.types-schema.json#/definitions/uuid",
                        "description": "A unique string that could be a UUIDv4"
                    }
                ],
                "description": "ID for the collection. Can be a value that is unique to the use context or a UUIDv4"
            },
            "di": {
                "$ref": "oic.types-schema.json#/definitions/uuid",
                "description": "The device ID which is an UUIDv4 string; used for
```

```
backward compatibility with Spec A definition of /oic/res"
                                    },
                                    "rts": {
                                        "$ref": "oic.core-
 schema.json#/definitions/oic.core/properties/rt",
                                        "description": "Defines the list of allowable resource types (for Target
and anchors) in links included in the collection; new links being created can only be from this list"        },
                                    "drel": {
                                        "type": "string",
                                        "description": "When specified this is the default relationship to
use when an OIC Link does not specify an explicit relationship with *rel* parameter"
                                    },
                                    "links": {
                                        "$ref": "#/definitions/oic.collection.alllinks"
                                    }
                                }
                            }
                        },
                        "type": "object",
                        "allOf": [
                            {"$ref": "oic.core-schema.json#/definitions/oic.core"},
                            {"$ref": "#/definitions/oic.collection"}
                        ]
                    }
```

2. 属性定义

属性定义如表 9-2 所示。

表 9-2 属性定义

属性名	属性值类型	是否强制	描　述
rt	数组	是	资源类型
di	多类型		设备唯一身份符号
eps	数组		目标资源的终端信息
title	字符串		链接关系的标题
pri(eps)	整型		多个终端的优先级
ep(eps)	字符串		传输协议套件 URI+定义终端位置信息
p	对象		规定从目标 URI 引用资源的框架策略
bm(p)	整型	是	规定为了可观察和可发现的目标 URI 引用资源的框架策略
ins	多类型		集合中使用链接数组中的 Web 实例标识符
href	字符串	是	目标 URI，它可以被指定为相对引用或完全限定的 URI。应该与 di 参数一起使用，使其能够唯一
rel	多类型		内容 URI 链接引用目标 URI 的关系
type	数组		提示由目标 URI 引用的资源表示。这表示用于接收和发送的媒体类型
anchor	字符串		用于覆盖内容 URI。例如，覆盖包含集合的 URI
if	数组	是	该资源支持的接口

3. CRUDN 行为

CRUDN 行为如表 9-3 所示。

表 9-3　CRUDN 行为

资　源	创建	检索	更新	删除	通知
/CollectionBaselineInterfaceURI		get	post		

4. 引用的 JSON 模式

```
oic.oic-link-schema.jason
{
  "$schema": "http://json-schema.org/draft-04/schema#",
  "description" : "Copyright (c) 2016, 2017 Open Connectivity Foundation, Inc. All rights reserved.",
  "id": "http://www.openconnectivity.org/ocf-apis/core/schemas/oic.oic-link-schema.json#",
  "definitions": {
    "oic.oic-link": {
      "type": "object",
      "properties": {
        "href": {
          "type": "string",
          "maxLength": 256,
          "description": "This is the target URI, it can be specified as a Relative Reference or fully-
qualified URI. Relative Reference should be used along with the di parameter to make it unique.",
          "format": "uri"
        },
        "rel": {
          "oneOf":[
            {
              "type": "array",
              "items": {
                  "type": "string",
                  "maxLength": 64
              },
              "minItems": 1,
              "default": ["hosts"]
            },
            {
              "type": "string",
              "maxLength": 64,
              "default": "hosts"
            }
          ]
         "description": "The relation of the target URI referenced by the link to the context URI"
        },
        "rt": {
          "type": "array",
          "items" : {
              "type" : "string",
              "maxLength": 64
            },
          "minItems" : 1,
          "description": "Resource Type"
```

```
        },
        "if": {
          "type": "array",
          "items": {
              "type" : "string",
              "enum" : ["oic.if.baseline", "oic.if.ll", "oic.if.b", "oic.if.rw", "oic.if.r",
"oic.if.a", "oic.if.s" ]
            },
          "minItems": 1,
          "description": "The interface set supported by this resource"
        },
        "di": {
           "$ref": "oic.types-schema.json#/definitions/uuid",
           "description": "Unique identifier for device (UUID)"
        },
        "p": {
          "description": "Specifies the framework policies on the Resource referenced by the target URI",
          "type": "object",
          "properties": {
            "bm": {
              "description": "Specifies the framework policies on the Resource referenced by the target
URI for e.g. observable and discoverable",
              "type": "integer"
            }
          }
          "required" : ["bm"]
        },
        "title": {
          "type": "string",
          "maxLength": 64,
          "description": "A title for the link relation. Can be used by the UI to provide a context"
        },
        "anchor": {
          "type": "string",
          "maxLength": 256,
          "description": "This is used to override the context URI e.g. override the URI of the containing
collection",
          "format": "uri"
        },
        "ins": {
          "oneOf": [
            {
              "type": "integer",
              "description": "An ordinal number that is not repeated - must be unique in the collection
context"
            },
            {
              "type": "string",
              "maxLength": 256,
              "format" : "uri",
              "description": "Any unique string including a URI"
            },
            {
              "$ref": "oic.types-schema.json#/definitions/uuid",
```

```
                "description": "Unique identifier (UUID)"
              }
            ],
            "description": "The instance identifier for this web link in an array of web links - used in
collections"
          },
          "type": {
            "type": "array",
            "description": "A hint at the representation of the resource referenced by the target URI. This
represents the media types that are used for both accepting and emitting",
            "items" : {
              "type": "string",
              "maxLength": 64
            },
            "minItems": 1,
            "default": "application/cbor"
          },
          "eps": {
            "type": "array",
            "description": "the Endpoint information of the target Resource",
            "items": {
              "type": "object",
              "properties": {
                "ep": {
                  "type": "string",
                  "format": "uri",
                  "description": "URI with Transport Protocol Suites + Endpoint Locator as specified"
                },
                "pri": {
                  "type": "integer",
                  "minimum": 1,
                  "description": "The priority among multiple Endpoints as specified"
                }
              }
            }
          }
        },
        "required": [ "href", "rt", "if" ]
      }
    },
    "type": "object",
    "allOf": [
      { "$ref": "#/definitions/oic.oic-link" }
    ]
}
```

9.2　设备配置

允许为设备配置特定信息的资源，URI 示例为“/example/DeviceConfigurationResURI”，资源类型为“oic. wk. con”。

1. RAML 定义

```
#%RAML 0.8
title: OCF Configuration
version: v1-20160622
traits:
  - interface-rw :
      queryParameters:
        if:
          enum: ["oic.if.rw"]
  - interface-all :
      queryParameters:
        if:
          enum: ["oic.if.rw", "oic.if.baseline"]
/example/DeviceConfigurationResURI:
  description: |
    Resource that allows for Device specific information to be configured.
  get:
    description: |
      Retrieves the current Device configuration settings
    is : ['interface-all']
    responses :
      200:
        body:
          application/json:
            schema: |
              {
                "id": "http://www.openconnectivity.org/ocf-apis/core/schemas/oic.wk.con-schema.json#",
                "$schema": "http://json-schema.org/draft-04/schema#",
                "description" : "Copyright (c) 2016, 2017 Open Connectivity Foundation, Inc. All rights reserved.",
                "definitions": {
                  "oic.wk.con": {
                    "type": "object",
                    "properties": {
                      "loc": {
                        "type": "array",
                        "description": "Location information",
                        "items": {
                          "type": "number"
                        },
                        "minItems": 2,
                        "maxItems": 2
                      },
                      "locn": {
                        "type": "string",
                        "maxLength": 64,
                        "description": "Human Friendly Name for location"
                      },
                      "c": {
                        "type": "string",
                        "maxLength": 64,
                        "description": "Currency"
```

```
                    },
                    "r": {
                      "type": "string",
                      "maxLength": 64,
                      "description": "Region"
                    },
                    "ln": {
                      "type": "array",
                      "items" :
                        {
                          "type": "object",
                          "properties": {
                            "language": {
                              "$ref": "oic.types-schema.json#/definitions/language-tag",
                              "description": "An RFC 5646 language tag."
                            },
                            "value": {
                              "type": "string",
                              "maxLength": 64,
                              "description": "Device description in the indicated language."
                            }
                          }
                        },
                      "minItems" : 1,
                      "description": "Localized names"
                    },
                    "dl": {
                      "$ref": "oic.types-schema.json#/definitions/language-tag",
                      "description": "Default Language"
                    }
                  }
                }
              },
              "type": "object",
              "allOf": [
                { "$ref": "oic.core-schema.json#/definitions/oic.core"},
                { "$ref": "#/definitions/oic.wk.con" }
              ],
              "required": ["n"]
            }
          example: |
            {
              "n": "My Friendly Device Name",
              "rt": ["oic.wk.con"],
              "loc": [32.777, -96.797],
              "locn": "My Location Name",
              "c": "USD",
              "r": "MyRegion",
              "dl": "en"
            }
  post:
    description: |
      Update the information about the Device
    is : ['interface-rw']
```

```
        body:
          application/json:
            schema: |
              {
                "id": "http://www.openconnectivity.org/ocf-apis/core/schemas/oic.wk.con-Update-schema.json#",
                "$schema": "http://json-schema.org/draft-04/schema#",
                "description" : "Copyright (c) 2016 Open Connectivity Foundation, Inc. All rights reserved.",
                "definitions": {
                  "oic.wk.con": {
                    "type": "object",
                    "properties": {
                      "loc": {
                        "type": "array",
                        "description": "Location information",
                        "items": {
                          "type": "number"
                        },
                        "minItems": 2,
                        "maxItems": 2
                      },
                      "locn": {
                        "type": "string",
                        "maxLength": 64,
                        "description": "Human Friendly Name for location"
                      },
                      "c": {
                        "type": "string",
                        "maxLength": 64,
                        "description": "Currency"
                      },
                      "r": {
                        "type": "string",
                        "maxLength": 64,
                        "description": "Region"
                      },
                      "ln": {
                        "type": "array",
                        "items" :
                          {
                            "type": "object",
                            "properties": {
                              "language": {
                                "$ref": "oic.types-schema.json#/definitions/language-tag",
                                "description": "An RFC 5646 language tag."
                              },
                              "value": {
                                "type": "string",
                                "maxLength": 64,
                                "description": "Device description in the indicated language."
                              }
                            }
                          },
```

```
                    "minItems" : 1,
                    "description": "Localized names"
                  },
                  "dl": {
                    " $ ref": "oic. types - schema. json # /definitions/language - tag",
                    "description": "Default Language"
                  }
                }
              }
            },
            "type": "object",
            "allOf": [
              { " $ ref": "oic. core - schema. rw. json # /definitions/oic. core"},
              { " $ ref": " # /definitions/oic. wk. con" }
            ],
            "required": ["n"]
          }
        example: |
          {
            "n": "Nuevo Nombre Amistoso",
            "r": "MyNewRegion",
            "ln": [ { "language": "es", "value": "Nuevo Nombre Amistoso" } ],
            "dl": "es"
          }
    responses :
      200:
        body:
          application/json:
            schema: |
              {
                " id": "http://www. openconnectivity. org/ocf - apis/core/schemas/oic. wk. con - Update -
schema. json # ",
                " $ schema": "http://json - schema. org/draft - 04/schema # ",
                "description" : "Copyright (c) 2016 Open Connectivity Foundation, Inc. All rights
reserved. ",
                "definitions": {
                  "oic. wk. con": {
                    "type": "object",
                    "properties": {
                      "loc": {
                        "type": "array",
                        "description": "Location information",
                        "items": {
                          "type": "number"
                        },
                        "minItems": 2,
                        "maxItems": 2
                      },
                      "locn": {
                        "type": "string",
                        "maxLength": 64,
                        "description": "Human Friendly Name for location"
                      },
                      "c": {
```

```
            "type": "string",
            "maxLength": 64,
            "description": "Currency"
          },
          "r": {
            "type": "string",
            "maxLength": 64,
            "description": "Region"
          },
          "ln": {
            "type": "array",
            "items" :
              {
                "type": "object",
                "properties": {
                  "language": {
                    "$ref": "oic.types-schema.json#/definitions/language-tag",
                    "description": "An RFC 5646 language tag."
                  },
                  "value": {
                    "type": "string",
                    "maxLength": 64,
                    "description": "Device description in the indicated language."
                  }
                }
              },
            "minItems" : 1,
            "description": "Localized names"
          },
          "dl": {
            "$ref": "oic.types-schema.json#/definitions/language-tag",
            "description": "Default Language"
          }
        }
      }
    },
    "type": "object",
    "allOf": [
      { "$ref": "oic.core-schema.rw.json#/definitions/oic.core"},
      { "$ref": "#/definitions/oic.wk.con" }
    ],
    "required": ["n"]
  }
example: |
  {
      "n": "Nuevo Nombre Amistoso",
      "r": "MyNewRegion",
      "ln": [ { "language": "es", "value": "Nuevo Nombre Amistoso" } ],
      "dl": "es"
  }
```

2. 属性定义

属性定义如表 9-4 所示。

表 9-4 属性定义

属性名	属性值类型	访问模式	是否强制	描 述
n	字符串	读写	是	只读,人性化的名称
di	字符串	只读	是	设备的唯一标识符(UUID)
icv	字符串	只读	是	OCF 服务器的版本
dmv	字符串	只读		垂直领域和/或资源规范的版本

3. CRUDN 行为

CRUDN 行为如表 9-5 所示。

表 9-5 CRUDN 行为

资 源	创建	检索	更新	删除	通知
/example/DeviceConfigurationResURI		get			

9.3 平台配置

允许平台配置指定信息的资源。URI 示例为"/example/PlatformConfigurationResURI",资源类型定义为"oic. wk. con. p"。

1. RAML 定义

```
# %RAML 0.8
  title: OCF Platform Configuration
  version: v1 - 20160622
  traits:
    - interface - rw :
      queryParameters:
        if:
          enum: ["oic.if.rw"]
    - interface - all :
      queryParameters:
        if:
          enum: ["oic.if.rw", "oic.if.baseline"]
 /example/PlatformConfigurationResURI:
   description: |
     Resource that allows for platform specific information to be configured.
   get:
     description: |
       Retrieves the current platform configuration settings
    is : ['interface - all']
     responses :
       200:
         body:
           application/json:
             schema: |
               {
                 "id": "http://www.openconnectivity.org/ocf - apis/core/schemas/oic.wk.con.p - schema
.json#",
```

```
                    "$schema": "http://json-schema.org/draft-04/schema#",
                    "description" : "Copyright (c) 2017 Open Connectivity Foundation, Inc. All rights reserved.",
                    "definitions": {
                      "oic.wk.con.p": {
                        "type": "object",
                        "properties": {
                          "mnpn": {
                            "type": "array",
                            "items" :
                              {
                                "type": "object",
                                "properties": {
                                  "language": {
                                    "$ref": "oic.types-schema.json#/definitions/language-tag",
                                    "description": "An RFC 5646 language tag."
                                  },
                                  "value": {
                                    "type": "string",
                                    "maxLength": 64,
                                    "description": "Platform description in the indicated language."
                                  }
                                }
                              },
                            "minItems" : 1,
                            "description": "Platform names"
                          }
                        }
                      }
                    },
                    "type": "object",
                    "allOf": [
                      { "$ref": "oic.core-schema.json#/definitions/oic.core"},
                      { "$ref": "#/definitions/oic.wk.con.p" }
                    ]
                  }
                example: |
                  {
                    "rt": ["oic.wk.con.p"],
                    "mnpn": [ { "language": "en", "value": "My Friendly Device Name" } ]
                  }
    post:
      description: |
        Update the information about the platform
      is : ['interface-rw']
      body:
        application/json:
          schema: |
            {
                "id": "http://www.openconnectivity.org/ocf-apis/core/schemas/oic.wk.con.p-Update-
schema.json#",
                "$schema": "http://json-schema.org/draft-04/schema#",
                "description" : "Copyright (c) 2017 Open Connectivity Foundation, Inc. All rights
reserved.",
                "definitions": {
```

```
              "oic.wk.con.p": {
                "type": "object",
                "properties": {
                  "mnpn": {
                    "type": "array",
                    "items" :
                      {
                        "type": "object",
                        "properties": {
                          "language": {
                            "$ref": "oic.types-schema.json#/definitions/language-tag",
                            "description": "An RFC 5646 language tag."
                          },
                          "value": {
                            "type": "string",
                            "maxLength": 64,
                            "description": "Platform description in the indicated language."
                          }
                        }
                      },
                    "minItems" : 1,
                    "description": "Platform names"
                  }
                }
              }
            },
            "type": "object",
            "allOf": [
              { "$ref": "oic.core-schema.rw.json#/definitions/oic.core"},
              { "$ref": "#/definitions/oic.wk.con.p" }
            ],
            "required": ["mnpn"]
          }
        example: |
          {
            "n": "Nuevo nombre",
            "mnpn": [ { "language": "es", "value": "Nuevo nombre de Plataforma Amigable" } ]
          }
    responses :
      200:
        body:
          application/json:
            schema: |
              {
                "id": "http://www.openconnectivity.org/ocf-apis/core/schemas/oic.wk.con.p-Update-
schema.json#",
                "$schema": "http://json-schema.org/draft-04/schema#",
                "description" : "Copyright (c) 2017 Open Connectivity Foundation, Inc. All rights reserved.",
                "definitions": {
                    "type": "object",
                    "properties": {
                      "mnpn": {
                        "type": "array",
                        "items" :
```

```
                {
                  "type": "object",
                  "properties": {
                    "language": {
                      "$ref": "oic.types-schema.json#/definitions/language-tag",
                      "description": "An RFC 5646 language tag."
                    },
                    "value": {
                      "type": "string",
                      "maxLength": 64,
                      "description": "Platform description in the indicated language."
                    }
                  }
                },
              "minItems" : 1,
              "description": "Platform names"
            }
          }
        }
      },
      "type": "object",
      "allOf": [
        { "$ref": "oic.core-schema.rw.json#/definitions/oic.core"},
        { "$ref": "#/definitions/oic.wk.con.p" }
      ],
      "required": ["mnpn"]
    }
  example: |
    {
      "n": "Nuevo nombre",
      "mnpn": [ { "language": "es", "value": "Nuevo nombre de Plataforma Amigable" } ]
    }
```

2. 属性定义

属性定义如表 9-6 所示。

表 9-6 属性定义

属 性 名	属性值类型	访问模式	是否强制	描 述
mnpn	数组			平台名称
value	字符串			指定语言下的平台描述
language	多类型			RFC 5646 语言标签

3. CRUDN 行为

CRUDN 行为如表 9-7 所示。

表 9-7 CRUDN 行为

资 源	创建	检索	更新	删除	通知
/example/PlatformConfigurationResURI		get	post		

9.4 设备

每个服务器端托管的已知资源允许发现逻辑设备特定信息。URI 示例为"/oic/d",资源类型定义为"oic. wk. d"。

1. RAML 定义

```
#%RAML 0.8
title: OIC Root Device
version: v1-20160622
traits:
 - interface :
     queryParameters:
       if:
         enum: ["oic.if.r", "oic.if.baseline"]
/oic/d:
  description: |
    Known resource that is hosted by every Server. Allows for logical device specific information to be discovered.
  is : ['interface']
  get:
    description: |
      Retrieve the information about the Device
    responses :
      200:
        body:
          application/json:
            schema: |
              {
                "$schema": "http://json-schemas.org/draft-04/schema#",
                "description" : "Copyright (c) 2016, 2017 Open Connectivity Foundation, Inc. All rights reserved.",
                "id": "http://www.openconnectivity.org/ocf-apis/core/schemas/oic.wk.d-schema.json#",
                "definitions": {
                  "oic.wk.d": {
                    "type": "object",
                    "properties": {
                      "di": {
                        "$ref": "oic.types-schema.json#/definitions/uuid",
                        "readOnly": true,
                        "description": "Unique identifier for device (UUID)"
                      },
                      "icv": {
                        "type": "string",
                        "maxLength": 64,
                        "readOnly": true,
                        "description": "The version of the OIC Server"
                      },
                      "dmv": {
                        "type": "string",
                        "maxLength": 256,
                        "readOnly": true,
```

```
                "description": "Spec versions of the Resource and Device Specifications to which
this device data model is implemented"
            },
            "ld": {
                "type": "array",
                "items" :
                  {
                    "type": "object",
                    "properties": {
                      "language": {
                        "$ref": "oic.types-schema.json#/definitions/language-tag",
                        "readOnly": true,
                        "description": "An RFC 5646 language tag."
                      },
                      "value": {
                        "type": "string",
                        "maxLength": 64,
                        "readOnly": true,
                        "description": "Device description in the indicated language."
                      }
                    }
                  },
                "minItems" : 1,
                "readOnly": true,
                "description": "Localized Description."
            },
            "sv": {
                "type": "string",
                "maxLength": 64,
                "readOnly": true,
                "description": "Software version."
            },
            "dmn": {
                "type": "array",
                "items" :
                  {
                    "type": "object",
                    "properties": {
                      "language": {
                        "$ref": "oic.types-schema.json#/definitions/language-tag",
                        "readOnly": true,
                        "description": "An RFC 5646 language tag."
                      }, 4
                      "value": {
                        "type": "string",
                        "maxLength": 64,
                        "readOnly": true,
                        "description": "Manufacturer name in the indicated language."
                      }
                    }
                  },
                "minItems" : 1,
                "readOnly": true,
                "description": "Manufacturer Name."
```

```
                    },
                    "dmno": {
                      "type": "string",
                      "maxLength": 64,
                      "readOnly": true,
                      "description": "Model number as designated by manufacturer."
                    },
                    "piid": {
                      "$ref": "oic.types-schema.json#/definitions/uuid",
                      "readOnly": true,
                      "description": "Protocol independent unique identifier for device (UUID) that is
immutable."
                    }
                  }
                }
              },
              "type": "object",
              "allOf": [
                { "$ref": "oic.core-schema.json#/definitions/oic.core"},
                { "$ref": "#/definitions/oic.wk.d" }
              ],
              "required": [ "n", "di", "icv", "dmv", "piid" ]
            }
          example: |
            {
              "n":     "Device 1",
              "rt":    ["oic.wk.d"],
              "di":    "54919CA5-4101-4AE4-595B-353C51AA983C",
              "icv":   "ocf.1.0.0",
              "dmv":   "ocf.res.1.0.0, ocf.sh.1.0.0",
              "piid":  "6F0AAC04-2BB0-468D-B57C-16570A26AE48"
            }
```

2. 属性定义

属性定义如表 9-8 所示。

表 9-8　属性定义

属　性　名	属性值类型	访问模式	是否强制	描　　述
id	数组	只读		本地描述
value(id)	字符串	只读		用指定语言进行设备描述
language	多类型	只读		RFC 5646 语言标签
piid	多类型	只读	是	不可变的、独立于协议的设备唯一标识符(UUID)
di	多类型	只读	是	设备唯一标识符(UUID)
dmno	字符串	只读		由制造商设计的型号
sv	字符串	只读		软件版本
dmn	数组	只读		制造商名称
value(dmn)	字符串	只读		制定语言下的制造商名称
language(dmn)	多类型	只读		RFC 5646 语言标签
icv	字符串	只读	是	OCF 服务器的版本
dmv	字符串	只读	是	该设备数据模型实现的资源和设备规格的版本

3. CRUDN 行为

CRUDN 行为如表 9-9 所示。

表 9-9 CRUDN 行为

资　源	创建	检索	更新	删除	通知
/oic/d		get			

9.5 维护

设备维护所用的资源,可用于诊断目的。出厂重置参数“fr”是一个布尔值,值 0 表示无操作(默认值);值 1 表示在出厂后复位。开始出厂复位后,此值应更改回默认值“rb”(重新启动),它也是一个布尔值,值 0 表示无操作(默认值),值 1 表示开始重新启动,之后此值应更改回默认值。URI 示例为“/oic/mnt”,资源类型定义为“oic. wk. mnt”。

1. RAML 定义

```
#%RAML 0.8
title: Maintenance
version: v1-20160622
traits:
- interface-rw :
    queryParameters:
      if:
        enum: ["oic.if.rw", "oic.if.baseline"]
  - interface-all :
    queryParameters:
      if:
        enum: ["oic.if.rw", "oic.if.r", "oic.if.baseline"]
/oic/mnt:
  description: |
    The resource through which a Device is maintained and can be used for diagnostic purposes.
    fr (Factory Reset) is a boolean.
    The value 0 means No action (Default), the value 1 means Start Factory Reset
    After factory reset, this value shall be changed back to the default value
    rb (Reboot) is a boolean.
    The value 0 means No action (Default), the value 1 means Start Reboot
    After Reboot, this value shall be changed back to the default value
get:
    is : ['interface-all']
    description: |
      Retrieve the maintenance action status
    responses :
      200:
        body:
          application/json:
            schema: |
              {
                "$schema": "http://json-schemas.org/draft-04/schema#",
                "description" : "Copyright (c) 2016, 2017 Open Connectivity Foundation, Inc. All rights
```

```
reserved.",
                "id": "http://www.openconnectivity.org/ocf-apis/core/schemas/oic.wk.mnt-
schema.json#",
                "definitions": {
                  "oic.wk.mnt": {
                    "type": "object",
                    "anyOf": [
                      {"required": ["fr"]},
                      {"required": ["rb"]}
                    ],
                    "properties": {
                      "fr":{
                        "type": "boolean",
                        "description": "Factory Reset"
                      },
                      "rb": {
                        "type": "boolean",
                       "description": "Reboot Action"
                      }
                    }
                  }
                },
                "type": "object",
                "allOf": [
                  { "$ref": "oic.core-schema.json#/definitions/oic.core"},
                  { "$ref": "#/definitions/oic.wk.mnt" }
                ]
              }
            example: |
              {
                "rt": ["oic.wk.mnt"],
                "fr": false,
                "rb": false
              }
post:
    is : ['interface-rw']
    description: |
      Set the maintenance action(s)
    body:
      application/json:
        schema: |
          {
            "$schema": "http://json-schemas.org/draft-04/schema#",
            "description" : "Copyright (c) 2016, 2017 Open Connectivity Foundation, Inc. All rights reserved.",
            "id": "http://www.openconnectivity.org/ocf-apis/core/schemas/oic.wk.mnt-schema.json#",
            "definitions": {
              "oic.wk.mnt": {
                "type": "object",
                "anyOf": [
                  {"required": ["fr"]},
                  {"required": ["rb"]}
                ],
                "properties": {
                  "fr":{
```

```
                  "type": "boolean",
                  "description": "Factory Reset"
                },
                "rb": {
                  "type": "boolean",
                  "description": "Reboot Action"
                }
              }
            }
          },
          "type": "object",
          "allOf": [
            { "$ref": "oic.core-schema.json#/definitions/oic.core"},
            { "$ref": "#/definitions/oic.wk.mnt" }
          ]
        }
      example: |
        {
          "fr": false,
          "rb": false
        }
  responses :
   200:
      body:
        application/json:
          schema: |
            {
              "$schema": "http://json-schemas.org/draft-04/schema#",
              "description" : "Copyright (c) 2016, 2017 Open Connectivity Foundation, Inc. All rights
reserved.",
              "id": "http://www.openconnectivity.org/ocf-apis/core/schemas/oic.wk.mnt-
schema.json#",
              "definitions": {
                "oic.wk.mnt": {
                  "type": "object",
                  "anyOf": [
                    {"required": ["fr"]},
                    {"required": ["rb"]}
                  ],
                  "properties": {
                    "fr":{
                      "type": "boolean",
                      "description": "Factory Reset"
                    },
                    "rb": {
                      "type": "boolean",
                      "description": "Reboot Action"
                    }
                  }
                }
              },
              "type": "object",
              "allOf": [
                { "$ref": "oic.core-schema.json#/definitions/oic.core"},
```

```
        { "$ref": "#/definitions/oic.wk.mnt" }
      ]
    }
  example: |
    {
      "fr": false,
      "rb": false
    }
```

2. 属性定义

属性定义如表 9-10 所示。

表 9-10 属性定义

属 性 名	属性值类型	访问模式	是否强制	描 述
fr	布尔型	只读	是	工厂重置
rb	布尔型	只读		重新启动操作

3. CRUDN 行为

CRUDN 行为如表 9-11 所示。

表 9-11 CRUDN 行为

资 源	创建	检索	更新	删除	通知
/oic/mnt		get	post		

9.6 平台

已知资源定义托管服务器端的平台,允许发现特定于平台的信息。URI 示例为"/oic/p",资源类型定义为"oic. wk. p"。

1. RAML 定义

```
# %RAML 0.8
title: Platform
version: v1 - 20160622
traits:
  - interface :
    queryParameters:
      if:
        enum: ["oic.if.r", "oic.if.baseline"]
/oic/p:
  description: |
    Known resource that is defines the platform on which an Server is hosted.
    Allows for platform specific information to be discovered.
  is : ['interface']
  get:
    description: |
      Retrieves the information about the Platform
    responses :
```

```
      200:
        body:
          application/json:
            schema: |
              {
                "$schema": "http://json-schema.org/draft-04/schema#",
                "description" : "Copyright (c) 2016 Open Connectivity Foundation, Inc. All rights reserved.",
                "id": "http://openconnectivityfoundation.org/core/schemas/oic.wk.p-schema.json#",
                "definitions": {
                  "oic.wk.p": {
                    "type": "object",
                    "properties": {
                      "di": {
                        "type": "string",
                        "pattern": "^[a-fA-F0-9]{8}-[a-fA-F0-9]{4}-[a-fA-F0-9]{4}-[a-
fA-F0-9]{4}-[a-fA-F0-9]{12}$",
                        "description": "ReadOnly, Platform Identifier as a UUID"
                      },
                      "mnmn": {
                        "type": "string",
                        "description": "ReadOnly, Manufacturer Name",
                        "maxLength": 64
                      },
                      "mnml": {
                        "type": "string",
                        "description": "ReadOnly, Manufacturer's URL",
                        "maxLength": 256,
                        "format": "uri"
                      },
                      "mnmo": {
                        "type": "string",
                        "maxLength": 64,
                        "description": "ReadOnly, Model number as designated by manufacturer"
                      },
                      "mndt": {
                        "type": "string",
                        "description": "ReadOnly, Manufacturing Date",
                        "format": "date"
                      },
                      "mnpv": {
                        "type": "string",
                        "maxLength": 64,
                        "description": "ReadOnly, Platform Version"
                      },
                      "mnos": {
                        "type": "string",
                        "maxLength": 64,
                        "description": "Readonly, Platform Resident OS Version"
                      },
                      "mnhw": {
                        "type": "string",
                        "maxLength": 64,
                        "description": "Readonly, Platform Hardware Version"
                      },
```

```
            "mnfv": {
              "type": "string",
              "maxLength": 64,
              "description": "ReadOnly, Manufacturer's firmware version"
            },
            "mnsl": {
              "type": "string",
              "description": "ReadOnly, Manufacturer's Support Information URL",
              "maxLength": 256,
              "format": "uri"
            }
            "st": {
              "type": "string",
              "description": "ReadOnly, Reference time for the device",
              "format": "date - time"
            },
            "vid": {
              "type": "string",
              "maxLength": 64,
                "description": "ReadOnly, Manufacturer's defined string for the platform. The
string is freeform and up to the manufacturer on what text to populate it"
            }
          }
        }
      },
      "type": "object",
      "allOf": [
        { "$ref": "oic.core - schema.json#/definitions/oic.core"},
        { "$ref": "#/definitions/oic.wk.p" }
      ],
      "required": [ "pi", "mnmn"]
    }
  example: |
    {
      "pi": "54919CA5 - 4101 - 4AE4 - 595B - 353C51AA983C",
      "rt": ["oic.wk.p"],
      "mnmn": "Acme, Inc"
    }
```

2. 属性定义

属性定义如表 9-12 所示。

表 9-12 属性定义

属 性 名	属性值类型	访问模式	是否强制	描 述
pi	字符串	只读	是	平台标识符作为 UUID
mnmn	字符串	只读	是	名称
mnml	字符串	只读		URL
mnmo	字符串	只读		由制造商指定的型号
mndt	字符串	只读		制造日期
mnpv	字符串	只读		平台版本
mnos	字符串	只读		只读,平台驻留操作系统版本

续表

属　性　名	属性值类型	访问模式	是否强制	描　　述
mnhw	字符串	只读		只读，平台硬件版本
mnfv	字符串	只读		制造商的固件版本
mnsl	字符串	只读		制造商的支持信息 URL
st	字符串	只读		设备的参考时间
vid	字符串	只读		制造商为平台定义的字符串。字符串是自由格式的，并且根据制造商填充文本

3. CRUDN 行为

CRUDN 行为如表 9-13 所示。

表 9-13　CRUDN 行为

资　　源	创建	检索	更新	删除	通知
/oic/p		get			

9.7　ping

此资源用于客户端保持其与服务器端连接处于活动状态。URI 示例为"/oic/ping"，资源类型定义为"oic. wk. ping"。

1. RAML 定义

```
#%RAML 0.8
title: Ping
version: v1-20160622
traits:
  - interface :
    queryParameters:
      if:
        enum: ["oic.if.rw", "oic.if.baseline"]
/oic/ping:
  description: |
    The resource using which a Client keeps its Connection with a Server active.
is : ['interface']
get:
  description: |
    Retrieve the ping information
  responses :
    200:
      body:
        application/json:
          schema: |
            {
              "$schema": "http://json-schema.org/draft-04/schema#",
              "description" : "Copyright (c) 2016 Open Connectivity Foundation, Inc. All rights reserved.",
              "id": "http://openconnectivityfoundation.org/core/schemas/oic.wk.ping-schema.json#",
              "definitions": {
```

```
            "oic.wk.ping": {
              "type": "object",
              "properties": {
                "in": {
                  "type": "integer",
                    "description": "ReadWrite, Indicates the interval for which connection shall be
kept alive"}
                }
              }
            },
            "type": "object",
            "allOf": [
              { "$ref": "oic.core-schema.json#/definitions/oic.core"},
              { "$ref": "#/definitions/oic.wk.ping" }
            ],
            "required": [ "in"]
          }
        example: |
          {
            "rt": ["oic.wk.ping"],
            "n": "Ping Information",
            "in": 16
          }
```

2. 属性定义

属性定义如表 9-14 所示。

表 9-14 属性定义

属性名	属性值类型	访问模式	是否强制	描述
in	整型	读写		读写，指示连接应保持活动的时间间隔

3. CRUDN 行为

CRUDN 行为如表 9-15 所示。

表 9-15 CRUDN 行为

资源	创建	检索	更新	删除	通知
/oic/ping		get			

9.8 可发现资源基准接口

可发现资源基准接口为"/oic/res"，的基准表示，可发现资源的列表。URI 示例为"/oic/res"，资源类型定义为"oic. wk. res"。

1. RAML 定义

```
#%RAML 0.8
title: Discoverable Resources
version: v1-20160622
traits:
```

```
  - interface - ll :
      queryParameters:
        if:
          enum: ["oic.if.ll"]
  - interface - baseline :
      queryParameters:
        if:
          enum: ["oic.if.baseline"]
/oic - res - BaselineInterfaceURI:
  description: |
   Baseline representation of /oic/res; list of discoverable resources
  is : ['interface - baseline']
  get:
    description: |
      Retrieve the discoverable resource set, baseline interface
      responses :
      200:
        body:
          application/json:
            schema: |
              {
                "$schema": "http://json - schema.org/draft - v4/schema#",
                "description" : "Copyright (c) 2016, 2017 Open Connectivity Foundation, Inc. All rights
reserved.",
                "id": "http://www.openconnectivity.org/ocf - apis/core/schemas/oic.wk.res -
schema.json#",
                "definitions": {
                  "oic.res - baseline": {
                    "type": "object",
                    "properties": {
                      "rt": {
                        "type": "array",
                        "items" : {
                            "type" : "string",
                            "maxLength": 64
                          },
                        "minItems" : 1,
                        "readOnly": true,
                        "description": "Resource Type"
                      },
                      "if": {
                        "type": "array",
                        "items": {
                            "type" : "string",
                            "enum" : ["oic.if.baseline", "oic.if.ll"]
                          },
                        "minItems": 1,
                        "readOnly": true,
                        "description": "The interface set supported by this resource"
                      },
                      "n": {
                        "type": "string",
                        "maxLength": 64,
                        "readOnly": true,
```

```
            "description": "Human friendly name"
          },
          "mpro": {
            "readOnly": true,
            "description": "Supported messaging protocols",
            "type": "string",
            "maxLength": 64
          },
          "links": {
            "type": "array",
            "items": {
              "$ref": "oic.oic-link-schema.json#/definitions/oic.oic-link"
            }
          }
        },
        "required": ["rt", "if", "links"]
      }
    },
    "description": "The list of resources expressed as OIC links",
    "type": "array",
    "items": {
      "$ref": "#/definitions/oic.res-baseline"
    }
  }
example: |
  [
    {
    "rt": ["oic.wk.res"],
    "if": ["oic.if.baseline", "oic.if.ll" ],
    "links":
      [
        {
          "href": "/humidity",
          "rt": ["oic.r.humidity"],
          "if": ["oic.if.s"],
          "p": {"bm": 3},
          "eps": [
              {"ep": "coaps://[fe80::b1d6]:1111", "pri": 2},
              {"ep": "coaps://[fe80::b1d6]:1122"},
              {"ep": "coap+tcp://[2001:db8:a::123]:2222", "pri": 3}
          ]
        },
        {
          "href": "/temperature",
          "rt": ["oic.r.temperature"],
          "if": ["oic.if.s"],
          "p": {"bm": 3},
          "eps": [
              {"ep": "coaps://[[2001:db8:a::123]:2222"}
          ]
        }
      ]
    }
  ]
```

2. 属性定义

属性定义如表 9-16 所示。

表 9-16 属性定义

属 性 名	属性值类型	访问模式	是否强制	描 述
n	字符串	只读		人性化名称
rt	数组	只读		资源类型
mpro	字符串	只读		支持的消息协议
links	数组		是	链接
if	数组	只读	是	该资源支持的接口集

3. CRUDN 行为

CRUDN 行为如表 9-17 所示。

表 9-17 CRUDN 行为

资源	创建	检索	更新	删除	通知
/oic/res		get			

9.9 可发现资源的链接表接口

可发现资源的链接表接口是"/oic/res"的链接表示,可发现资源的列表。URI 示例为"/oic/res",资源类型定义为"oic. wk. res"。

1. RAML 定义

```
#%RAML 0.8
title: Discoverable Resources
version: v1-20160622
traits:
  - interface-ll :
    queryParameters:
      if:
        enum: ["oic.if.ll"]
  - interface-baseline :
    queryParameters:
      if:
        enum: ["oic.if.baseline"]
/oic-res-llInterfaceURI:
  description: |
    Link list representation of /oic/res; list of discoverable resources
  is : ['interface-ll']
  get:
    description: |
      Retrieve the discoverable resource set, link list interface
    responses :
      200:
        body:
```

```
        application/json:
          schema: |
            {
              "$schema": "http://json-schema.org/draft-v4/schema#",
              "description" : "Copyright (c) 2016, 2017 Open Connectivity Foundation, Inc. All rights
reserved.",
              "id": "http://www.openconnectivity.org/ocf-apis/core/schemas/oic.wk.res-schema-
ll.json#",
              "description": "The list of resources expressed as OCF links without di",
              "definitions": {
                "oic.res-ll": {
                  "$ref": "oic.oic-link-schema.json#/definitions/oic.oic-link"
                }
              },
              "type": "array",
              "items": {
                "$ref": "#/definitions/oic.res-ll"
              }
            }
          example: |
            [
              {
                "href": "/humidity",
                "rt": ["oic.r.humidity"],
                "if": ["oic.if.s"],
                "p": {"bm": 3},
                "eps": [
                  {"ep": "coaps://[fe80::b1d6]:1111", "pri": 2},
                  {"ep": "coaps://[fe80::b1d6]:1122"},
                 {"ep": "coaps+tcp://[2001:db8:a::123]:2222", "pri": 3}
                ]
              },
              {
                "href": "/temperature",
                "rt": ["oic.r.temperature"],
                "if": ["oic.if.s"],
                "p": {"bm": 3},
                "eps": [
                  {"ep": "coaps://[[2001:db8:a::123]:2222"}
                ]
              }
            ]
```

2. 属性定义

属性定义如表 9-18 所示。

表 9-18 属性定义

属性名	属性值类型	是否强制	描述
rt	数组	是	资源类型
di	多类型		设备唯一标识符(UUID)
title	字符串		链接关系标题,可为界面提供内容
eps	数组		目标资源的终端信息

续表

属　性　名	属性值类型	是否强制	描　　述
pri(eps)	整型		多终端的优先级
ep(eps)	字符串		传输协议套件 URI+终端位置信息
p	对象		规定从目标 URI 引用资源的框架策略
bm(p)	整型	是	规定为了可观察和可发现目标 URI 引用资源的框架策略
ins	多类型		集合中使用链接数组中的 Web 实例标识符
href	字符串	是	目标 URI,它可以被指定为相对引用或完全限定的 URI。应该与 di 参数一起使用使其能够唯一
rel	多类型		内容 URI 链接引用目标 URI 的关系
type	数组		提示由目标 URI 引用的资源表示。这表示用于接收和发送的媒体类型
anchor	字符串		用于覆盖内容 URI,例如,覆盖包含集合的 URI
if	数组	是	该资源支持的接口

3. CRUDN 行为

CRUDN 行为如表 9-19 所示。

表 9-19　CRUDN 行为

资源	创建	检索	更新	删除	通知
/oic/res		get			

4. 引用 JSON 方案

```
oic.oic-link-schema.json
{
  "$schema": "http://json-schema.org/draft-04/schema#",
  "description" : "Copyright (c) 2016, 2017 Open Connectivity Foundation, Inc. All rights reserved.",
  "id": "http://www.openconnectivity.org/ocf-apis/core/schemas/oic.oic-link-schema.json#",
  "definitions": {
    "oic.oic-link": {
      "type": "object",
      "properties": {
        "href": {
          "type": "string",
          "maxLength": 2
          "description": "This is the target URI, it can be specified as a Relative Reference orfully-
qualified URI. Relative Reference should be used along with the di parameter to make it unique.",
          "format": "uri"
        },
        "rel": {
          "oneOf":[
            {
              "type": "array",
              "items": {
                "type": "string",
                "maxLength": 64
              },
```

```
                "minItems": 1,
                "default": ["hosts"]
              },
              {
                "type": "string",
                "maxLength": 64,
                "default": "hosts"
              }
            ],
          "description": "The relation of the target URI referenced by the link to the context URI"
          },
          "rt": {
            "type": "array",
            "items" : {
              "type" : "string",
              "maxLength": 64
            },
            "minItems" : 1,
            "description": "Resource Type"
          },
          "if": {
            "type": "array",
            "items": {
              "type" : "string",
              "enum" : ["oic.if.baseline", "oic.if.ll", "oic.if.b", "oic.if.rw", "oic.if.r",
"oic.if.a", "oic.if.s" ]
            },
            "minItems": 1,
            "description": "The interface set supported by this resource"
          },
          "di": {
            "$ref": "oic.types-schema.json#/definitions/uuid",
            "description": "Unique identifier for device (UUID)"
          },
          "p": {
            "description": "Specifies the framework policies on the Resource referenced by the target URI",
            "type": "object",
            "properties": {
              "bm": {
                "description": "Specifies the framework policies on the Resource referenced by the target
URI for e.g. observable and discoverable",
                "type": "integer"
              }
            },
            "required" : ["bm"]
          },
          "title": {
            "type": "string",
            "maxLength": 64,
            "description": "A title for the link relation. Can be used by the UI to provide a context"
          },
          "anchor": {
            "type": "string",
            "maxLength": 256,
```

```
            "description": "This is used to override the context URI e.g. override the URI of the containing collection",
            "format": "uri"
        },
        "ins": {
            "oneOf": [
                {
                    "type": "integer",
                    "description": "An ordinal number that is not repeated - must be unique in the collection context"
                },
                {
                    "type": "string",
                    "maxLength": 256,
                    "format" : "uri",
                    "description": "Any unique string including a URI"
                },
                {
                    "$ref": "oic.types-schema.json#/definitions/uuid",
                    "description": "Unique identifier (UUID)"
                }
            ],
            "description": "The instance identifier for this web link in an array of web links - used in collections"
        },
        "type": {
            "type": "array",
            "description": "A hint at the representation of the resource referenced by the target URI. This represents the media types that are used for both accepting and emitting",
            "items" : {
                "type": "string",
                "maxLength": 64
            },
            "minItems": 1,
            "default": "application/cbor"
        },
        "eps": {
            "type": "array",
            "description": "the Endpoint information of the target Resource",
            "items": {
                "type": "object",
                "properties": {
                    "ep": {
                        "type": "string",
                        "format": "uri",
                        "description": "URI with Transport Protocol Suites + Endpoint Locator as specified"
                    },
                    "pri": {
                        "type": "integer",
                        "minimum": 1,
                        "description": "The priority among multiple Endpoints as specified"
                    }
                }
            }
```

```
            }
          },
          "required": [ "href", "rt", "if" ]
        }
      },
      "type": "object",
      "allOf": [
        { "$ref": "#/definitions/oic.oic-link" }
      ]
}
```

9.10 场景(顶层)

顶层场景资源是通用集合资源,"rts"值应包含"oic. sceneCollection"资源类型。URI 示例为"/SceneListResURI",资源类型定义为"oic. wk. sceneList"。

1. RAML 定义

```
#%RAML 0.8
title: Scene
version: v1-20160622
traits:
  - interface :
    queryParameters:
      if:
        enum: ["oic.if.a", "oic.if.ll", "oic.if.baseline"]
/SceneListResURI:
  description: |
    Toplevel Scene resource. This resource is a generic collection resource. The rts value shall contain oic.
    sceneCollection resource types.
  get:
    description: |
      Provides the current list of web links pointing to scenes
    responses :
      200:
        body:
          application/json:
            schema: |
              {
                "$schema": "http://json-schema.org/draft-04/schema#",
                "description" : "Copyright (c) 2016 Open Connectivity Foundation, Inc. All rights reserved.",
                "id": "http://openconnectivityfoundation.org/core/schemas/oic.collection-schema.json#",
                "title": "Collection",
                "definitions": {
                  "oic.collection.setoflinks": {
                    "type": "object",
                    "description": "A set (array) of simple or individual OIC Links",
                    "properties": {
                      "links": {
                        "type": "array",
                        "description": "Array of OIC Links. In addition to properties required for an OIC
Link, the identifier for that link in this set is also required",
```

```
                "items": {
                  "allOf": [
                    {
                      "$ref": "oic.oic-link-schema.json#"
                    }
                  ],
                  "required": ["ins"]
                }
              }
            }
          },
          "oic.collection.tagged-setoflinks": {
            "type": "object",
            "description": "A tagged link is a set (array) of links that are tagged with one or more
key-value pairs usually either an ID or Name or both",
            "allOf": [
              {
            "$ref":"#/definitions/oic.collection.setoflinks"
              },
              {
                "properties": {
                  "n": {
                    "type": "string",
                    "description": "Used to name i.e. tag the set of links",
                    "format": "UTF8"
                  },
                  "id": {
                    "oneOf": [
                      {
                        "type": "integer",
                         "description": "A number that is unique to that collection; like an
ordinal number that is not repeated"
                      },
                      {
                        "type": "string",
                        "description": "A unique string that could be a hash or similarly unique"
                      },
                      {
                        "type": "string",
                        "format": "UUID",
                        "description": "A unique string that could be a UUIDv4"
                      }
                    ],
                    "description": "Id for each set of links i.e. tag. Can be an value that is
unique to the use context or a UUIDv4"
                  },
                  "di": {
                    "type": "string",
                     "description": "The device ID which is an UUIDv string; used for backward
compatibility with Spec A defintion of /oic/res",
                    "format": "UUID"
                  }
                }
              }
```

```
              ],
              "required": [ "links" ]
            },
            "oic.collection.setof - tagged - setoflinks": {
              "type": "array",
              "items": {
                "$ref": "#/definitions/oic.collection.tagged - setoflinks"
              }
            },
            "oic.collection.alllinks": {
              "description": "All forms of links in a collection",
              "oneOf": [
                {
                  "$ref": "#/definitions/oic.collection.setof - tagged - setoflinks"
                },
                {
              "$ref":"#/definitions/oic.collection.setoflinks"
                }
              ],
              "required": [ "links" ]
            },
            "oic.collection": {
              "type": "object",
               "description": "A collection is a set (array) of tagged - link or set (array) of
simple links along with additional properties to describe the collection itself",
              "allOf": [
                {
              "$ref":"#/definitions/oic.collection.alllinks"
                },
                {
                  "properties": {
                    "n": {
                      "type": "string",
                      "description": "User friendly name of the collection",
                      "format": "UTF"
                    },
                    "id": {
                      "oneOf": [
                      {
                        "type": "integer",
                        "description": " A number that is unique to that collection; like an
ordinal number that is not repeated"
                      },
                      {
                        "type": "string",
                        "description": "A unique string that could be a hash or similarly unique"
                      },
                      {
                        "type": "string",
                        "format": "UUID", "description": "A unique string that could be a UUIDv4"
                      }
                    ],
                    "description": "ID for the collection. Can be an value that is unique to the
use context or a UUIDv4"
```

```
                    },
                    "rts": {
                      "type": "string",
                      "description": "Defines the list of allowable resource types (for Target and
anchors) in links included in the collection; new links being created can only be from this list",
                      "format": "UTF8"
                    },
                    "drel": {
                      "type": "string",
                      "description": "When specified this is the default relationship to use when an
OIC Link does not specify an explicit relationship with * rel * parameter"
                    }
                  }
                }
              ]
            }
          },
          "type": "object",
          "allOf": [
            {
              "$ref": "oic.core-schema.json#/definitions/oic.core"
            },
            {
              "$ref": "#/definitions/oic.collection"
            }
          ]
        }
      example: |
        {
          "rt": "oic.wk.sceneList",
          "n": "list of scene Collections",
          "rts": "oic.wk.sceneCollection",
          "links": [  ]
        }
```

2. 属性定义

属性定义如表 9-20 所示。

表 9-20 属性定义

属 性 名	属性值类型	是否强制	描 述
links	多类型		
di	多类型		设备 ID 是一个 UUIDv4 字符串；用于与规范中"/oic/res"的定义向后兼容
id	多类型		集合 ID
rts	多类型		定义了集合中包含的链接中允许的资源类型(对于目标和锚点)的列表，正在创建的新链接只能来自此列表
drel	字符串		在 OCF 链接未通过"rel"参数指明一个显式关系时，指定使用该默认关系

3. CRUDN 行为

CRUDN 行为如表 9-21 所示。

表 9-21 CRUDN 行为

资　源	创建	检索	更新	删除	通知
/SceneListResURI		get			

9.11 场景集合

场景集合是一组场景建模的集合。该资源是具有其他参数的通用集合资源。"rts"值应包含"oic. sceneMember"资源类型,附加参数是"lastScene",是最后由任何 OCF 客户端"sceneValueList"设置的场景值,这是可用场景的列表,"lastScene"应该列在"sceneValueList"中。URI 示例为"/SceneCollectionResURI",资源类型定义为"oic. wk. sceneCollection"。

1. RAML 定义

```
#%RAML 0.8
title: Scene
version: v1-20160622
traits:
  - interface :
    queryParameters:
      if:
        enum: ["oic.if.a", "oic.if.ll", "oic.if.baseline"]
/SceneCollectionResURI:
  description: |
    Collection that models a set of Scenes. This resource is a generic collection resource with additional
    parameters. The rts value shall contain oic.sceneMember resource types. The additional parameters are
    lastScene, this is the scene value last set by any OIC Client sceneValueList, this is the list of
    available scenes lastScene shall be listed in sceneValueList.
  get:
    description: |
      Provides the current list of web links pointing to scenes
    responses :
      200:
        body:
          application/json:
            schema: |
              {
                "$schema": "http://json-schema.org/draft-04/schema#",
                "description" : "Copyright (c) 2016 Open Connectivity Foundation, Inc. All rights reserved.",
                "id": "http://openconnectivityfoundation.org/core/schemas/oic.collection-schema.json#",
                "title": "Scene Collection",
                "definitions": {
                  "oic.scenecollection": {
                    "type": "object",
                    "properties": {
                      "lastScene": {
                        "type": "string",
                        "description": "Last selected Scene, shall be part of sceneValues",
                        "format": "UTF8"
                      },
```

```
                "sceneValues": {
                  "type": "string",
                  "description": "ReadOnly, All available scene values",
                  "format": "CSV"
                },
                "n": {
                  "type": "string",
                  "description": "Used to name the Scene collection",
                  "format": "UTF8"
                },
                "id": {
                  "type": "string",
                  "description" : "A unique string that could be a hash or similarly unique"
                },
                "rts": {
                  "type": "string",
                  "description": "ReadOnly, Defines the list of allowable resource types in links
included in the collection; new links being created can only be from this list",
                  "format": "UTF8"
                },
                "links": {
                  "type": "array",
                  "description": "Array of OIC web links that are reference from this collection",
                  "items" : {
                    "allOf": [
                      { "$ref": "oic.oic-link-schema.json#/definitions/oic.oic-link" },
                      { "required" : [ "ins" ] }
                    ]
                  }
                }
              },
              "required":["lastScene","sceneValues","rts","id" ]
            }
          },
          "type": "object",
          "allOf" : [
            { "$ref": "oic.core-schema.json#/definitions/oic.core" },
            { "$ref": "#/definitions/oic.sceneCollection" }
          ]
        }
      example: |
        {
          "lastScene": "off",
          "sceneValues": "off,Reading,TVWatching",
          "rt": "oic.wk.sceneCollection",
          "n": "My Scenes for my living room",
          "id": "B-F-F-BEC-ECBDADC",
          "rts": "oic.wk.sceneMember",
          "links": [ ]
        }
  post:
    description: |
      Provides the action to create a new sceneMember in the SceneCollection resource. The only resource type
that is allowed to be created is "oic.wk.sceneMember". The id of the resource will be generated by the
```

```
implementation. As example the mappings of the scenes are mapped to different states of a binary switch.
    body:
      application/json:
        schema: |
          {
            "$schema": "http://json-schema.org/draft-04/schema#",
            "description" : "Copyright (c) 2016 Open Connectivity Foundation, Inc. All rights reserved.",
            "id":"http://openconnectivityfoundation.org/core/schemas/oic.sceneMember- schema.json#",
            "title" : "Scene Member",
            "definitions": {
              "oic.sceneMember": {
                "type": "object",
                "properties": {
                  "n": {
                    "type": "string",
                    "description": "Used to name the Scene collection",
                    "format": "UTF8"
                  },
                  "id": {
                    "type": "string",
                    "description": "Can be an value that is unique to the use context or a UUIDv4"
                  },
"SceneMappings" : {
                    "type": "array",
                    "description":"array of mappings per scene,can be",
                    "items": [
                      {
                        "type": "object",
                        "properties": {
                          "scene": {
                            "type": "string",
                            "description": "Specifies a scene value that will acted upon"
                          },
                          "memberProperty": {
                            "type": "string",
                            "description": "ReadOnly, property name that will be mapped"
                          },
                          "memberValue": {
                            "type": "string",
                            "description": "ReadOnly, value of the Member Property"
                          }
                        },
                        "required": [ "scene", "memberProperty", "memberValue" ]
                      }
                    ]
                  },
                  "link": {
                    "type": "string",
                    "description": "web link that points at an resource",
                    "$ref": "oic.oic-link-schema.json#"
                  }
                },
                "required": [ "link" ]
              }
```

```
          },
          "type": "object",
          "allOf" : [
            { "$ref": "oic.core-schema.json#/definitions/oic.core" },
            { "$ref": "#/definitions/oic.sceneMember" } ] }
        example: |
          {
            "link": { "href":"coap://mydevice/mybinaryswitch",
                      "if": "oic.if.a",
                      "rt": "oic.r.switch.binary" },
            "n": "my binary switch (for light bulb) mappings",
            "sceneMappings": [
              {
                "scene": "off",
                "memberProperty": "value",
                "memberValue": true
              },
              {
                "scene": "Reading",
                "memberProperty": "value",
                "memberValue": false
              },
              {
                "scene": "TVWatching",
                "memberProperty": "value",
                "memberValue": true
              }
            ]
          }
  responses :
    200:
      description: |
        Indicates that the target resource was created. The new resource attributes are provided in the
response.
      body:
        application/json:
          schema: |
            {
              "$schema": "http://json-schema.org/draft-04/schema#",
              "description" : "Copyright (c) 2016 Open Connectivity Foundation, Inc. All rights reserved.",
              "id":"http://openconnectivityfoundation.org/core/schemas/oic.sceneMember-schema.json#",
              "title" : "Scene Member",
              "definitions":{
                "oic.sceneMember": {
                  "type": "object",
                  "properties": {
                    "n": {
                      "type": "string",
                      "description": "Used to name the Scene collection",
                      "format": "UTF8"
                    },
                    "id": {
                      "type": "string",
                      "description": "Can be an value that is unique to the use context or a UUIDv4"
```

```
              },
"SceneMappings" : {
                "type": "array",
                "description": "array of mappings per scene, can be 1",
                "items": [
                  {
                    "type": "object",
                    "properties": {
                      "scene": {
                        "type": "string",
                        "description": "Specifies a scene value that will acted upon"
                      },
                      "memberProperty": {
                        "type": "string",
                        "description": "ReadOnly, property name that will be mapped"
                      },
                      "memberValue": {
                        "type": "string",
                        "description": "ReadOnly, value of the Member Property"
                      }
                    },
                    "required": [ "scene", "memberProperty", "memberValue" ]
                  }
                ]
              },
              "link": {
                "type": "string",
                "description": "web link that points at an resource",
                "$ref": "oic.oic-link-schema.json#"
              }
            },
            "required": [ "link" ]
          }
        },
        "type": "object",
        "allOf" : [
          { "$ref": "oic.core-schema.json#/definitions/oic.core" },
          { "$ref": "#/definitions/oic.sceneMember" }
        ]
      }
    example: |
      {
        "id": "0685B960-FFFF-46F7-BEC0-9E6234671ADC1",
        "n": "my binary switch (for light bulb) mappings",
        "link": { "href":"coap://mydevice/mybinaryswitch",
                  "if": "oic.if.a",
                  "rt": "oic.r.switch.binary" },
        "sceneMappings": [
          {
            "scene": "off",
            "memberProperty": "value",
            "memberValue": true
          },
          {
```

```
                    "scene": "Reading",
                    "memberProperty": "value",
                    "memberValue": false
                  },
                  {
                    "scene": "TVWatching",
                    "memberProperty": "value",
                    "memberValue": true
                  }
                ]
              }
put:
  description: |
    Provides the action to change the last settted scene selection. Calling this method shall update of all
sceneMembers to the prescribed membervalue.
    When this method is called with the same value as the current lastScene value then all sceneMembers shall
be updated.
  body:
    application/json:
      schema: |
        {
          "$schema": "http://json-schema.org/draft-04/schema#",
          "description" : "Copyright (c) 2016 Open Connectivity Foundation, Inc. All rights reserved.",
          "id":"http://openconnectivityfoundation.org/core/schemas/oic.sceneCollection- schema.json#",
          "title" : "Scene Collection",
          "definitions": {
            "oic.sceneCollection": {
              "type": "object",
              "properties": {
                "lastScene": {
                "type": "string",
                "description": "Last selected Scene, shall be part of sceneValues",
                "format": "UTF8"
              },
              "sceneValues": {
                "type": "string",
                "description": "ReadOnly, All available scene values",
                "format": "CSV"
              },
              "n": {
                "type": "string",
                "description": "Used to name the Scene collection",
                "format": "UTF8"
              },
              "id": {
                "type": "string",
                "description" : "A unique string that could be a hash or similarly unique"
              },
              "rts": {
                "type": "string",
                "description": "ReadOnly, Defines the list of allowable resource types in links included in
the collection; new links being created can only be from this list",
                "format": "UTF8"
              },
              "links": {
```

```
                "type": "array",
                "description": "Array of OIC web links that are reference from this collection",
                "items" : {
                  "allOf": [
                    { "$ref": "oic.oic-link-schema.json#/definitions/oic.oic-link" },
                    { "required" : [ "ins" ] }
                  ]
                }
              }
            },
            "required": [ "lastScene" ]
          }
        },
        "type": "object",
        "allOf" : [
          { "$ref": "oic.core-schema.json#/definitions/oic.core" },
          { "$ref": "#/definitions/oic.sceneCollection" }
        ]
      }
    example: |
      {
        "lastScene": "Reading"
      }
```

2. 属性定义

属性定义如表 9-22 所示。

表 9-22 属性定义

属 性 名	属性值类型	访问模式	是否强制	描 述
lastScene	字符串	读写	是	最后选择的场景，应该是"sceneValues"的一部分
sceneValues	字符串	只读	是	所有可用的场景值
n	字符串	读写		用于命名场景集合
id	对象	读写	是	一个唯一的字符串，可以是散列或类似唯一的
rts	字符串	读写	是	定义了集合中包含的链接中允许的资源类型(对于目标和锚点)列表，正在创建的新链接只能来自此列表
links	数组	读写		从该集合中引用的 OCF 链接数组

3. CRUDN 行为

CRUDN 行为如表 9-23 所示。

表 9-23 CRUDN 行为

资 源	创建	检索	更新	删除	通知
/SceneCollectionResURI		get	post		

9.12 场景成员

场景成员建模为一个"sceneMember"的集合。URI 示例为"/SceneMemberResURI"，资源类型定义为"oic. r. switch. binary"。

1. RAML 定义

```
#%RAML 0.8
title: Scene
version: v1-20160622
traits:
  - interface :
    queryParameters:
      if:
        enum: ["oic.if.a", "oic.if.ll", "oic.if.baseline"]
/SceneMemberResURI:
  description: |
    Collection that models a sceneMenber.
  get:
    description: |
      Provides the scene member
    responses :
      200:
        body:
          application/json:
            schema: |
              {
                "$schema": "http://json-schema.org/draft-04/schema#",
                "description" : "Copyright (c) 2016 Open Connectivity Foundation, Inc. All rights reserved.",
                "id": "http://openconnectivityfoundation.org/core/schemas/oic.collection-schema.json#",
                "title": "Scene Collection",
                "definitions": {
                  "oic.sceneMember": {
                    "type": "object",
                    "properties": {
                      "n": {
                        "type": "string",
                        "description": "Used to name the Scene collection",
                        "format": "UTF8"
                      },
                      "id": {
                        "type": "string",
                        "description" : "Can be an value that is unique to the use context or a UUIDv4"
                      },
                    "SceneMappings" : {
                    "type": "array",
                    "description": "array of mappings per scene, can be 1",
                    "items": [
                      {
                        "type": "object",
                        "properties": {
                          "scene": {
                            "type": "string",
                            "description": "Specifies a scene value that will acted upon"
                          },
                          "memberProperty": {
                            "type": "string",
                            "description": "ReadOnly, property name that will be mapped"
                          },
```

```
                "memberValue": {
                  "type": "string",
                  "description": "ReadOnly, value of the Member Property"
                }
              },
              "required": [ "scene", "memberProperty", "memberValue" ]
            }
          ]
        },
        "link": {
          "type": "string",
          "description": "web link that points at an resource",
          "$ref": "oic.oic-link-schema.json#"
        }
      },
      "required": [ "link" ]
    }
  },
  "type": "object",
  "allOf" : [
    { "$ref": "oic.core-schema.json#/definitions/oic.core" },
    { "$ref": "#/definitions/oic.sceneMember" }
  ]
}
example: |
  {
    "id": "0685B960-FFFF-46F7-BEC0-9E6234671ADC1",
    "n": "my binary switch (for light bulb) mappings",
    "link": { "href":"coap://mydevice/mybinaryswitch",
              "if": "oic.if.a",
              "rt": "oic.r.switch.binary" },
    "sceneMappings": [
      {
        "scene": "off",
        "memberProperty": "value",
        "memberValue": true
      },
      {
        "scene": "Reading",
        "memberProperty": "value",
        "memberValue": false
      },
      {
        "scene": "TVWatching",
        "memberProperty": "value",
        "memberValue": true
      }
    ]
  }
```

2. 属性定义

属性定义如表 9-24 所示。

表 9-24　属性定义

属　性　名	属性值类型	访问模式	是否强制	描　　述
n	字符串	读写		用于命名场景集合
id	字符串	读写		可以是对使用上下文或 UUIDv4 唯一的值
sceneMappings	数组	读写		每个场景的映射数组,可以是 1
scene	字符串	读写	是	指定执行操作的一个场景
memberProperty	字符串	只读	是	被映射的属性名
memberValue	字符串	只读	是	成员属性的值
link	字符串	读写	是	指向一个资源链接

3. CRUDN 行为

CRUDN 行为如表 9-25 所示。

表 9-25　CRUDN 行为

资　　源	创建	检索	更新	删除	通知
/SceneMemberResURI		get			

9.13　资源目录资源

资源目录资源可以作为资源目录,是被设备公开的资源。使用 GET 请求提供选择器标准(例如整数);使用 POST 请求在"/oic/res"中发布或更新链接;使用 DELETE 请求删除"/oic/res"中的链接。URI 示例为"/oic/rd",资源类型定义为"oic. wk. rd"。

1. RAML 定义

```
#%RAML 0.8
title: Resource Directory
version: v1-20160622
traits:
  - rddelete-di :
      queryParameters:
        di:
          description: This is used to determine which set of links to operata on. (Need authentication to
ensure that there is no spoofing). If instance is ommitted then the entire set of links from this device ID is
deleted.
Example: DELETE /oic/rd?di="0685B960-736F-46F7-BEC0-9E6CBD671ADC1"
  - rddelete-ins :
      queryParameters:
        ins:
          description: Instance of the link to delete Value of parameter is a string where instance to be
deleted are comma separated.
Example: DELETE /oic/rd?di="0685B960-736F-46F7-BEC0-9E6CBD671ADC1";ins="20"
  - rdgetinterface :
      queryParameters:
        if:
          enum: ["oic.if.baseline"]
          description: Interface is optional since there is only one interface supported for the Resource
```

```
Type. Both for RD selection and for publish.
Example: GET /oic/rd?if = oic.if.baseline
   - rdpostinterface :
      queryParameters:
         rt:
            enum: ["oic.wk.rdpub"]
            description: Used in POST request to ask the RD to add the Links in payload to /oic/res.
Example: POST /oic/rd?rt = oic.wk.rdpub
/oic/rd:
   description: |
      Resource to be exposed by any Device that can act as a Resource Directory.
      1) Provides selector criteria (e.g., integer) with GET request
      2) Publish or Update a Link in /oic/res with POST request
      3) Delete a Link in /oic/res with DELETE request
   get:
      description: |
        Get the attributes of the Resource Directory for selection purposes.
      is : ['rdgetinterface']
      responses :
        200:
           description: |
             Respond with the selector criteria - either the set of attributes or the bias factor
           body:
             application/json:
                schema: |
                   {
                      "$schema": "http://json-schema.org/draft-04/schema#",
                      "description" : "Copyright (c) 2016, 2017 Open Connectivity Foundation, Inc. All rights
reserved.",
                      "id": "http://www.openconnectivity.org/ocf-apis/core/schemas/oic.rd.selection-
schema.json#",
                      "title" : "RD Selection",
                      "definitions": {
                        "oic.rd.attributes": {
                            "type": "object",
                            "oneOf": [
                              {
                                "properties": {
                                  "sel": {
                                    "type": "integer",
                                   "minimum": 0,
                                    "maximum": 100,
                                     "description": "A bias factor calculated by the Resource directory - the
value is in the range of 0 to 100 - 0 implies that RD is not to be selected. Client chooses RD with highest
bias factor or randomly between RDs that have same bias factor"
                                  }
                                },
                                "required": ["sel"]
                              },
                              {
                                "properties": {
                                  "sel": {
                                     "description": "Selection criteria that a device wanting to publish to any RD
can use to choose this Resource Directory over others that are discovered",
```

```
                              "type": "object",
                              "properties": {
                                "pwr": {
                                  "type": "string",
                                  "enum": [ "ac", "batt", "safe" ],
                                  "description": "A hint about how the RD is powered. If AC then this is
stronger than battery powered. If source is reliable (safe) then appropriate mechanism for managing power
failure exists"
                                },
                                "conn": {
                                  "type": "string",
                                  "enum": [ "wrd", "wrls" ],
                                  "description": "A hint about the networking connectivity of the RD.
*wrd* if wired connected and *wrls* if wireless connected."
                                },
                                "bw": {
                                  "type": "string",
                                  "description": "Qualitative bandwidth of the connection",
                                  "enum": [ "high", "low", "lossy" ]
                                },
                                "mf": {
                                  "type": "integer",
                                  "description": "Memory factor - Ratio of available memory to total
memory expressed as a percentage"
                                },
                                "load": {
                                  "type": "array",
                                  "items": {
                                    "type": "number"
                                  },
                                  "minItems": 3,
                                  "maxItems": 3,
                                  "description": "Current load capacity of the RD. Expressed as a load
factor 3-tuple (upto two decimal points each). Load factor is based on request processed in a 1 minute, 5
minute window and 15 minute window"
                                }
                              }
                            }
                          },
                          "required": ["sel"]
                        }
                      ]
                    }
                  },
                  "type": "object",
                  "allOf": [
                    { "$ref": "oic.core-schema.json#/definitions/oic.core" },
                    { "$ref": "#/definitions/oic.rd.attributes" }
                  ]
                }
              example: |
                {
                  "rt": ["oic.wk.rd"],
                  "if": ["oic.if.baseline"],
```

```
                "sel": 50
              }
    post:
      description: |
        Publish the resource information for the first time or Update the existing one in /oic/res.
        Appropriates parts of the information, i.e., Links of the published Resources will be discovered
through /oic/res.
        1) When a Device first publishes a Link, the request payload to RD may include the Links without "ins"
Parameter.
        2) Upon granting the request, the RD assigns a unique instance value identifying the Link among all
the Links it advertises and sends back the instance value in "ins" Parameter in the Link to the publishing
Device.
        3) When later the publishing Device updates the existing Link, i.e., changing its Endpoint
information, the request payload to RD needs to include the instance value in "ins" Parameter to identify the
Link to update.
      is : ['rdpostinterface']
      body:
        application/json:
          schema: |
            {
              "$schema": "http://json-schema.org/draft-04/schema#",
              "description": "Copyright (c) 2016, 2017 Open Connectivity Foundation, Inc. All rights
reserved.",
              "id": "http://www.openconnectivity.org/ocf-apis/core/schemas/oic.rd.publish-schema.
json#",
              "title": "RD Publish & Update",
              "definitions": {
                "oic.rd.publish": {
                  "description": "Publishes resources as OIC Links into the resource directory",
                  "properties": {
                    "di": {
                      "$ref": "oic.types-schema.json#/definitions/uuid",
                      "description": "A unique identifier for the publishing Device, i.e., its device ID"
                    },
                    "links": {
                      "$ref": "oic.collection-schema.json#/definitions/oic.collection.setoflinks"
                    },
                    "ttl": {
                      "type": "integer",
                      "description": "Time to indicate a RD, how long to keep this published item. After
this time (in seconds) elapses, the RD invalidates the links. To keep link alive the publishing device updates
the ttl using the update schema"
                    }
                  }
                }
              },
              "type": "object",
              "allOf": [
                {
                  "$ref": "oic.core-schema.json#/definitions/oic.core"
                },
                {
                  "$ref": "#/definitions/oic.rd.publish"
                }
```

```
          ],
          "required": [
            "di",
            "links",
            "ttl"
          ]
        }
      example: |
        {
          "di": "e61c3e6b-9c54-4b81-8ce5-f9039c1d04d9",
          "links": [
            {
              "anchor": "ocf://e61c3e6b-9c54-4b81-8ce5-f9039c1d04d9",
              "href":  "/myLightSwitch",
              "rt":    ["oic.r.switch.binary"],
              "if":    ["oic.if.a", "oic.if.baseline"],
              "p":     {"bm": 3},
              "eps": [
                  {"ep":"coaps://[2001:db8:a::b1d6]:1111", "pri": 2},
                  {"ep": "coaps://[2001:db8:a::b1d6]:1122"},
                  {"ep": "coaps+tcp://[2001:db8:a::123]:2222", "pri": 3}
              ]
            },
            {
              "anchor": "ocf://e61c3e6b-9c54-4b81-8ce5-f9039c1d04d9",
              "href":  "/myLightBrightness",
              "rt":    ["oic.r.brightness"],
              "if":    ["oic.if.a", "oic.if.baseline"],
              "p":     {"bm": 3},
              "eps": [
                  {"ep": "coaps://[[2001:db8:a::123]:2222"}
              ]
            }
          ],
          "ttl": 600
        }
    responses :
      200:
        description: |
          Respond with the same schema as publish but, when a Link is first published, with the additional "
ins" Parameter in the Link. This value is used by the receiver to manage that OCF Link instance.
        body:
          application/json:
            schema: |
              {
                "$schema": "http://json-schema.org/draft-04/schema#",
                "description": "Copyright (c) 2016, 2017 Open Connectivity Foundation, Inc. All rights
reserved.",
                "id": "http://www.openconnectivity.org/ocf-apis/core/schemas/oic.rd.publish-
schema.json#",
                "title": "RD Publish & Update",
                "definitions": {
                  "oic.rd.publish": {
                    "description": "Publishes resources as OIC Links into the resource directory",
```

```
                "properties": {
                  "di": {
                    "$ref": "oic.types-schema.json#/definitions/uuid",
                    "description": "A unique identifier for the publishing Device, i.e., its device ID"
                  },
                  "links": {
                        "$ref": "oic.collection-schema.json#/definitions/oic.collection.
setoflinks"
                  },
                  "ttl": {
                    "type": "integer",
                    "description": "Time to indicate a RD, how long to keep this published item. After
this time (in seconds) elapses, the RD invalidates the links. To keep link alive the publishing device updates
the ttl using the update schema"
                  }
                }
              }
            },
            "type": "object",
            "allOf": [
              {
                "$ref": "oic.core-schema.json#/definitions/oic.core"
              },
              {
                "$ref": "#/definitions/oic.rd.publish"
              }
            ],
            "required": [
              "di",
              "links",
              "ttl"
            ]
          }
        example: |
          {
            "di": "e61c3e6b-9c54-4b81-8ce5-f9039c1d04d9",
            "links": [
              {
                "anchor": "ocf://e61c3e6b-9c54-4b81-8ce5-f9039c1d04d9",
                "href":   "/myLightSwitch",
                "rt":     ["oic.r.switch.binary"],
                "if":     ["oic.if.a", "oic.if.baseline"],
                "p":      {"bm": 3},
                "eps": [
                  {"ep": "coaps://[2001:db8:a::b1d6]:1111", "pri": 2},
                  {"ep": "coaps://[2001:db8:a::b1d6]:1122"},
                  {"ep": "coaps+tcp://[2001:db8:a::123]:2222", "pri": 3}
                ],
                "ins": "11235"
              },
              {
                "anchor": "ocf://e61c3e6b-9c54-4b81-8ce5-f9039c1d04d9",
                "href":   "/myLightBrightness",
                "rt":     ["oic.r.brightness"],
                "if":     ["oic.if.a", "oic.if.baseline"],
```

```
                    "p":      {"bm": 3},
                    "eps": [
                        {"ep": "coaps://[2001:db8:a::123]:2222"}
                    ],
                    "ins": "112358"
                }
            ],
            "ttl": 600
        }
delete:
  description: |
    Delete a particular OIC Link - the link may be a simple link or a link in a tagged set.
  is : ['rddelete-di','rddelete-ins']
  responses :
    200:
      description: |
        The delete succeeded
```

2. 属性定义

属性定义如表 9-26 所示。

表 9-26 属性定义

属 性 名	属性值类型	访问模式	是否强制	描 述
sel	对象		是	一种选择器标准,能用此选择在其他可发现资源上的目录
pwr	字符串	读写		关于 RD 如何供电的提示。如果改为交流电,那么这比电池供电更强。如果电源可靠(安全),则存在用于管理电源故障的适当机制
conn	字符串	读写		关于 RD 的网络连接的提示。如果是有线连接,则为"wrd";如果是无线连接,则为"wrls"
bw	字符串	读写		连接的带宽
mf	整型	读写		内存因子——可用内存与总内存之比,以百分比表示
load	数组			当前 RD 的加载能力

3. CRUDN 行为

CRUDN 行为如表 9-27 所示。

表 9-27 CRUDN 行为

资 源	创建	检索	更新	删除	通知
/oic/rd		get	post	delete	

9.14 图标

此资源描述与图标相关联的属性。URI 示例为"/IconResURI",资源类型定义为"oic. r. icon"。

1. RAML 定义

```
#%RAML 0.8
title: OICIcon
```

```
version: v1.1.0 - 20161107
traits:
  - interface :
      queryParameters:
        if:
          enum: ["oic.if.r", "oic.if.baseline"]
/IconResURI:
   description: |
     This resource describes the attributes associated with an Icon.
   is : ['interface']
   get:
     description: |
       Retrieves the current icon properties.
     responses :
       200:
         body:
           application/json:
             schema: |
               {
                 "id": "http://www.openconnectivity.org/ocf - apis/core/schemas/oic.r.icon.json#",
                 "$schema": "http://json - schema.org/draft - 04/schema#",
                 "description" : "Copyright (c) 2017 Open Connectivity Foundation, Inc. All rights reserved.",
                 "title": "Icon",
                 "definitions": {
                   "oic.r.icon": {
                     "properties": {
                       "mimetype": {
                         "type": "string",
                         "maxLength": 64,
                         "readOnly": true,
                         "description": "Specifies the format of the MIME Type"
                       },
                       "width": {
                         "type": "integer",
                         "minimum": 1,
                         "readOnly": true,
                         "description": "Specifies the width in pixels"
                       },
                       "height": {
                         "type": "integer",
                         "minimum": 1,
                         "readOnly": true,
                       "description": "Specifies the height in pixels"
                       },
                       "media": {
                         "type": "string",
                         "maxLength": 256,
                         "format" : "uri",
                         "readOnly": true,
                         "description": "Specifies the media URL to icon"
                       }
                     }
                   }
                 },
```

```
        "type": "object",
        "allOf": [
          { "$ref": "oic.core-schema.json#/definitions/oic.core"},
          { "$ref": "#/definitions/oic.r.icon"}
        ]
        "required": ["mimetype","width","height","media"]
      }
    example: |
      {
      "rt": ["oic.r.icon"],
      "id": "unique_example_id",
      "mimetype": "image/png",
      "width": 256,
      "height": 256,
      "media": "http://findbetter.ru/public/uploads/1481662800/2043.png"
      }
```

2. 属性定义

属性定义如表 9-28 所示。

表 9-28 属性定义

属性名	属性值类型	访问模式	是否强制	描述
mimetype	字符串	只读	是	规定 MIME 类型的形式
width	整型	只读	是	规定宽度像素
media	字符串	只读	是	规定媒体 URL 为图标
height	字符串	只读	是	规定高度像素

3. CRUDN 行为

CRUDN 行为如表 9-29 所示。

表 9-29 CRUDN 行为

资源	创建	检索	更新	删除	通知
/IconResURI		get			

9.15 内省资源

该资源提供一种方法，去获取所有终端设备的内省数据。该资源托管的 URL 既不是内部 URL 也不是外部 URL。URI 示例为“/IntrospectionResURI”，资源类型为“oic. wk. introspection”。

1. RAML 定义

```
#%RAML 0.8
title: OICIntrospection
version: v1.0.0-20160707
traits:
  - interface :
      queryParameters:
        if:
```

```
        enum: ["oic.if.r", "oic.if.baseline"]
/IntrospectionResURI:
  description: |
    This resource provides the means to get the device introspection data specifiying all the
endpoints of the device. The url hosted by this resource is either a local or an external url.
  is : ['interface']
  get:
    responses :
      200:
        body:
          application/json:
            schema: |
              {
                "id": "http://www.openconnectivity.org/ocf - apis/core/schemas/oic.wk.introspectionInfo.
json#",
                "$schema": "http://json - schema.org/draft - 04/schema#",
                "description" : "Copyright (c) 2017 Open Interconnect Consortium, Inc. All rights
reserved.",
                "title": "introspection resource",
                "definitions": {
                  "oic.wk.introspectionInfo": {
                    "type": "object",
                    "properties": {
                      "urlInfo": {
                        "type": "array",
                        "description": "The valid range for the value Property",
                        "readOnly": true,
                        "minItems": 1,
                        "items": {
                          "type" : "object",
                          "properties": {
                            "url": {
                              "type": "string",
                              "format": "uri",
                              "description" : "url to download the description"
                            },
                            "protocol": {
                              "type": "string",
                              "enum": [ "coap", "coaps", "http", "https", "coap + tcp", "coaps + tcp" ],
                              "description" : "protocol to be used to download the introspection"
                            },
                            "content - type": {
                              "type": "string",
                              "enum": [ "application/json", "application/cbor" ],
                              "default" : "application/cbor",
                                "description" : "content - type of the introspection data"
                            },
                            "version": {
                              "type": "integer",
                              "enum": [ 1 ],
                              "default" : 1,
                              "description" : "version the introspection data that can be downloaded"
                            }
                          },
```

```
                "required" : [ "url","protocol"]
              }
            }
          },
          "required" : ["urlInfo"]
        }
      },
      "type": "object",
      "allOf": [
        {"$ref": "#/definitions/oic.wk.introspectionInfo"},
        {"$ref": "oic.core-schema.json#/definitions/oic.core"}
      ]
    }
  example: |
    {
      "rt" : ["oic.wk.introspection"],
      "urlInfo" : [
        {
          "content-type" : "application/cbor",
          "protocol" : "coap",
          "url" : "coap://[fe80::1]:1234/IntrospectionExampleURI"
        }
      ]
    }
```

2. 属性定义

属性定义如表 9-30 所示。

表 9-30　属性定义

属　性　名	属性值类型	访问模式	是否强制	描　述
urlinfo	字符串	只读	是	值属性的可获取范围
url(urlinfo)	整型		是	下载描述的 URL
content-type(urlinfo)	字符串			内省数据的内容类型
version(urlinfo)	整型			内省数据的版本
protocol(urlinfo)	字符串		是	下载的内省数据所使用的协议

3. CRUDN 行为

CRUDN 行为如表 9-31 所示。

表 9-31　CRUDN 行为

资　源	创建	检索	更新	删除	通知
/IntrospectionResURI		get			

第 10 章 Swagger 定义核心资源类型

CHAPTER 10

Swagger 是一款 RESTful 接口的文档在线自动生成、功能测试软件。Swagger 可以生成一个具有互动性的 API 控制台,开发者可以用来快速学习和尝试 API。Swagger 可以生成客户端 SDK 代码用于各种平台上的实现,本章介绍 OCF 对 Swagger 2.0 的支持代码。

10.1 图标

此资源描述与图标相关联的属性,检索当前的图标属性。URI 示例为"/IconResURI",被定义资源类型为"oic.r.icon"。

1. Swagger 2.0 定义

```
{
  "swagger": "2.0",
  "info": {
    "title": "Icon",
    "version": "v1.1.0 - 20161107",
    "license": {
      "name": "copyright 2016 - 2017 Open Connectivity Foundation, Inc. All rights reserved.",
      "x - description": "Redistribution and use in source and binary forms, with or without modification, are permitted provided that the following conditions are met:\n 1. Redistributions of source code must retain the above copyright notice, this list of conditions and the following disclaimer. \n 2. Redistributions in binary form must reproduce the above copyright notice, this list of conditions and the following disclaimer in the documentation and/or other materials provided with the distribution. \n\n THIS SOFTWARE IS PROVIDED BY THE Open Connectivity Foundation, INC. \"AS IS\" AND ANY EXPRESS OR IMPLIED WARRANTIES, INCLUDING, BUT NOT LIMITED TO, THE IMPLIED WARRANTIES OF MERCHANTABILITY AND FITNESS FOR A PARTICULAR PURPOSE OR WARRANTIES OF NON - INFRINGEMENT, ARE DISCLAIMED. \ n IN NO EVENT SHALL THE Open Connectivity Foundation, INC. OR CONTRIBUTORS BE LIABLE FOR ANY DIRECT, INDIRECT, INCIDENTAL, SPECIAL, EXEMPLARY, OR CONSEQUENTIAL DAMAGES (INCLUDING, BUT NOT LIMITED TO, PROCUREMENT OF SUBSTITUTE GOODS OR SERVICES; LOSS OF USE, DATA, OR PROFITS; OR BUSINESS INTERRUPTION) \n HOWEVER CAUSED AND ON ANY THEORY OF LIABILITY, WHETHER IN CONTRACT, STRICT LIABILITY, OR TORT (INCLUDING NEGLIGENCE OR OTHERWISE) ARISING IN ANY WAY OUT OF THE USE OF THIS SOFTWARE, EVEN IF ADVISED OF THE POSSIBILITY OF SUCH DAMAGE. \n"
    }
  },
  "schemes": ["http"],
  "consumes": ["application/json"],
  "produces": ["application/json"],
  "paths": {
    "/IconResURI" : {
      "get": {
```

```
        "description": "This resource describes the attributes associated with an Icon. \nRetrieves the
current icon properties. \n",
        "parameters": [
          {"$ref": "#/parameters/interface"}
        ],
        "responses": {
          "200": {
            "description" : "",
            "x-example":
              {
              "rt": ["oic.r.icon"],
              "id": "unique_example_id",
              "mimetype": "image/png",
              "width": 256,
             "height": 256,
              "media": "http://findbetter.ru/public/uploads/1481662800/2043.png"
              }
              ,
            "schema": { "$ref": "#/definitions/Icon" }
          }
        }
      }
    }
  },
  "parameters": {
    "interface" : {
      "in" : "query",
      "name" : "if",
      "type" : "string",
      "enum" : ["oic.if.r", "oic.if.baseline"]
    }
  },
  "definitions": {
    "Icon" :
          {
      "properties": {
        "height": {
          "description": "Specifies the height in pixels",
          "minimum": 1,
          "readOnly": true,
          "type": "integer"
        }
        "id": {
          "description": "Instance ID of this specific resource",
          "maxLength": 64,
          "readOnly": true,
          "type": "string"
        }
        "if": {
          "description": "The interface set supported by this resource",
          "items": {
            "enum": [
              "oic.if.baseline",
              "oic.if.ll",
```

```
        "oic.if.b",
        "oic.if.lb",
        "oic.if.rw",
        "oic.if.r",
        "oic.if.a",
        "oic.if.s"
      ],
      "type": "string"
    },
    "minItems": 1,
    "readOnly": true,
    "type": "array"
  },
  "media": {
    "description": "Specifies the media URL to icon",
    "format": "uri",
    "maxLength": 256,
    "readOnly": true,
    "type": "string"
  },
  "mimetype": {
    "description": "Specifies the format of the MIME Type",
    "maxLength": 64,
    "readOnly": true,
    "type": "string"
  },
  "n": {
    "description": "Friendly name of the resource",
    "maxLength": 64,
    "readOnly": true,
    "type": "string"
  },
  "rt": {
    "description": "Resource Type",
    "items": {
      "maxLength": 64,
      "type": "string"
    },
    "minItems": 1,
    "readOnly": true,
    "type": "array"
  },
  "width": {
    "description": "Specifies the width in pixels",
    "minimum": 1,
    "readOnly": true,
    "type": "integer"
  }
},
"required": [
  "mimetype",
  "width",
  "height",
  "media"
```

```
            ]
          }
        }
      }
```

2. 属性定义

属性定义如表 10-1 所示。

表 10-1 属性定义

属 性 名	属性值类型	访问模式	是否强制	描 述
width	整型	只读	是	指定以像素为单位的宽度
rt	数组	只读		资源类型
id	字符串	只读		规定资源的示例 ID
height	整型	只读	是	指定以像素为单位的高度
mimetype	字符串	只读	是	指定 MIME 类型的格式
n	字符串	只读		资源的友好名称
if	数组	只读		该资源支持的接口
media	字符串	只读	是	指定媒体 URL 为图标

3. CRUDN 行为

CRUDN 行为如表 10-2 所示。

表 10-2 CRUDN 行为

资 源	创建	检索	更新	删除	通知
/IconResURI		get			

10.2 内省资源

该资源提供一种方法，获取所有终端设备的内省数据。该资源托管的 URL 既不是内部 URL，也不是外部 URL。URI 示例为"/IntrospectionResURI"，资源类型为"oic.wk.introspection"。

1. Swagger 2.0 定义

```
{
  "swagger": "2.0",
  "info": {
    "title": "Introspection Resource",
    "version": "v1.0.0 - 20160707",
    "license": {
      "name": "copyright 2016 - 2017 Open Connectivity Foundation, Inc. All rights reserved.",
      "x - description": "Redistribution and use in source and binary forms, with or without
modification, are permitted provided that the following conditions are met:\n 1. Redistributions of source
code must retain the above copyright notice, this list of conditions and the following disclaimer. \n 2.
Redistributions in binary form must reproduce the above copyright notice, this list of conditions and the
following disclaimer in the documentation and/orother materials provided with the distribution. \n\n THIS
SOFTWARE IS PROVIDED BY THE Open Connectivity Foundation, INC. \"AS IS\" AND ANY EXPRESS OR IMPLIED
WARRANTIES, INCLUDING, BUT NOT LIMITED TO, THE IMPLIED WARRANTIES OF MERCHANTABILITY AND FITNESS FOR A
```

```
PARTICULAR PURPOSE OR WARRANTIES OF NON - INFRINGEMENT, ARE DISCLAIMED. \n IN NO EVENT SHALL THE Open Connectivity Foundation, INC. OR CONTRIBUTORS BE LIABLE FOR ANY DIRECT, INDIRECT, INCIDENTAL, SPECIAL, EXEMPLARY, OR CONSEQUENTIAL DAMAGES (INCLUDING, BUT NOT LIMITED TO, PROCUREMENT OF SUBSTITUTE GOODS OR SERVICES; LOSS OF USE, DATA, OR PROFITS; OR BUSINESS INTERRUPTION)\n HOWEVER CAUSED AND ON ANY THEORY OF LIABILITY, WHETHER IN CONTRACT, STRICT LIABILITY, OR TORT (INCLUDING NEGLIGENCE OR OTHERWISE) ARISING IN ANY WAY OUT OF THE USE OF THIS SOFTWARE, EVEN IF ADVISED OF THE POSSIBILITY OF SUCH DAMAGE.\n"
    }
  },
  "schemes": ["http"],
  "consumes": ["application/json"],
  "produces": ["application/json"],
  "paths": {
    "/IntrospectionResURI" : {
      "get": {
        "description": "This resource provides the means to get the device introspection data specifiying all the endpoints of the device.\nThe url hosted by this resource is either a local or an external url.\n",
        "parameters": [
          {"$ref": "#/parameters/interface"}
        ],
        "responses": {
            "200": {
              "description" : "",
              "x-example":
                {
                  "rt" : ["oic.wk.introspection"],
                  "urlInfo" : [
                    {
                      "content-type" : "application/cbor",
                      "protocol" : "coap",
                      "url" : "coap://[fe80::1]:1234/IntrospectionExampleURI"
                    }
                  ]
                }
                ,
              "schema": { "$ref": "#/definitions/oic.wk.introspectionInfo" }
            }
        }
      }
    }
  },
  "parameters": {
    "interface" : {
      "in" : "query",
      "name" : "if",
      "type" : "string",
      "enum" : ["oic.if.r", "oic.if.baseline"]
    }
  },
  "definitions": {
    "oic.wk.introspectionInfo" :
              {
        "properties": {
          "id": {
            "description": "Instance ID of this specific resource",
```

```
        "maxLength": 64,
        "readOnly": true,
        "type": "string"
      },
      "if": {
        "description": "The interface set supported by this resource",
        "items": {
          "enum": [
            "oic.if.baseline",
            "oic.if.ll",
            "oic.if.b",
            "oic.if.lb",
            "oic.if.rw",
            "oic.if.r",
            "oic.if.a",
            "oic.if.s"
          ],
          "type": "string"
        }, 6869
        "minItems": 1,
        "readOnly": true,
        "type": "array"
      },
      "n": {
        "description": "Friendly name of the resource",
        "maxLength": 64,
        "readOnly": true,
        "type": "string"
      },
      "rt": {
        "description": "Resource Type",
        "items": {
          "maxLength": 64,
          "type": "string"
        },
        "minItems": 1,
        "readOnly": true,
        "type": "array"
      },
      "urlInfo": {
        "description": "The valid range for the value Property",
        "items": {
          "properties": {
            "content - type": {
              "default": "application/cbor",
              "description": "content - type of the introspection data",
              "enum": [
                "application/json",
                "application/cbor"
              ],
              "type": "string"
            },
            "protocol": {
              "description": "protocol to be used to download the introspection",
```

```
                    "enum": [
                      "coap",
                      "coaps",
                      "http",
                      "https",
                      "coap + tcp",
                      "coaps + tcp"
                    ],
                    "type": "string"
                  },
                  "url": {
                    "description": "url to download the description",
                    "format": "uri",
                    "type": "string"
                  },
                  "version": {
                    "default": 1,
                    "description": "version the introspection data that can be downloaded",
                    "enum": [
                      1
                    ],
                    "type": "integer"
                  }
                },
                "required": [
                  "url",
                  "protocol"
                ],
                "type": "object"
              },
              "minItems": 1,
              "readOnly": true,
              "type": "array"
            }
          },
          "required": [
            "urlInfo"
          ],
          "type": "object"
        }
      }
    }
```

2. 属性定义

属性定义如表 10-3 所示。

表 10-3 属性定义

属 性 名	属性值类型	访问模式	是否强制	描 述
if	数组	只读		该资源支持的接口集合
id	字符串	只读		此特定资源的实例 ID
urlinfo	数组	只读	是	值属性的可获取范围
rt	数组	只读		资源类型
n	字符串	只读		资源的友好名称

3. CRUDN 行为

CRUDN 行为如表 10-4 所示。

表 10-4 CRUDN 行为

资　源	创建	检索	更新	删除	通知
/IntrospectionResURI		get			

10.3 OCF 集合

OCF 集合资源类型包含属性和链接。"oic. if. baseline"接口暴露了集合资源本身的链接和属性表示。URI 示例为"/CollectionBaselineInterfaceURI",资源类型为"oic. wk. col"。

1. Swagger 2.0 定义

```
{
  "swagger": "2.0",
  "info": {
    "title": "OCF Collection",
    "version": "1.0",
    "license": {
      "name": "copyright 2016 - 2017 Open Connectivity Foundation, Inc. All rights reserved.",
"x - description": "Redistribution and use in source and binary forms, with or without modification, are permitted provided that the following conditions are met:\n 1. Redistributions of source code must retain the above copyright notice, this list of conditions and the following disclaimer.\n 2. Redistributions in binary form must reproduce the above copyright notice, this list of conditions and the following disclaimer in the documentation and/or other materials provided with the distribution.\n\n THIS SOFTWARE IS PROVIDED BY THE Open Connectivity Foundation, INC. \"AS IS\" AND ANY EXPRESS OR IMPLIED WARRANTIES, INCLUDING, BUT NOT LIMITED TO, THE IMPLIED WARRANTIES OF MERCHANTABILITY AND FITNESS FOR A PARTICULAR PURPOSE OR WARRANTIES OF NON - INFRINGEMENT, ARE DISCLAIMED.\n IN NO EVENT SHALL THE Open Connectivity Foundation, INC. OR CONTRIBUTORS BE LIABLE FOR ANY DIRECT, INDIRECT, INCIDENTAL, SPECIAL, EXEMPLARY, OR CONSEQUENTIAL DAMAGES (INCLUDING, BUT NOT LIMITED TO, PROCUREMENT OF SUBSTITUTE GOODS OR SERVICES; LOSS OF USE, DATA, OR PROFITS; OR BUSINESS INTERRUPTION)\n HOWEVER CAUSED AND ON ANY THEORY OF LIABILITY, WHETHER IN CONTRACT, STRICT LIABILITY, OR TORT (INCLUDING NEGLIGENCE OR OTHERWISE) ARISING IN ANY WAY OUT OF THE USE OF THIS SOFTWARE, EVEN IF ADVISED OF THE POSSIBILITY OF SUCH DAMAGE.\n"
    }
  },
  "schemes": ["http"],
  "consumes": ["application/json"],
  "produces": ["application/json"],
  "paths": {
    "/CollectionBaselineInterfaceURI" : {
      "get": {
        "description": "OCF Collection Resource Type contains properties and links.\nThe oic.if.baseline interface exposes a representation of\nthe links and the properties of the collection resource itself\nRetrieve on Baseline Interface\n",
        "parameters": [
          {"$ref": "#/parameters/interface - baseline"}
        ],
        "responses": {
            "200": {
```

```
            "description" : "",
            "x-example":
              {
                "rt": ["oic.wk.col"],
                "id": "unique_example_id",
                "rts": [ "oic.r.switch.binary", "oic.r.airflow" ],
                "links": [
                  {
                    "href": "switch",
                    "rt": ["oic.r.switch.binary"],
                   "if": ["oic.if.a", "oic.if.baseline"],
                    "eps": [
                        {"ep": "coap://[fe80::b1d6]:1111", "pri": 2},
                        {"ep": "coaps://[fe80::b1d6]:1122"},
                        {"ep": "coap+tcp://[2001:db8:a::123]:2222", "pri": 3}
                    ]
                  },
                  {
                    "href": "airFlow",
                    "rt": ["oic.r.airflow"],
                    "if": ["oic.if.a", "oic.if.baseline"],
                    "eps": [
                        {"ep": "coap://[fe80::b1d6]:1111", "pri": 2},
                        {"ep": "coaps://[fe80::b1d6]:1122"},
                        {"ep": "coap+tcp://[2001:db8:a::123]:2222", "pri": 3}
                    ]
                  }
                ]
              }
              ,
            "schema": { "$ref": "#/definitions/sbaseline" }
          }
        }
      },
      "post": {
        "description": "Update on Baseline Interface\n",
        "parameters": [
          {"$ref": "#/parameters/interface-baseline"},
          {
            "name": "body",
            "in": "body",
            "required": true,
            "schema": { "$ref": "#/definitions/sbaseline" }
          }
        ],
        "responses": {
            "200": {
              "description" : "",
              "schema": { "$ref": "#/definitions/sbaseline" }
            }
        }
      }
    },
    "/CollectionBatchInterfaceURI" : {
```

```
"get": {
    "description": "OCF Collection Resource Type contains properties and links.\nThe oic.if.b interfacce exposes a composite representation of the\nresources pointed to by the links\nRetrieve on Batch Interface\n",
    "parameters": [
      {"$ref": "#/parameters/interface-b"}
    ],
    "responses": {
        "200": {
          "description" : "All targets returned OK status (HTTP 200 or CoAP 2.05 Content)",
          "x-example":
            [
              {
                "href": "switch",
                "rep":
                  {
                    "value": true
                  }
              },
              {
                "href": "airFlow",
                "rep":
                  {
                    "direction": "floor",
                    "speed": 3
                  }
              }
            ]
            ,
          "schema": { "$ref": "#/definitions/sbatch-retrieve" }
        },
        "404": {
          "description" : "One or more targets did not return an OK status, return a representation containing returned properties from the targets that returned OK",
          "x-example":
            [
              {
                "href": "switch",
                "rep":
                  {
                    "value": true
                  }
              }
            ]
            ,
          "schema": { "$ref": "#/definitions/sbatch-retrieve" }
        }
    }
},
"post": {
  "description": "Update on Batch Interface\n",
  "parameters": [
    {"$ref": "#/parameters/interface-b"},
    {
```

```
            "name": "body",
            "in": "body",
            "required": true,
            "schema": { "$ref": "#/definitions/sbatch-update" },
            "x-example":
              [
                {
                  "href": "switch",
                  "rep":
                    {
                      "value": true
                    }
                },
                {
                  "href": "airFlow",
                  "rep":
                    {
                      "direction": "floor",
                      "speed": 3
                    }
                }
              ]
          }
        ],
        "responses": {
            "200": {
              "description" : "all targets returned OK status (HTTP 200 or CoAP 2.04 Changed) return a representation of the current state of all targets",
              "x-example":
                [
                  {
                    "href": "switch",
                    "rep":
                      {
                        "value": true
                      }
                  },
                  {
                    "href": "airFlow",
                    "rep":
                      {
                        "direction": "demist",
                        "speed": 5
                      }
                  }
                ]
                ,
              "schema": { "$ref": "#/definitions/sbatch-retrieve" }
            },
            "403": {
              "description" : " one or more targets did not return OK status; return a retrieve representation of the current state of all targets in the batch",
              "x-example":
                [
```

```
                {
                  "href": "switch",
                  "rep":
                    {
                      "value": true
                    }
                },
                {
                  "href": "airFlow",
                  "rep":
                    {
                      "direction": "floor",
                      "speed": 3
                    }
                }
              ]
              ,
            "schema": { "$ref": "#/definitions/sbatch-retrieve" }
          }
        }
      }
    },
    "/CollectionLinkListInterfaceURI" : {
      "get": {
        "description": "OCF Collection Resource Type contains properties and links.\nThe oic.if.ll interface exposes a representation of the links\nRetrieve on Link List Interface\n",
        "parameters": [
          {"$ref": "#/parameters/interface-ll"}
        ],
        "responses": {
          "200": {
            "description" : "",
            "x-example":
              [
                {
                  "href": "switch",
                  "rt": ["oic.r.switch.binary"],
                  "if": ["oic.if.a", "oic.if.baseline"],
                  "eps": [
                    {"ep": "coap://[fe80::b1d6]:1111", "pri": 2},
                    {"ep": "coaps://[fe80::b1d6]:1122"},
                    {"ep": "coap+tcp://[2001:db8:a::123]:2222", "pri": 3}
                  ]
                },
                {
                  "href": "airFlow",
                  "rt": ["oic.r.airflow"],
                  "if": ["oic.if.a", "oic.if.baseline"],
                  "eps": [
                    {"ep": "coap://[fe80::b1d6]:1111", "pri": 2},
                    {"ep": "coaps://[fe80::b1d6]:1122"},
                    {"ep": "coap+tcp://[2001:db8:a::123]:2222", "pri": 3}
                  ]
                }
```

```
              ]
             ,
             "schema": { "$ref": "#/definitions/slinks" }
           }
         }
       }
     }
   },
   "parameters": {
     "interface-ll" : {
       "in" : "query",
       "name" : "if",
       "type" : "string",
       "enum" : ["oic.if.ll"]
     },
     "interface-b" : {
       "in" : "query",
       "name" : "if",
       "type" : "string",
       "enum" : ["oic.if.b"]
     },
     "interface-baseline" : {
       "in" : "query",
       "name" : "if",
       "type" : "string",
       "enum" : ["oic.if.baseline"]
     },
     "interface-all" : {
       "in" : "query",
       "name" : "if",
       "type" : "string",
       "enum" : ["oic.if.ll", "oic.if.baseline", "oic.if.b"]
     }
   },
   "definitions": {
     "sbaseline" :
           {
       "description": "A set (array) of simple or individual OIC Links. In addition to properties required
for an OIC Link, the identifier for that link in this set is also required",
         "items": {
           "properties": {
             "anchor": {
               "description": "This is used to override the context URI e.g. override the URI of the
containing collection",
               "format": "uri",
               "maxLength": 256,
               "type": "string"
             },
             "di": {
               "description": "Unique identifier for device (UUID)",
               "pattern": "^[a-fA-F0-9]{8}-[a-fA-F0-9]{4}-[a-fA-F0-9]{4}-[a-fA-F0-9]
{4}-[a-fA-F0-9]{12}$",
               "type": "string"
             },
```

```
"drel": {
  "description": "When specified this is the default relationship to use when an OIC Link does
not specify an explicit relationship with * rel * parameter",
  "type": "string"
},
"eps": {
  "description": "the Endpoint information of the target Resource",
  "items": {
    "properties": {
      "ep": {
        "description": "URI with Transport Protocol Suites + Endpoint Locator as specified",
        "format": "uri",
        "type": "string"
      },
      "pri": {
        "description": "The priority among multiple Endpoints as specified",
        "minimum": 1,
        "type": "integer"
      }
    },
    "type": "object"
  },
  "type": "array"
},
"href": {
  "description": "This is the target URI, it can be specified as a Relative Reference or fully
- qualified URI. Relative Reference should be used along with the di parameter to make it unique.",
  "format": "uri",
  "maxLength": 256,
  "type": "string"
},
"id": {
  "description": "Instance ID of this specific resource",
  "maxLength": 64,
  "readOnly": true,
  "type": "string"
},
"if": {
  "description": "The interface set supported by this resource",
  "items": {
    "enum": [
      "oic.if.baseline",
      "oic.if.ll",
      "oic.if.b",
      "oic.if.rw",
      "oic.if.r",
      "oic.if.a",
      "oic.if.s"
    ],
    "type": "string"
  },
  "minItems": 1,
  "type": "array"
},
```

```
            "ins": {
              "description": "The instance identifier for this web link in an array of web links - used in
collections",
              "oneOf": [
                {
                  "description": " An ordinal number that is not repeated - must be unique in the
collection context",
                  "type": "integer"
                },
                {
                  "description": "Any unique string including a URI",
                  "format": "uri",
                  "maxLength": 256,
                  "type": "string"
                },
                {
                  "description": "Unique identifier (UUID)",
                  "pattern": "^[a-fA-F0-9]{8}-[a-fA-F0-9]{4}-[a-fA-F0-9]{4}-[a-fA-F0
-9]{4}-[a-fA-F0-9]{12}$",
                  "type": "string"
                }
              ]
            },
            "links": {
              "description": "All forms of links in a collection",
              "oneOf": [
                {
                  "description": " A set (array) of simple or individual OIC Links. In addition to
properties required for an OIC Link, the identifier for that link in this set is also required",
                  "items": {
                    "properties": {
                      "anchor": {
                        "description": "This is used to override the context URI e.g. override the URI of
the containing collection",
                        "format": "uri",
                        "maxLength": 256,
                        "type": "string"
                      },
                      "di": {
                        "description": "Unique identifier for device (UUID)",
                        "pattern": "^[a-fA-F0-9]{8}-[a-fA-F0-9]{4}-[a-fA-F0-9]{4}-[a-
fA-F0-9]{4}-7367[a-fA-F0-9]{12}$",
                        "type": "string"
                      },
                      "eps": {
                        "description": "the Endpoint information of the target Resource",
                        "items": {
                          "properties": {
                            "ep": {
                              "description": "URI with Transport Protocol Suites + Endpoint Locator as
specified ",
                              "format": "uri",
                              "type": "string"
                            },
```

```
                    "pri": {
                      "description": "The priority among multiple Endpoints as specified",
                      "minimum": 1,
                      "type": "integer"
                    }
                  },
                  "type": "object"
                },
                "type": "array"
              },
              "href": {
                "description": " This is the target URI, it can be specified as a Relative
Reference or fully-qualified URI. Relative Reference should be used along with the di parameter to make it
unique.",
                "format": "uri",
                "maxLength": 256,
                "type": "string"
              }
              "if": {
                "description": "The interface set supported by this resource",
                "items": {
                  "enum": [
                    "oic.if.baseline",
                    "oic.if.ll",
                    "oic.if.b",
                    "oic.if.rw",
                    "oic.if.r",
                    "oic.if.a",
                    "oic.if.s"
                  ],
                  "type": "string"
                },
                "minItems": 1,
                "type": "array"
              },
              "ins": {
                "description": "The instance identifier for this web link in an array of web links
 - used in collections",
                "oneOf": [
                  {
                    "description": "An ordinal number that is not repeated - must be unique in
the collection context",
                    "type": "integer"
                  },
                  {
                    "description": "Any unique string including a URI",
                    "format": "uri",
                    "maxLength": 256,
                    "type": "string"
                  },
                  {
                    "description": "Unique identifier (UUID)",
                    "pattern": "^[a-fA-F0-9]{8}-[a-fA-F0-9]{4}-[a-fA-F0-9]{4}-
[a-fA-F0-9]{4}-[a-fA-F0-9]{12}$",
```

```
                        "type": "string"
                      }
                    ]
                  },
                  "p": {
                    "description": "Specifies the framework policies on the Resource referenced by
the target URI",
                    "properties": {
                      "bm": {
                        "description": "Specifies the framework policies on the Resource referenced
by the target URI for e.g. observable and discoverable",
                        "type": "integer"
                      }
                    },
                    "required": [
                      "bm"
                    ],
                    "type": "object"
                  },
                  "rel": {
                    "description": "The relation of the target URI referenced by the link to the
context URI",
                    "oneOf": [
                      {
                        "default": [
                          "hosts"
                        ],
                        "items": {
                          "maxLength": 64,
                          "type": "string"
                        },
                        "minItems": 1,
                        "type": "array"
                      },
                      {
                        "default": "hosts",
                        "maxLength": 64,
                        "type": "string"
                      }
                    ]
                  },
                  "rt": {
                    "description": "Resource Type",
                    "items": {
                      "maxLength": 64,
                      "type": "string"
                    },
                    "minItems": 1,
                    "type": "array"
                  },
                  "title": {
                    "description": "A title for the link relation. Can be used by the UI to provide a
context",
                    "maxLength": 64,
```

```
                    "type": "string"
                  },
                  "type": {
                    "default": "application/cbor",
                    "description": "A hint at the representation of the resource referenced by the
target URI. This represents the media types that are used for both accepting and emitting",
                    "items": {
                      "maxLength": 64,
                      "type": "string"
                    },
                    "minItems": 1,
                    "type": "array"
                  }
                },
                "required": [
                  "href",
                  "rt",
                  "if"
                ],
                "type": "object"
              },
              "type": "array"
            }
          ]
        },
        "n": {
          "description": "Friendly name of the resource",
          "maxLength": 64,
          "readOnly": true,
          "type": "string"
        },
        "p": {
          "description": "Specifies the framework policies on the Resource referenced by the target URI",
          "properties": {
            "bm": {
              "description": "Specifies the framework policies on the Resource referenced by the
target URI for e.g. observable and discoverable",
              "type": "integer"
            }
          },
          "required": [
            "bm"
          ],
          "type": "object"
        },
        "rel": {
          "description": "The relation of the target URI referenced by the link to the context URI",
          "oneOf": [
            {
              "default": [
                "hosts"
              ],
              "items": {
                "maxLength": 64,
```

```
                "type": "string"
              },
              "minItems": 1,
              "type": "array"
            },
            {
              "default": "hosts",
              "maxLength": 64,
              "type": "string"
            }
          ]
        },
        "rt": {
          "description": "Resource Type",
          "items": {
            "maxLength": 64,
            "type": "string"
          },
          "minItems": 1,
          "type": "array"
        },
        "rts": {
          "description": "Defines the list of allowable resource types (for Target and anchors) in
links included in the collection; new links being created can only be from this list",
          "items": {
            "maxLength": 64,
            "type": "string"
          },
          "minItems": 1,
          "readOnly": true,
          "type": "array"
        },
        "title": {
          "description": "A title for the link relation. Can be used by the UI to provide a context",
          "maxLength": 64,
          "type": "string"
        },
        "type": {
          "default": "application/cbor",
          "description": "A hint at the representation of the resource referenced by the target URI.
This represents the media types that are used for both accepting and emitting",
          "items": {
            "maxLength": 64,
            "type": "string"
          },
          "minItems": 1,
          "type": "array"
        }
      },
      "required": [
        "href",
        "rt",
        "if"
      ],
```

```
            "type": "object"
          },
          "type": "array"
        }
      "sbatch - retrieve" :
          {
          "items": {
            "additionalProperties": true,
            "properties": {
              "href": {
                "description": "URI of the target resource relative assuming the collection URI as anchor",
                "format": "uri",
                "maxLength": 256,
                "type": "string"
              },
              "rep": {
                "oneOf": [
                  {
                    "description": "The response payload from a single resource",
                    "type": "object"
                  },
                  {
                    "description": " The response payload from a collection (batch) resource",
                    "type": "array"
                  }
                ]
              }
            },
            "required": [
              "href",
              "rep"
            ],
            "type": "object"
          },
          "minItems": 1,
          "type": "array"
        }
      "sbatch - update" :
          {
        "description": "array of resource representations to apply to the batch collection, using href to
indicate which resource(s) in the batch to update. If the href property is empty, effectively making the URI
reference to the collection itself, the representation is to be applied to all resources in the batch",
        "items": {
          "additionalProperties": true,
           "properties": {
             "href": {
               "description": "URI of the target resource relative assuming the collection URI as anchor",
               "format": "uri",
               "maxLength": 256,
               "type": "string"
             },
             "rep": {
               "oneOf": [
                 {
```

```
                "description": "The response payload from a single resource",
                "type": "object"
              },
              {
                "description": " The response payload from a collection (batch) resource",
                "type": "array"
              }
            ]
          }
        },
        "required": [
          "href",
          "rep"
        ],
        "type": "object"
      },
      "minItems": 1,
      "type": "array"
    }
  "slinks" :
        {
      "description": "All forms of links in a collection",
      "oneOf": [
        {
          "description": "A set (array) of simple or individual OIC Links. In addition to properties
required for an OIC Link, the identifier for that link in this set is also required",
          "items": {
            "properties": {
              "anchor": {
                "description": "This is used to override the context URI e.g. override the URI of the
containing collection",
                "format": "uri",
                "maxLength": 256,
                "type": "string"
              },
              "di": {
                "description": "Unique identifier for device (UUID)",
                "pattern": "^[a-fA-F0-9]{8}-[a-fA-F0-9]{4}-[a-fA-F0-9]{4}-[a-fA-
F0-9]{4}-[a-fA-F0-9]{12}$",
                "type": "string"
              },
              "eps": {
                "description": "the Endpoint information of the target Resource",
                "items": {
                  "properties": {
                    "ep": {
                      "description": "URI with Transport Protocol Suites + Endpoint Locator as specified",
                      "format": "uri",
                      "type": "string"
                    },
                    "pri": {
                      "description":"The priority among multiple Endpoints as specified ",
                      "minimum": 1,
                      "type": "integer"
```

```
                }
              },
                "type": "object"
              },
              "type": "array"
            },
            "href": {
              "description": "This is the target URI, it can be specified as a Relative Reference or
fully - qualified URI. Relative Reference should be used along with the di parameter to make it unique.",
              "format": "uri",
              "maxLength": 256,
              "type": "string"
            },
            "if": {
              "description": "The interface set supported by this resource",
              "items": {
                "enum": [
                  "oic.if.baseline",
                  "oic.if.ll",
                  "oic.if.b",
                  "oic.if.rw",
                  "oic.if.r",
                  "oic.if.a",
                  "oic.if.s"
                ],
                "type": "string"
              },
              "minItems": 1,
              "type": "array"
            },
            "ins": {
              "description": "The instance identifier for this web link in an array of web links -
used in collections",
              "oneOf": [
                {
                  "description": "An ordinal number that is not repeated - must be unique in the
collection context",
                  "type": "integer"
                },
                {
                  "description": "Any unique string including a URI",
                  "format": "uri",
                  "maxLength": 256,
                  "type": "string"
                },
                {
                  "description": "Unique identifier (UUID)",
                  "pattern": "^[a-fA-F0-9]{8}-[a-fA-F0-9]{4}-[a-fA-F0-9]{4}-[a-
fA-F0-9]{4}-[a-fA-F0-9]{12}$",
                  "type": "string"
                }
              ]
            },
            "p": {
```

```
                "description": "Specifies the framework policies on the Resource referenced by the target URI",
                "properties": {
                  "bm": {
                    "description": "Specifies the framework policies on the Resource referenced by the target URI for e.g. observable and discoverable",
                    "type": "integer"
                  }
                },
                "required": [
                  "bm"
                ],
                "type": "object"
              },
              "rel": {
                "description": "The relation of the target URI referenced by the link to the context URI",
                "oneOf": [
                  {
                    "default": [
                      "hosts"
                    ],
                    "items": {
                      "maxLength": 64,
                      "type": "string"
                    },
                    "minItems": 1, 7
                    "type": "array" 7
                  },
                  {
                    "default": "hosts",
                    "maxLength": 64,
                    "type": "string"
                  }
                ]
              },
              "rt": {
                "description": "Resource Type",
                "items": {
                  "maxLength": 64,
                  "type": "string"
                },
                "minItems": 1,
                "type": "array"
              },
              "title": {
                "description": "A title for the link relation. Can be used by the UI to provide a context",
                "maxLength": 64,
                "type": "string"
              },
              "type": {
                "default": "application/cbor",
                "description": "A hint at the representation of the resource referenced by the target URI. This represents the media types that are used for both accepting and emitting",
                "items": {
```

```
                    "maxLength": 64,
                    "type": "string"
                  },
                  "minItems": 1,
                  "type": "array"
                }
              },
              "required": [
                "href",
                "rt",
                "if"
              ],
              "type": "object"
            },
            "type": "array"
          }
        ]
      }
    }
  }
```

2. 属性定义

属性定义如表 10-5 所示。

表 10-5　属性定义

属性名	属性值类型	访问模式	是否强制	描述
links	多类型			集合中链接的所有形式
rt	数组		是	资源类型
n	字符串	只读		资源的友好名称
rts	数组	只读		定义在集合中的链接里允许的资源列表
id	字符串	只读		指定资源的示例 ID
di	多类型			设备唯一身份符号
drel	字符串			当指定时，这是 OCF 链接未指定与"rel"参数的显式关系时使用的默认关系
eps	数组			目标资源的终端信息
title	字符串			链接关系的标题
rep	多类型		是	
p	对象			规定从目标 URI 引用资源的框架策略
ins	多类型			集合中使用链接数组的 Web 实例标识符
href	字符串		是	这是目标 URI，它可以被指定为相对引用或完全限定的 URI。应该与"di"参数一起使用使其能够唯一
rel	多类型			内容 URI 的链接引用目标 URI 的关系
type	数组			提示由目标 URI 引用的资源表示，这表示用于接收和发送的媒体类型
anchor	字符串			用于覆盖内容 URI。例如，覆盖包含集合的 URI
if	数组		是	该资源支持的接口

3. CRUDN 行为

CRUDN 行为如表 10-6 所示。

表 10-6 CRUDN 行为

资 源	创建	检索	更新	删除	通知
/CollectionBaselineInterfaceURI		get	post		

10.4 平台配置

该资源允许平台配置指定信息，检索当前平台配置信息。URI 示例为“/example/PlatformConfigurationResURI”，资源类型为“oic. wk. con. p”。

1. Swagger 2.0 定义

```
{
  "swagger": "2.0",
  "info": {
    "title": "Platform Configuration",
    "version": "v1 - 20160622",
    "license": {
      "name": "copyright 2016 - 2017 Open Connectivity Foundation, Inc. All rights reserved.",
      "x - description": "Redistribution and use in source and binary forms, with or without modification, are permitted provided that the following conditions are met:\n 1. Redistributions of source code must retain the above copyright notice, this list of conditions and the following disclaimer. \n 2. Redistributions in binary form must reproduce the above copyright notice, this list of conditions and the following disclaimer in the documentation and/or other materials provided with the distribution. \n\n THIS SOFTWARE IS PROVIDED BY THE Open Connectivity Foundation, INC. \"AS IS\" AND ANY EXPRESS OR IMPLIED WARRANTIES, INCLUDING, BUT NOT LIMITED TO, THE IMPLIED WARRANTIES OF MERCHANTABILITY AND FITNESS FOR A PARTICULAR PURPOSE OR WARRANTIES OF NON - INFRINGEMENT, ARE DISCLAIMED. \ n IN NO EVENT SHALL THE Open Connectivity Foundation, INC. OR CONTRIBUTORS BE LIABLE FOR ANY DIRECT, INDIRECT, INCIDENTAL, SPECIAL, EXEMPLARY, OR CONSEQUENTIAL DAMAGES (INCLUDING, BUT NOT LIMITED TO, PROCUREMENT OF SUBSTITUTE GOODS OR SERVICES; LOSS OF USE, DATA, OR PROFITS; OR BUSINESS INTERRUPTION) \ n HOWEVER CAUSED AND ON ANY THEORY OF LIABILITY, WHETHER IN CONTRACT, STRICT LIABILITY, OR TORT (INCLUDING NEGLIGENCE OR OTHERWISE) ARISING IN ANY WAY OUT OF THE USE OF THIS SOFTWARE, EVEN IF ADVISED OF THE POSSIBILITY OF SUCH DAMAGE. \n"
    }
  },
  "schemes": ["http"],
  "consumes": ["application/json"],
  "produces": ["application/json"],
  "paths": {
    "/example/PlatformConfigurationResURI" : {
      "get": {
          "description": "Resource that allows for platform specific information to be configured. \nRetrieves the current platform configuration settings\n",
        "parameters": [
          {"$ref": "#/parameters/interface - all"}
        ],
        "responses": {
            "200": {
              "description" : "",
              "x - example":
                {
                  "rt": ["oic.wk.con.p"],
```

```
                    "mnpn": [ { "language": "en", "value": "My Friendly Device Name" } ]
                  }
                  ,
                "schema": { "$ref": "#/definitions/Conf_Platform" }
              }
            }
          },
          "post": {
            "description": "Update the information about the platform\n",
            "parameters": [
              {"$ref": "#/parameters/interface-rw"},
            {
                "name": "body",
                "in": "body",
                "required": true,
                "schema": { "$ref": "#/definitions/Update_Platform" },
                "x-example":
                  {
                    "n": "Nuevo nombre",
                    "mnpn": [ { "language": "es", "value": "Nuevo nombre de Plataforma Amigable" } ]
                  }
              }
            ],
            "responses": {
              "200": {
                "description": "",
                "x-example":
                  {
                    "n": "Nuevo nombre",
                    "mnpn": [ { "language": "es", "value": "Nuevo nombre de Plataforma Amigable" } ]
                  }
                  ,
                "schema": { "$ref": "#/definitions/Update_Platform" }
              }
            }
          }
        }
      },
      "parameters": {
        "interface-rw" : {
          "in" : "query",
          "name" : "if",
          "type" : "string",
          "enum" : ["oic.if.rw"]
        },
        "interface-all" : {
          "in" : "query",
          "name" : "if",
          "type" : "string",
          "enum" : ["oic.if.rw", "oic.if.baseline"]
        }
      },
      "definitions": {
        "Conf_Platform" :
```

```
    {
"properties": {
  "id": {
    "description": "Instance ID of this specific resource",
    "maxLength": 64,
    "readOnly": true,
    "type": "string"
  },
  "if": {
    "description": "The interface set supported by this resource",
    "items": {
      "enum": [
        "oic.if.baseline",
        "oic.if.ll",
        "oic.if.b",
        "oic.if.lb",
        "oic.if.rw",
        "oic.if.r",
        "oic.if.a",
        "oic.if.s"
      ],
      "type": "string"
    },
    "minItems": 1,
    "readOnly": true,
    "type": "array"
  },
  "mnpn": {
    "description": "Platform names",
    "items": {
      "properties": {
        "language": {
          "description": "An RFC 5646 language tag.",
          "pattern": "^[A-Za-z]{1,8}(-[A-Za-z0-9]{1,8})*$",
          "type": "string"
        },
        "value": {
          "description": "Platform description in the indicated language.",
          "maxLength": 64,
          "type": "string"
        }
      },
      "type": "object"
    },
    "minItems": 1,
    "type": "array"
  },
  "n": {
    "description": "Friendly name of the resource",
    "maxLength": 64,
    "readOnly": true,
    "type": "string"
  },
  "rt": {
```

```
            "description": "Resource Type",
            "items": {
              "maxLength": 64,
              "type": "string"
            },
            "minItems": 1,
            "readOnly": true,
            "type": "array"
          }
        },
        "type": "object"
      }
    ,
    "Update_Platform" :
            {
        "properties": {
          "id": {
            "description": "Instance ID of this specific resource",
            "maxLength": 64,
            "readOnly": true,
            "type": "string"
          },
          "if": {
            "description": "The interface set supported by this resource",
            "items": {
              "enum": [
                "oic.if.baseline",
                "oic.if.ll",
                "oic.if.b",
                "oic.if.lb",
                "oic.if.rw",
                "oic.if.r",
                "oic.if.a",
                "oic.if.s"
              ],
              "type": "string"
            },
            "minItems": 1,
            "readOnly": true,
            "type": "array"
          },
          "mnpn": {
            "description": "Platform names",
            "items": {
              "properties": {
                "language": {
                  "description": "An RFC 5646 language tag.",
                  "pattern": "^[A - Za - z]{1,8}( - [A - Za - z0 - 9]{1,8}) * $ ",
                  "type": "string"
                },
                "value": {
                  "description": "Platform description in the indicated language.",
                  "maxLength": 64,
                  "type": "string"
```

```
            }
          },
          "type": "object"
        },
        "minItems": 1,
        "type": "array"
      },
      "n": {
        "description": "Friendly name of the resource",
        "maxLength": 64,
        "type": "string"
      },
      "rt": {
        "description": "Resource Type",
        "items": {
          "maxLength": 64,
          "type": "string"
        },
        "minItems": 1,
        "readOnly": true,
        "type": "array"
      }
    },
    "required": [
      "mnpn"
    ],
    "type": "object"
  }
 }
}
```

2. 属性定义

属性定义如表 10-7 所示。

表 10-7 属性定义

属 性 名	属性值类型	访问模式	是否强制	描 述
mnpn	数组			平台名称
if	数组	只读		资源支持的接口集合
rt	数组	只读		资源类型
id	字符串	只读		指定资源的 ID 示例
n	字符串	只读		资源的友好名称

3. CRUDN 行为

CRUDN 行为如表 10-8 所示。

表 10-8 CRUDN 行为

资 源	创建	检索	更新	删除	通知
/example/PlatformConfigurationResURI		get	post		

10.5 设备配置

该资源允许配置设备特定信息，检索当前设备配置信息。URI 示例为“/example/DeviceConfigurationResURI”，资源类型为“oic. wk. com”。

1. Swagger 2.0 定义

```
{
  "swagger": "2.0",
  "info": {
    "title": "Device Configuration",
    "version": "v1-20160622",
    "license": {
      "name": "copyright 2016-2017 Open Connectivity Foundation, Inc. All rights reserved.",
      "x-description": "Redistribution and use in source and binary forms, with or without modification, are permitted provided that the following conditions are met:\n 1. Redistributions of source code must retain the above copyright notice, this list of conditions and the following disclaimer.\n 2. Redistributions in binary form must reproduce the above copyright notice, this list of conditions and the following disclaimer in the documentation and/or other materials provided with the distribution.\n\n THIS SOFTWARE IS PROVIDED BY THE Open Connectivity Foundation, INC. \"AS IS\" AND ANY EXPRESS OR IMPLIED WARRANTIES, INCLUDING, BUT NOT LIMITED TO, THE IMPLIED WARRANTIES OF MERCHANTABILITY AND FITNESS FOR A PARTICULAR PURPOSE OR WARRANTIES OF NON-INFRINGEMENT, ARE DISCLAIMED.\n IN NO EVENT SHALL THE Open Connectivity Foundation, INC. OR CONTRIBUTORS BE LIABLE FOR ANY DIRECT, INDIRECT, INCIDENTAL, SPECIAL, EXEMPLARY, OR CONSEQUENTIAL DAMAGES (INCLUDING, BUT NOT LIMITED TO, PROCUREMENT OF SUBSTITUTE GOODS OR SERVICES; LOSS OF USE, DATA, OR PROFITS; OR BUSINESS INTERRUPTION)\n HOWEVER CAUSED AND ON ANY THEORY OF LIABILITY, WHETHER IN CONTRACT, STRICT LIABILITY, OR TORT (INCLUDING NEGLIGENCE OR OTHERWISE) ARISING IN ANY WAY OUT OF THE USE OF THIS SOFTWARE, EVEN IF ADVISED OF THE POSSIBILITY OF SUCH DAMAGE.\n"
    }
  },
  "schemes": ["http"],
  "consumes": ["application/json"],
  "produces": ["application/json"],
  "paths": {
    "/example/DeviceConfigurationResURI" : {
      "get": {
        "description": "Resource that allows for Device specific information to be configured.\nRetrieves the current Device configuration settings\n",
        "parameters": [
          {"$ref": "#/parameters/interface-all"}
        ],
        "responses": {
          "200": {
            "description" : "",
            "x-example":
              {
                "n": "My Friendly Device Name",
                "rt": ["oic.wk.con"],
                "loc": [32.777, -96.797],
                "locn": "My Location Name",
                "c": "USD",
                "r": "MyRegion",
                "dl": "en"
```

```
            }
            ,
          "schema": { " $ ref": " # /definitions/Configuration" }
        }
      }
    },
    "post": {
      "description": "Update the information about the Device\n",
      "parameters": [
        {" $ ref": " # /parameters/interface - rw"},
        {
          "name": "body",
          "in": "body",
          "required": true,
          "schema": { " $ ref": " # /definitions/Update" },
          "x - example":
            {
              "n": "Nuevo Nombre Amistoso",
              "r": "MyNewRegion",
              "ln": [ { "language": "es", "value": "Nuevo Nombre Amistoso" } ],
              "dl": "es"
            }
        }
      ],
      "responses": {
          "200": {
            "description" : "",
            "x - example":
              {
                  "n": "Nuevo Nombre Amistoso",
                  "r": "MyNewRegion",
                  "ln": [ { "language": "es", "value": "Nuevo Nombre Amistoso" } ],
                  "dl": "es"
              }
              ,
            "schema": { " $ ref": " # /definitions/Update" }
          }
      }
    }
  }
},
"parameters": {
  "interface - rw" : {
    "in" : "query",
    "name" : "if",
    "type" : "string",
    "enum" : ["oic. if. rw"]
  },
  "interface - all" : {
    "in" : "query",
    "name" : "if",
    "type" : "string",
    "enum" : ["oic. if. rw", "oic. if. baseline"]
  }
```

```
},
"definitions": {
  "Configuration" :
        {
    "properties": {
      "c": {
        "description": "Currency",
        "maxLength": 64,
        "type": "string"
      },
      "dl": {
       "description": "Default Language",
        "pattern": "^[A-Za-z]{1,8}(-[A-Za-z0-9]{1,8})*$",
        "type": "string"
      },
      "id": {
        "description": "Instance ID of this specific resource",
        "maxLength": 64,
        "readOnly": true,
        "type": "string"
      },
      "if": {
        "description": "The interface set supported by this resource",
        "items": {
          "enum": [
            "oic.if.baseline",
            "oic.if.ll",
            "oic.if.b",
            "oic.if.lb",
            "oic.if.rw",
            "oic.if.r",
            "oic.if.a",
            "oic.if.s"
          ],
          "type": "string"
        },
        "minItems": 1,
        "readOnly": true,
        "type": "array"
      },
      "ln": {
        "description": "Localized names",
        "items": {
          "properties": {
            "language": {
              "description": "An RFC 5646 language tag.",
              "pattern": "^[A-Za-z]{1,8}(-[A-Za-z0-9]{1,8})*$",
              "type": "string"
            },
            "value": {
              "description": "Device description in the indicated language.",
              "maxLength": 64,
              "type": "string"
            }
```

```
          },
          "type": "object"
        },
        "minItems": 1,
        "type": "array"
      },
      "loc": {
        "description": "Location information",
        "items": {
          "type": "number"
        },
        "maxItems": 2,
        "minItems": 2,
        "type": "array"
      },
      "locn": {
        "description": "Human Friendly Name for location",
        "maxLength": 64,
        "type": "string"
      },
      "n": {
        "description": "Friendly name of the resource",
        "maxLength": 64,
        "readOnly": true,
        "type": "string"
      },
      "r": {
        "description": "Region",
        "maxLength": 64,
        "type": "string"
      },
      "rt": {
        "description": "Resource Type",
        "items": {
          "maxLength": 64,
          "type": "string"
        },
        "minItems": 1,
        "readOnly": true,
        "type": "array"
      }
    },
    "required": [
      "n"
    ],
    "type": "object"
  }
,
"Update" :
        {
    "properties": {
      "c": {
        "description": "Currency",
        "maxLength": 64,
```

```
            "type": "string"
        },
        "dl": {
          "description": "Default Language",
          "pattern": "^[A-Za-z]{1,8}(-[A-Za-z0-9]{1,8})*$",
          "type": "string"
        },
        "id": {
          "description": "Instance ID of this specific resource",
          "maxLength": 64,
          "readOnly": true,
          "type": "string"
        },
        "if": {
          "description": "The interface set supported by this resource",
          "items": {
            "enum": [
              "oic.if.baseline",
              "oic.if.ll",
              "oic.if.b",
              "oic.if.lb",
              "oic.if.rw",
              "oic.if.r",
              "oic.if.a",
              "oic.if.s"
            ],
            "type": "string"
          },
          "minItems": 1,
          "readOnly": true,
          "type": "array"
        },
        "ln": {
          "description": "Localized names",
          "items": {
            "properties": {
              "language": {
                "description": "An RFC 5646 language tag.",
                "pattern": "^[A-Za-z]{1,8}(-[A-Za-z0-9]{1,8})*$",
                "type": "string"
              },
              "value": {
                "description": "Device description in the indicated language.",
                "maxLength": 64,
                "type": "string"
              }
            },
            "type": "object"
          },
          "minItems": 1,
          "type": "array"
        },
        "loc": {
          "description": "Location information",
```

```
            "items": {
              "type": "number"
            },
            "maxItems": 2,
            "minItems": 2,
            "type": "array"
          },
          "locn": {
            "description": "Human Friendly Name for location",
            "maxLength": 64,
            "type": "string"
          },
          "n": {
            "description": "Friendly name of the resource",
            "maxLength": 64,
            "type": "string"
          },
          "r": {
            "description": "Region",
            "maxLength": 64,
            "type": "string"
          },
          "rt": {
            "description": "Resource Type",
            "items": {
              "maxLength": 64,
              "type": "string"
            },
            "minItems": 1,
            "readOnly": true,
            "type": "array"
          }
        },
        "required": [
          "n"
        ],
        "type": "object"
      }
    }
  }
```

2. 属性定义

属性定义如表 10-9 所示。

表 10-9 属性定义

属 性 名	属性值类型	是否强制	描 述
icon	字符串		位置的人性化名称
dl	字符串		默认语言
c	字符串		货币
rt	数组		资源类型
id	字符串		指定资源的 ID

续表

属 性 名	属性值类型	是否强制	描 述
ln	数组		本地名称
n	字符串	是	资源的友好名称
loc	数组		位置信息
r	字符串		地区
if	数组		资源支持的接口集合

3. CRUDN 行为

CRUDN 行为如表 10-10 所示。

表 10-10 CRUDN 行为

资 源	创建	检索	更新	删除	通知
/example/DeviceConfigurationResURI		get	post		

10.6 设备

每个服务器端托管的已知资源,允许发现逻辑设备特定信息,检索设备信息。URI 示例为"/oic/d",资源类型为"oic. wk. d"。

1. Swagger 2.0 定义

```
{
  "swagger": "2.0",
  "info": {
    "title": "Device",
    "version": "v1 - 20160622",
    "license": {
      "name": "copyright 2016 - 2017 Open Connectivity Foundation, Inc. All rights reserved.",
      "x - description": "Redistribution and use in source and binary forms, with or without modification, are permitted provided that the following conditions are met:\n 1. Redistributions of source code must retain the above copyright notice, this list of conditions and the following disclaimer. \n 2. Redistributions in binary form must reproduce the above copyright notice, this list of conditions and the following disclaimer in the documentation and/or other materials provided with the distribution. \n\n THIS SOFTWARE IS PROVIDED BY THE Open Connectivity Foundation, INC. \"AS IS\" AND ANY EXPRESS OR IMPLIED WARRANTIES, INCLUDING, BUT NOT LIMITED TO, THE IMPLIED WARRANTIES OF MERCHANTABILITY AND FITNESS FOR A PARTICULAR PURPOSE OR WARRANTIES OF NON - INFRINGEMENT, ARE DISCLAIMED. \ n IN NO EVENT SHALL THE Open Connectivity Foundation, INC. OR CONTRIBUTORS BE LIABLE FOR ANY DIRECT, INDIRECT, INCIDENTAL, SPECIAL, EXEMPLARY, OR CONSEQUENTIAL DAMAGES (INCLUDING, BUT NOT LIMITED TO, PROCUREMENT OF SUBSTITUTE GOODS OR SERVICES; LOSS OF USE, DATA, OR PROFITS; OR BUSINESS INTERRUPTION) \ n HOWEVER CAUSED AND ON ANY THEORY OF LIABILITY, WHETHER IN CONTRACT, STRICT LIABILITY, OR TORT (INCLUDING NEGLIGENCE OR OTHERWISE) ARISING IN ANY WAY OUT OF THE USE OF THIS SOFTWARE, EVEN IF ADVISED OF THE POSSIBILITY OF SUCH DAMAGE. \n"
    }
  },
  "schemes": ["http"],
  "consumes": ["application/json"],
  "produces": ["application/json"],
  "paths": {
```

```
    "/oic/d" : {
      "get": {
        "description": "Known resource that is hosted by every Server.\nAllows for logical device specific
information to be discovered.\nRetrieve the information about the Device\n",
        "parameters": [
          {"$ref": "#/parameters/interface"}
        ],
        "responses": {
            "200": {
              "description" : "",
              "x-example":
                {
                  "n":    "Device 1",
                  "rt":   ["oic.wk.d"],
                  "di":   "54919CA5-4101-4AE4-595B-353C51AA983C",
                  "icv":  "ocf.1.0.0",
                  "dmv":  "ocf.res.1.0.0, ocf.sh.1.0.0",
                  "piid": "6F0AAC04-2BB0-468D-B57C-16570A26AE48"
                }
              ,
              "schema": { "$ref": "#/definitions/Device" }
            }
          }
        }
      }
    },
    "parameters": {
      "interface" : {
        "in" : "query",
        "name" : "if",
        "type" : "string",
        "enum" : ["oic.if.r", "oic.if.baseline"]
      }
    },
    "definitions": {
      "Device" :
            {
        "properties": {
          "di": {
            "description": "Unique identifier for device (UUID)", 8499
            "pattern": "^[a-fA-F0-9]{8}-[a-fA-F0-9]{4}-[a-fA-F0-9]{4}-[a-fA-F0-9]
{4}-[a-fA-F0-9]{12}$",
            "readOnly": true,
            "type": "string"
          },
          "dmn": {
            "description": "Manufacturer Name.",
            "items": {
              "properties": {
              "language": {
                "description": "An RFC 5646 language tag.",
                "pattern": "^[A-Za-z]{1,8}(-[A-Za-z0-9]{1,8})*$",
                "readOnly": true,
                "type": "string"
```

```
              },
              "value": {
                "description": "Manufacturer name in the indicated language.",
                "maxLength": 64,
                "readOnly": true,
                "type": "string"
              }
            },
            "type": "object"
          },
          "minItems": 1,
          "readOnly": true,
          "type": "array"
        },
        "dmno": {
          "description": "Model number as designated by manufacturer.",
          "maxLength": 64,
          "readOnly": true,
          "type": "string"
        },
        "dmv": {
          "description": "Spec versions of the Resource and Device Specifications to which this device
data model is implemented",
          "maxLength": 256,
          "readOnly": true,
          "type": "string"
        },
        "icv": {
          "description": "The version of the OIC Server",
          "maxLength": 64,
          "readOnly": true,
          "type": "string"
        },
        "id": {
          "description": "Instance ID of this specific resource",
          "maxLength": 64,
          "readOnly": true,
          "type": "string"
        },
        "if": {
          "description": "The interface set supported by this resource",
          "items": {
            "enum": [
              "oic.if.baseline",
              "oic.if.ll",
              "oic.if.b",
              "oic.if.lb",
              "oic.if.rw",
              "oic.if.r",
              "oic.if.a",
              "oic.if.s"
            ],
            "type": "string"
          },
```

```
        "minItems": 1,
        "readOnly": true,
        "type": "array"
      },
      "ld": {
        "description": "Localized Description.",
        "items": {
          "properties": {
            "language": {
              "description": "An RFC 5646 language tag.",
              "pattern": "^[A-Za-z]{1,8}(-[A-Za-z0-9]{1,8})*$",
              "readOnly": true,
              "type": "string"
            },
            "value": {
              "description": "Device description in the indicated language.",
              "maxLength": 64,
              "readOnly": true,
              "type": "string"
            }
          },
          "type": "object"
        },
        "minItems": 1,
        "readOnly": true,
        "type": "array"
      },
      "n": {
        "description": "Friendly name of the resource",
        "maxLength": 64,
        "readOnly": true,
        "type": "string"
      },
      "piid": {
        "description": "Protocol independent unique identifier for device (UUID) that is immutable.",
        "pattern": "^[a-fA-F0-9]{8}-[a-fA-F0-9]{4}-[a-fA-F0-9]{4}-[a-fA-F0-9]
{4}-[a-fA-F0-9]{12}$",
        "readOnly": true,
        "type": "string"
      },
      "rt": {
        "description": "Resource Type",
        "items": {
          "maxLength": 64,
          "type": "string"
        },
        "minItems": 1,
        "readOnly": true,
        "type": "array"
      },
      "sv": {
        "description": "Software version.",
        "maxLength": 64,
        "readOnly": true,
```

```
          "type": "string"
        }
      },
      "required": [
        "n",
        "di",
        "icv",
        "dmv",
        "piid"
      ],
      "type": "object"
    }
  }
}
```


2. 属性定义

属性定义如表 10-11 所示。

表 10-11 属性定义

属 性 名	属性值类型	访问模式	是否强制	描 述
id	数组	只读		指定资源的 ID 示例
piid	多类型	只读	是	不可变的独立于协议的设备唯一标识符(UUID)
di	多类型	只读	是	设备唯一标识符
dmno	字符串	只读		由制造商设计的型号
sv	字符串	只读		软件版本
dmn	数组	只读		制造商名称
icv	字符串	只读	是	OCF 服务器端的版本
dmv	字符串	只读	是	该设备数据模型实现的资源和规格版本
if	数组	只读		资源支持的接口集合
ld	数组	只读		本地描述
rt	数组	只读		资源类型
n	字符串	只读	是	资源的友好名称

3. CRUDN 行为

CRUDN 行为如表 10-12 所示。

表 10-12 CRUDN 行为

资 源	创建	检索	更新	删除	通知
/oic/d		get			

10.7 维护

设备维护所用的资源可用于诊断目的。“fr”参数(出厂重置)是一个布尔值,值 0 表示无操作(默认值),值 1 表示在出场复位后。开始出厂复位之后,此值应更改回默认值。“rb”参数(重新启动)也是一个布尔值,值 0 表示无操作(默认值),值 1 表示重新启动,然后此值应更改回默认值。“ssc”参数(开始统计集合）是一个布尔值，值 0 表示无统计信息收集，值 1 表示开始收集统计信息。URI 示例为“/oic/mnt”，资源类型为“oic.wk.mnt”。

1. Swagger 2.0 定义

```
{
  "swagger": "2.0",
  "info": {
    "title": "Maintenance",
    "version": "v1 - 20160622",
    "license": {
      "name": "copyright 2016 - 2017 Open Connectivity Foundation, Inc. All rights reserved.",
      "x - description": "Redistribution and use in source and binary forms, with or without modification, are permitted provided that the following conditions are met:\n 1. Redistributions of source code must retain the above copyright notice, this list of conditions and the following disclaimer. \n 2. Redistributions in binary form must reproduce the above copyright notice, this list of conditions and the following disclaimer in the documentation and/or other materials provided with the distribution. \n\n THIS SOFTWARE IS PROVIDED BY THE Open Connectivity Foundation, INC. \" AS IS \" AND ANY EXPRESS OR IMPLIED WARRANTIES, INCLUDING, BUT NOT LIMITED TO, THE IMPLIED WARRANTIES OF MERCHANTABILITY AND FITNESS FOR A PARTICULAR PURPOSE OR WARRANTIES OF NON - INFRINGEMENT, ARE DISCLAIMED. \ n IN NO EVENT SHALL THE Open Connectivity Foundation, INC. OR CONTRIBUTORS BE LIABLE FOR ANY DIRECT, INDIRECT, INCIDENTAL, SPECIAL, EXEMPLARY, OR CONSEQUENTIAL DAMAGES ( INCLUDING, BUT NOT LIMITED TO, PROCUREMENT OF SUBSTITUTE GOODS OR SERVICES; LOSS OF USE, DATA, OR PROFITS; OR BUSINESS INTERRUPTION) \n HOWEVER CAUSED AND ON ANY THEORY OF LIABILITY, WHETHER IN CONTRACT, STRICT LIABILITY, OR TORT ( INCLUDING NEGLIGENCE OR OTHERWISE) ARISING IN ANY WAY OUT OF THE USE OF THIS SOFTWARE, EVEN IF ADVISED OF THE POSSIBILITY OF SUCH DAMAGE. \n"
    }
  },
  "schemes": ["http"],
  "consumes": ["application/json"],
  "produces": ["application/json"],
  "paths": {
    "/oic/mnt" : {
      "get": {
        "description": "The resource through which a Device is maintained and can be used for diagnostic purposes. \nfr (Factory Reset) is a boolean. \n The value 0 means No action (Default), the value 1 means Start Factory Reset\nAfter factory reset, this value shall be changed back to the default value\nrb (Reboot) is a boolean. \n The value 0 means No action (Default), the value 1 means Start Reboot\nAfter Reboot, this value shall be changed back to the default value\nRetrieve the maintenance action status",
        "parameters": [
          {" $ ref": " # /parameters/interface - all"}
        ],
        "responses": {
            "200": {
              "description" : "",
              "x - example":
                {
                  "rt": ["oic.wk.mnt"],
                  "fr": false,
                  "rb": false
                }
                ,
              "schema": { " $ ref": " # /definitions/MNT" }
            }
        }
```

```
        },
        "post": {
          "description": "Set the maintenance action(s)\n",
          "parameters": [
            {"$ref": "#/parameters/interface-rw"},
            {
              "name": "body",
              "in": "body",
              "required": true,
              "schema": { "$ref": "#/definitions/MNT" },
              "x-example":
                {
                  "fr": false,
                  "rb": false
                }
            }
          ],
          "responses": {
            "200": {
              "description" : "",
              "x-example":
                {
                  "fr": false,
                  "rb": false
                }
                ,
              "schema": { "$ref": "#/definitions/MNT" }
            }
          }
        }
      }
    },
    "parameters": {
      "interface-rw" : {
        "in" : "query",
        "name" : "if",
        "type" : "string",
        "enum" : ["oic.if.rw", "oic.if.baseline"]
      },
      "interface-all" : {
        "in" : "query",
        "name" : "if",
        "type" : "string",
        "enum" : ["oic.if.rw", "oic.if.r", "oic.if.baseline"]
      }
    },
    "definitions": {
      "MNT" :
            {
        "anyOf": [
          {
            "required": [
              "fr"
            ]
```

```
        },
        {
          "required": [
            "rb"
          ]
        }
      ],
      "properties": {
        "fr": {
          "description": "Factory Reset",
          "type": "boolean"
        },
        "id": {
          "description": "Instance ID of this specific resource",
          "maxLength": 64,
          "readOnly": true,
          "type": "string"
        },
        "if": {
          "description": "The interface set supported by this resource",
          "items": {
            "enum": [
              "oic.if.baseline",
              "oic.if.ll",
              "oic.if.b",
              "oic.if.lb",
              "oic.if.rw",
              "oic.if.r",
              "oic.if.a",
              "oic.if.s"
            ],
            "type": "string"
          },
          "minItems": 1,
          "readOnly": true,
          "type": "array"
        },
        "n": {
          "description": "Friendly name of the resource",
          "maxLength": 64,
          "readOnly": true,
          "type": "string"
        },
        "rb": {
          "description": "Reboot Action",
          "type": "boolean"
        },
        "rt": {
          "description": "Resource Type",
          "items": {
            "maxLength": 64,
            "type": "string"
          },
          "minItems": 1,
```

```
                "readOnly": true,
                "type": "array"
              }
            },
            "type": "object"
          }
      }
  }
```

2. 属性定义

属性定义如表 10-13 所示。

表 10-13 属性定义

属性名	属性值类型	访问模式	是否强制	描述
id	数组	只读		本地描述
fr	布尔	只读	是	工厂重置
rb	布尔	只读		重新启动操作
rt	数组	只读		资源类型
n	字符串	只读	是	资源的友好名称
if	数组	只读		资源支持的接口集合

3. CRUDN 行为

CRUDN 行为如表 10-14 所示。

表 10-14 CRUDN 行为

资源	创建	检索	更新	删除	通知
/oic/mnt		get	post		

10.8 平台

已知资源，定义托管服务器端的平台，允许发现特定于平台的信息，检索平台有关信息。URI 示例为“/oic/p”，该资源类型被定义为“oic. wk. p”。

1. Swagger 2.0 定义

```
{
  "swagger": "2.0",
  "info": {
    "title": "Platform",
    "version": "v1 - 20160622",
    "license": {
      "name": "copyright 2016 - 2017 Open Connectivity Foundation, Inc. All rights reserved.",
      "x - description": "Redistribution and use in source and binary forms, with or without modification, are permitted provided that the following conditions are met:\n 1. Redistributions of source code must retain the above copyright notice, this list of conditions and the following disclaimer. \n 2. Redistributions in binary form must reproduce the above copyright notice, this list of conditions and the following disclaimer in the documentation and/or other materials provided with the distribution. \n\n THIS SOFTWARE IS PROVIDED BY THE Open Connectivity Foundation, INC. \"AS IS\" AND ANY EXPRESS OR IMPLIED WARRANTIES, INCLUDING, BUT NOT
```

```
LIMITED TO, THE IMPLIED WARRANTIES OF MERCHANTABILITY AND FITNESS FOR A PARTICULAR PURPOSE OR WARRANTIES OF NON - INFRINGEMENT, ARE DISCLAIMED. \n IN NO EVENT SHALL THE Open Connectivity Foundation, INC. OR CONTRIBUTORS BE LIABLE FOR ANY DIRECT, INDIRECT, INCIDENTAL, SPECIAL, EXEMPLARY, OR CONSEQUENTIAL DAMAGES (INCLUDING, BUT NOT LIMITED TO, PROCUREMENT OF SUBSTITUTE GOODS OR SERVICES; LOSS OF USE, DATA, OR PROFITS; OR BUSINESS INTERRUPTION) \n HOWEVER CAUSED AND ON ANY THEORY OF LIABILITY, WHETHER IN CONTRACT, STRICT LIABILITY, OR TORT (INCLUDING NEGLIGENCE OR OTHERWISE) ARISING IN ANY WAY OUT OF THE USE OF THIS SOFTWARE, EVEN IF ADVISED OF THE POSSIBILITY OF SUCH DAMAGE.\n"
      }
    },
    "schemes": ["http"],
    "consumes": ["application/json"],
    "produces": ["application/json"],
    "paths": {
      "/oic/p" : {
        "get": {
          "description": "Known resource that is defines the platform on which an Server is hosted.\nAllows for platform specific information to be discovered.\nRetrieve the information about the Platform\n",
          "parameters": [
            {"$ref": "#/parameters/interface"}
          ],
          "responses": {
            "200": {
              "description" : "",
              "x-example":
                {
                  "pi": "54919CA5-4101-4AE4-595B-353C51AA983C",
                  "rt": ["oic.wk.p"],
                  "mnmn": "Acme, Inc"
                }
                ,
              "schema": { "$ref": "#/definitions/Platform" }
            }
          }
        }
      }
    },
    "parameters": {
      "interface" : {
        "in" : "query",
        "name" : "if",
        "type" : "string",
        "enum" : ["oic.if.r", "oic.if.baseline"]
      }
    },
    "definitions": {
      "Platform" :
            {
        "properties": {
          "id": {
            "description": "Instance ID of this specific resource",
            "maxLength": 64,
            "readOnly": true,
            "type": "string"
          },
```

```
"if": {
  "description": "The interface set supported by this resource",
  "items": {
    "enum": [
      "oic.if.baseline",
      "oic.if.ll",
      "oic.if.b",
      "oic.if.lb",
      "oic.if.rw",
      "oic.if.r",
      "oic.if.a",
      "oic.if.s"
    ],
    "type": "string"
  },
  "minItems": 1,
  "readOnly": true,
  "type": "array"
},
"mndt": {
  "description": "Manufacturing Date.",
  "pattern": "^([0-9]{4})-(1[0-2]|0[1-9])-(3[0-1]|2[0-9]|1[0-9]|0[1-9])$",
  "readOnly": true,
  "type": "string"
},
"mnfv": {
  "description": "Manufacturer's firmware version",
  "maxLength": 64,
  "readOnly": true,
  "type": "string"
},
"mnhw": {
  "description": "Platform Hardware Version",
  "maxLength": 64,
  "readOnly": true,
  "type": "string"
},
"mnml": {
  "description": "Manufacturer's URL",
  "format": "uri",
  "maxLength": 256,
  "readOnly": true,
  "type": "string"
},
"mnmn": {
  "description": "Manufacturer Name",
  "maxLength": 64,
  "readOnly": true,
  "type": "string"
},
"mnmo": {
  "description": "Model number as designated by manufacturer",
  "maxLength": 64,
  "readOnly": true,
  "type": "string"
```

```
        },
        "mnos": {
          "description": "Platform Resident OS Version",
          "maxLength": 64,
          "readOnly": true,
          "type": "string"
        },
        "mnpv": {
          "description": "Platform Version",
          "maxLength": 64,
          "readOnly": true,
          "type": "string"
        },
        "mnsl": {
          "description": "Manufacturer's Support Information URL",
          "format": "uri",
          "maxLength": 256,
          "readOnly": true,
          "type": "string"
        },
        "n": {
          "description": "Friendly name of the resource",
          "maxLength": 64,
          "readOnly": true,
          "type": "string"
        },
        "pi": {
          "description": "Platform Identifier as a UUID",
          "pattern": "^[a-fA-F0-9]{8}-[a-fA-F0-9]{4}-[a-fA-F0-9]{4}-[a-fA-F0-9]
{4}-[a-fA-F0-9]{12}$",
          "readOnly": true,
          "type": "string"
        },
        "rt": {
          "description": "Resource Type",
          "items": {
            "maxLength": 64,
            "type": "string"
          },
          "minItems": 1,
          "readOnly": true,
          "type": "array"
        },
        "st": {
          "description": "Reference time for the device as defined in ISO 8601, where concatenation of
'date' and 'time' with the 'T' as a delimiter between 'date' and 'time'.",
          "format": "date-time",
          "readOnly": true,
          "type": "string"
        },
        "vid": {
          "description": "Manufacturer's defined string for the platform. The string is freeform and up
to the manufacturer on what text to populate it",
          "maxLength": 64,
          "readOnly": true,
          "type": "string"
```

```
            }
          },
          "required": [
            "pi",
            "mnmn"
          ],
          "type": "object"
        }
    }
}
```

2. 属性定义

属性定义如表10-15所示。

表10-15 属性定义

属 性 名	属性值类型	访问模式	是否强制	描 述
pi	字符串	只读	是	平台标识符作为UUID
mnmn	字符串	只读	是	厂商名称
mnml	字符串	只读		厂商的URL
mnmo	字符串	只读		由厂商指定的型号
mndt	字符串	只读		制造日期
mnpv	字符串	只读		平台版本
mnos	字符串	读写		只读,平台驻留操作系统版本
mnhw	字符串	读写		只读,平台硬件版本
mnfv	字符串	只读		制造商的固件版本
mnsl	字符串	只读		制造商的支持信息URL
st	字符串	只读		设备的参考时间
vid	字符串	只读		制造商为平台定义的字符串。字符串是自由格式的,并且根据制造商填充文本
n	字符串	只读		资源的友好名称
if	数组	只读		资源支持的接口集合
rt	数组	只读		资源类型
id	字符串	只读		指定资源的ID示例

3. CRUDN 行为

CRUDN行为如表10-16所示。

表10-16 CRUDN 行为

资 源	创建	检索	更新	删除	通知
/oic/p		get			

10.9 ping

客户端保持其与服务器端连接的资源处于活动状态,检索ping的信息。URI示例为"/oic/ping",资源类型为"oic. wk. ping"。

1. Swagger 2.0 定义

```
{
  "swagger": "2.0",
  "info": {
    "title": "Ping",
    "version": "v1 - 20160622",
    "license": {
      "name": "copyright 2016 - 2017 Open Connectivity Foundation, Inc. All rights reserved.",
      "x - description": "Redistribution and use in source and binary forms, with or without modification,
are permitted provided that the following conditions are met:\n 1. Redistributions of source code must retain
the above copyright notice, this list of conditions and the following disclaimer. \n 2. Redistributions in
binary form must reproduce the above copyright notice, this list of conditions and the following disclaimer
in the documentation and/or other materials provided with the distribution. \n\n THIS SOFTWARE IS PROVIDED BY
THE Open Connectivity Foundation, INC. \"AS IS\" AND ANY EXPRESS OR IMPLIED WARRANTIES, INCLUDING, BUT NOT
LIMITED TO, THE IMPLIED WARRANTIES OF MERCHANTABILITY AND FITNESS FOR A PARTICULAR PURPOSE OR WARRANTIES OF
NON - INFRINGEMENT, ARE DISCLAIMED. \ n IN NO EVENT SHALL THE Open Connectivity Foundation, INC. OR
CONTRIBUTORS BE LIABLE FOR ANY DIRECT, INDIRECT, INCIDENTAL, SPECIAL, EXEMPLARY, OR CONSEQUENTIAL DAMAGES
(INCLUDING, BUT NOT LIMITED TO, PROCUREMENT OF SUBSTITUTE GOODS OR SERVICES; LOSS OF USE, DATA, OR PROFITS;
OR BUSINESS INTERRUPTION) \ n HOWEVER CAUSED AND ON ANY THEORY OF LIABILITY, WHETHER IN CONTRACT, STRICT
LIABILITY, OR TORT (INCLUDING NEGLIGENCE OR OTHERWISE) ARISING IN ANY WAY OUT OF THE USE OF THIS SOFTWARE,
EVEN IF ADVISED OF THE POSSIBILITY OF SUCH DAMAGE. \n"
    }
  },
  "schemes": ["http"],
  "consumes": ["application/json"],
  "produces": ["application/json"],
  "paths": {
    "/oic/ping" : {
      "get": {
        "description": "The resource using which an Client keeps its Connection with an Server active. \
nRetrieve the ping information",
        "parameters": [
          {"$ref": "#/parameters/interface"}
        ],
        "responses": {
            "200": {
              "description" : "",
              "x - example":
                {
                  "rt": ["oic.wk.ping"],
                  "n": "Ping Information",
                  "in": 16
                }
                ,
              "schema": { "$ref": "#/definitions/PING" }
            }
        }
      },
      "post": {
        "description": "Update or reset the alive interval",
        "parameters": [
          {"$ref": "#/parameters/interface"},
          {
```

```
                "name": "body",
                "in": "body",
                "required": true,
                "schema": { "$ref": "#/definitions/PING" },
                "x-example":
                  {
                    "in": 16
                  }
              }
            ],
            "responses": {
              "203": {
                "description" : "Successfully updated & restarted alive interval timer.",
                "x-example":
                  {
                    "in": 16
                  }
                  ,
                "schema": { "$ref": "#/definitions/PING" }
              }
            }
          }
        }
      },
      "parameters": {
        "interface" : {
          "in" : "query",
          "name" : "if",
          "type" : "string",
          "enum" : ["oic.if.rw", "oic.if.baseline"]
        }
      },
      "definitions": {
        "PING" :
            {
          "properties": {
            "id": {
              "description": "Instance ID of this specific resource",
              "maxLength": 64,
              "readOnly": true,
              "type": "string"
            },
            "if": {
              "description": "The interface set supported by this resource",
              "items": {
                "enum": [
                  "oic.if.baseline",
                  "oic.if.ll",
                  "oic.if.b",
                  "oic.if.lb",
                  "oic.if.rw",
                  "oic.if.r",
                  "oic.if.a",
                  "oic.if.s"
```

```
            ],
            "type": "string"
          },
          "minItems": 1,
          "readOnly": true,
          "type": "array"
        },
        "in": {
          "description": "Indicates the interval for which connection shall be kept alive",
          "readOnly": false,
          "type": "integer"
        },
        "n": {
          "description": "Friendly name of the resource",
          "maxLength": 64,
          "readOnly": true,
          "type": "string"
        },
        "rt": {
          "description": "Resource Type",
          "items": {
            "maxLength": 64,
            "type": "string"
          }
          "minItems": 1
         "readOnly": true,
          "type": "array"
        }
      },
      "required": [
        "in"
      ],
      "type": "object"
    }
  }
}
```

2. 属性定义

属性定义如表 10-17 所示。

表 10-17　属性定义

属 性 名	属性值类型	访问模式	是否强制	描　述
in	整型	读写	是	说明连接应该保持激活的间隔
n	字符串	只读		资源的友好名称
if	数组	只读		资源支持的接口集合
rt	数组	只读		资源类型
id	字符串	只读		指定资源的 ID 示例

3. CRUDN 行为

CRUDN 行为如表 10-18 所示。

表 10-18 CRUDN 行为

资 源	创建	检索	更新	删除	通知
/oic/ping		get	post		

10.10 资源目录资源

任何可以作为资源目录的设备公开资源：1)使用 GET 请求提供选择标准(例如整数)；2)使用 POST 请求在"/oic/res"中发布或更新链接；3)使用 DELETE 请求删除"/oic/res"中的链接。获取资源目录属性以进行选择。URI 示例为"/oic/rd",资源类型为"oic. wk. rd"。

1. Swagger 2.0 定义

```
{
  "swagger": "2.0",
  "info": {
    "title": "Resource directory resource",
    "version": "v1 - 20160622",
    "license": {
      "name": "copyright 2016 - 2017 Open Connectivity Foundation, Inc. All rights reserved.",
      "x - description": "Redistribution and use in source and binary forms, with or withoutmodification, are permitted provided that the following conditions are met:\n 1. Redistributions of source code must retain the above copyright notice, this list of conditions and the following disclaimer. \n 2. Redistributions in binary form must reproduce the above copyright notice, this list of conditions and the following disclaimer in the documentation and/or other materials provided with the distribution. \n\n THIS SOFTWARE IS PROVIDED BY THE Open Connectivity Foundation, INC. \"AS IS\" AND ANY EXPRESS OR IMPLIED WARRANTIES, INCLUDING, BUT NOT LIMITED TO, THE IMPLIED WARRANTIES OF MERCHANTABILITY AND FITNESS FOR A PARTICULAR PURPOSE OR WARRANTIES OF NON - INFRINGEMENT, ARE DISCLAIMED. \ n IN NO EVENT SHALL THE Open Connectivity Foundation, INC. OR CONTRIBUTORS BE LIABLE FOR ANY DIRECT, INDIRECT, INCIDENTAL, SPECIAL, EXEMPLARY, OR CONSEQUENTIAL DAMAGES (INCLUDING, BUT NOT LIMITED TO, PROCUREMENT OF SUBSTITUTE GOODS OR SERVICES; LOSS OF USE, DATA, OR PROFITS; OR BUSINESS INTERRUPTION) \ n HOWEVER CAUSED AND ON ANY THEORY OF LIABILITY, WHETHER IN CONTRACT, STRICT LIABILITY, OR TORT (INCLUDING NEGLIGENCE OR OTHERWISE) ARISING IN ANY WAY OUT OF THE USE OF THIS SOFTWARE, EVEN IF ADVISED OF THE POSSIBILITY OF SUCH DAMAGE. \n"
    }
  },
  "schemes": ["http"],
  "consumes": ["application/json"],
  "produces": ["application/json"],
  "paths": {
    "/oic/rd" : {
      "get": {
        "description": "Resource to be exposed by any Device that can act as a Resource Directory. \n1) Provides selector criteria (e.g., integer) with GET request\n2) Publish or Update a Link in /oic/res with POST request\n3) Delete a Link in /oic/res with DELETE request\nGet the attributes of the Resource Directory for selection purposes. \n",
        "parameters": [
          {"$ ref": "#/parameters/rdgetinterface"}
        ],
        "responses": {
          "200": {
            "description" : "Respond with the selector criteria - either the set of attributes or the
```

```
bias factor\n",
                "x-example":
                  {
                    "rt": ["oic.wk.rd"],
                    "if": ["oic.if.baseline"],
                    "sel": 50
                  }
                ,
                "schema": { "$ref": "#/definitions/rdSelection"}
              }
            }
          },
          "post": {
            "description": "Publish the resource information for the first time or Update the existing one in /
oic/res.\nAppropriates parts of the information, i.e., Links of the published Resources will be discovered
through /oic/res.\n1) When a Device first publishes a Link, the request payload to RD may include the Links
without \"ins\" Parameter.\n2) Upon granting the request, the RD assigns a unique instance value identifying
the Link among all the Links it advertises\n and sends back the instance value in \"ins\" Parameter in the
Link to the publishing Device.\n3) When later the publishing Device updates the existing Link, i.e., changing
its Endpoint information, \n the request payload to RD needs to include the instance value in \"ins\"
Parameter to identify the Link to update.\n",
            "parameters": [
              {"$ref": "#/parameters/rdpostinterface"},
              {
                "name": "body",
                "in": "body",
                "required": true,
                "schema": { "$ref": "#/definitions/rdPublish" },
                "x-example":
                  {
                    "di": "e61c3e6b-9c54-4b81-8ce5-f9039c1d04d9",
                    "links": [
                      {
                        "anchor": "ocf://e61c3e6b-9c54-4b81-8ce5-f9039c1d04d9",
                        "href":   "/myLightSwitch",
                        "rt":     ["oic.r.switch.binary"],
                        "if":     ["oic.if.a", "oic.if.baseline"],
                        "p":      {"bm": 3},
                        "eps": [
                            {"ep": "coaps://[2001:db8:a::b1d6]:1111", "pri": 2},
                            {"ep": "coaps://[2001:db8:a::b1d6]:1122"},
                            {"ep": "coaps+tcp://[2001:db8:a::123]:2222", "pri": 3}
                        ]
                      },
                      {
                        "anchor": "ocf://e61c3e6b-9c54-4b81-8ce5-f9039c1d04d9",
                        "href":   "/myLightBrightness",
                        "rt":     ["oic.r.brightness"],
                        "if":     ["oic.if.a", "oic.if.baseline"],
                        "p":      {"bm": 3},
                        "eps": [
                            {"ep": "coaps://[[2001:db8:a::123]:2222"}
                        ]
                      }
```

```
                ],
                "ttl": 600
              }
            }
          ],
          "responses": {
              "200": {
                "description" : "Respond with the same schema as publish but, when a Link is first published,
\nwith the additional \"ins\" Parameter in the Link.\nThis value is used by the receiver to manage that OCF
Link instance.\n",
                "x-example":
                  {
                    "di": "e61c3e6b-9c54-4b81-8ce5-f9039c1d04d9",
                    "links": [
                      {
                        "anchor": "ocf://e61c3e6b-9c54-4b81-8ce5-f9039c1d04d9",
                        "href":   "/myLightSwitch",
                        "rt":     ["oic.r.switch.binary"],
                        "if":     ["oic.if.a", "oic.if.baseline"],
                        "p":      {"bm": 3},
                        "eps": [
                            {"ep": "coaps://[2001:db8:a::b1d6]:1111", "pri": 2},
                            {"ep": "coaps://[2001:db8:a::b1d6]:1122"},
                            {"ep": "coaps+tcp://[2001:db8:a::123]:2222", "pri": 3}
                        ],
                        "ins": "11235"
                      },
                      {
                        "anchor": "ocf://e61c3e6b-9c54-4b81-8ce5-f9039c1d04d9",
                        "href":   "/myLightBrightness",
                        "rt":     ["oic.r.brightness"],
                        "if":     ["oic.if.a", "oic.if.baseline"],
                        "p":      {"bm": 3},
                        "eps": [
                            {"ep": "coaps://[2001:db8:a::123]:2222"}
                        ],
                        "ins": "112358"
                      }
                    ],
                    "ttl": 600
                  }
                ,
                "schema": { "$ref": "#/definitions/rdPublish" }
              }
          }
        },
        "delete": {
          "description":"Delete a particular OIC Link - the link may be a simple link or a link in a tagged set.\n",
          "parameters": [
            {"$ref": "#/parameters/rddelete-di"},
            {"$ref": "#/parameters/rddelete-ins"}
          ],
          "responses": {
              "200": {
```

```
                "description" : "The delete succeeded",
                "x-example":
                  {}
              }
            }
          }
        }
      },
      "parameters": {
        "rddelete-di" : {
          "in" : "query",
          "name" : "di",
          "type" : "string",
          "description" : "description"
        },
        "rddelete-ins" : {
          "in" : "query",
          "name" : "ins",
          "type" : "string",
          "description" : "description"
        },
        "rdgetinterface" : {
          "in" : "query",
          "name" : "if",
          "type" : "string",
          "enum" : ["oic.if.baseline"],
          "description" : "enumdescription"
        },
        "rdpostinterface" : {
          "in" : "query",
          "name" : "rt",
          "type" : "string",
          "enum" : ["oic.wk.rdpub"],
          "description" : "enumdescription"
        }
      },
      "definitions": {
        "rdSelection" :
            {
          "oneOf": [
            {
              "properties": {
                "sel": {
                  "description": "A bias factor calculated by the Resource directory - the value is in the
range of 0 to 100 - 0 implies that RD is not to be selected. Client chooses RD with highest bias factor or
randomly between RDs that have same bias factor",
                  "maximum": 100,
                  "minimum": 0,
                  "type": "integer"
                }
              },
              "required": [
                "sel"
              ]
```

```
            },
            {
              "properties": {
                "id": {
                  "description": "Instance ID of this specific resource",
                  "maxLength": 64,
                  "readOnly": true,
                  "type": "string"
                },
                "if": {
                  "description": "The interface set supported by this resource",
                  "items": {
                    "enum": [
                      "oic.if.baseline",
                      "oic.if.ll",
                      "oic.if.b",
                      "oic.if.lb",
                      "oic.if.rw",
                      "oic.if.r",
                      "oic.if.a",
                      "oic.if.s"
                    ],
                    "type": "string"
                  },
                  "minItems": 1,
                  "readOnly": true,
                  "type": "array"
                },
                "n": {
                  "description": "Friendly name of the resource",
                  "maxLength": 64,
                  "readOnly": true,
                  "type": "string"
                },
                "rt": {
                  "description": "Resource Type",
                  "items": {
                    "maxLength": 64,
                    "type": "string"
                  },
                  "minItems": 1,
                  "readOnly": true,
                  "type": "array"
                },
                "sel": {
                  "description": "Selection criteria that a device wanting to publish to any RD can use to
choose this Resource Directory over others that are discovered",
                  "properties": {
                    "bw": {
                      "description": "Qualitative bandwidth of the connection",
                      "enum": [
                        "high",
                        "low",
                        "lossy"
```

```
                        ],
                        "type": "string"
                    },
                    "conn": {
                        "description": "A hint about the networking connectivity of the RD. * wrd * if wired
connected and * wrls * if wireless connected.",
                        "enum": [
                            "wrd",
                            "wrls"
                        ],
                        "type": "string"
                    },
                    "load": {
                        "description": "Current load capacity of the RD. Expressed as a load factor 3 - tuple
(upto two decimal points each). Load factor is based on request processed in a 1 minute, 5 minute window and 15
minute window",
                        "items": {
                            "type": "number"
                        },
                        "maxItems": 3,
                        "minItems": 3,
                        "type": "array"
                    },
                    "mf": {
                        "description": "Memory factor - Ratio of available memory to total memory expressed
as a percentage",
                        "type": "integer"
                    },
                    "pwr": {
                        "description": "A hint about how the RD is powered. If AC then this is stronger than
battery powered. If source is reliable (safe) then appropriate mechanism for managing power failure exists",
                        "enum": [
                            "ac",
                            "batt",
                            "safe"
                        ],
                        "type": "string"
                    }
                },
                "type": "object"
            }
        },
        "required": [
            "sel"
        ]
    }
    ],
    "type": "object"
}

,
"rdPublish" :
    {
    "description": "Publishes resources as OIC Links into the resource directory",
```

```
            "properties": {
              "di": {
                "description": "A unique identifier for the publishing Device, i.e., its device ID",
                "pattern": "^[a-fA-F0-9]{8}-[a-fA-F0-9]{4}-[a-fA-F0-9]{4}-[a-fA-F0-9]
{4}-[a-fA-F0-9]{12}$",
                "type": "string"
              },
              "id": {
                "description": "Instance ID of this specific resource",
                "maxLength": 64,
                "readOnly": true,
                "type": "string"
              },
              "if": {
                "description": "The interface set supported by this resource",
                "items": {
                  "enum": [
                    "oic.if.baseline",
                    "oic.if.ll",
                    "oic.if.b",
                    "oic.if.lb",
                    "oic.if.rw",
                    "oic.if.r",
                    "oic.if.a",
                    "oic.if.s"
                  ],
                  "type": "string"
                },
                "minItems": 1,
                "readOnly": true,
                "type": "array"
              }
              "links": {
                "description": "A set (array) of simple or individual OIC Links. In addition to properties
required for an OIC Link, the identifier for that link in this set is also required",
                "items": {
                  "properties": {
                    "anchor": {
                      "description": "This is used to override the context URI e.g. override the URI of the
containing collection",
                      "format": "uri",
                      "maxLength": 256,
                      "type": "string"
                    },
                    "di": {
                      "description": "Unique identifier for device (UUID)",
                      "pattern": "^[a-fA-F0-9]{8}-[a-fA-F0-9]{4}-[a-fA-F0-9]{4}-[a-fA-
F0-9]{4}-[a-fA-F0-9]{12}$",
                      "type": "string"
                    },
                    "eps": {
                      "description": "the Endpoint information of the target Resource",
                      "items": {
                        "properties": {
```

```
                "ep": {
                  "description": "URI with Transport Protocol Suites + Endpoint Locator as specified",
                  "type": "string"
                },
                "pri": {
                  "description":"The priority among multiple Endpoints as specified ",
                  "minimum": 1,
                  "type": "integer"
                }
              },
              "type": "object"
            },
            "type": "array"
          },
          "href": {
            "description": "This is the target URI, it can be specified as a Relative Reference or
fully-qualified URI. Relative Reference should be used along with the di parameter to make it unique.",
            "format": "uri",
            "maxLength": 256,
            "type": "string"
          },
          "if": {
            "description": "The interface set supported by this resource",
            "items": {
              "enum": [
                "oic.if.baseline", 9
                "oic.if.ll",
                "oic.if.b",
                "oic.if.rw",
                "oic.if.r",
                "oic.if.a",
                "oic.if.s"
              ],
              "type": "string"
            },
            "minItems": 1,
            "type": "array"
          },
          "ins": {
            "description": "The instance identifier for this web link in an array of web links -
used in collections",
            "oneOf": [
              {
                "description": "An ordinal number that is not repeated - must be unique in the
collection context",
                "type": "integer"
              }, 9
              {
                "description": "Any unique string including a URI",
                "format": "uri",
                "maxLength": 256,
                "type": "string"
              },
              {
```

```
                    "description": "Unique identifier (UUID)",
                    "pattern": "^[a-fA-F0-9]{8}-[a-fA-F0-9]{4}-[a-fA-F0-9]{4}-[a-fA
-F0-9]{4}-[a- fA-F0-9]{12}$",
                    "type": "string"
                  }
                ]
              },
              "p": {
                "description": "Specifies the framework policies on the Resource referenced by the
target URI",
                "properties": {
                  "bm": {
                    "description": "Specifies the framework policies on the Resource referenced by the
target URI for e.g. observable and discoverable",
                    "type": "integer"
                  }
                },
                "required": [
                  "bm"
                ],
                "type": "object"
              },
              "rel": {
                "description": "The relation of the target URI referenced by the link to the context
URI",
                "oneOf": [
                  {
                    "default": [
                      "hosts"
                    ],
                    "items": {
                      "maxLength": 64,
                      "type": "string"
                    },
                    "minItems": 1,
                    "type": "array"
                  },
                  {
                    "default": "hosts",
                    "maxLength": 64,
                    "type": "string"
                  }
                ]
              },
              "rt": {
                "description": "Resource Type",
                "items": {
                  "maxLength": 64,
                  "type": "string"
                },
                "minItems": 1,
                "type": "array"
              },
              "title": {
```

```
                    "description": "A title for the link relation. Can be used by the UI to provide a
context",
                    "maxLength": 64,
                    "type": "string"
                  },
                  "type": {
                    "default": "application/cbor",
                    "description": "A hint at the representation of the resource referenced by the target
URI. This represents the media types that are used for both accepting and emitting",
                    "items": {
                      "maxLength": 64,
                      "type": "string"
                    },
                    "minItems": 1,
                    "type": "array"
                  }
                },
                "required": [
                  "href",
                  "rt",
                  "if"
                ],
                "type": "object"
              },
              "type": "array"
            },
            "n": {
              "description": "Friendly name of the resource",
              "maxLength": 64,
              "readOnly": true,
              "type": "string"
            },
            "rt": {
              "description": "Resource Type",
              "items": {
                "maxLength": 64,
                "type": "string"
              },
              "minItems": 1,
              "readOnly": true,
              "type": "array"
            },
            "ttl": {
              "description": "Time to indicate a RD, how long to keep this published item. After this time
(in seconds) elapses, the RD invalidates the links. To keep link alive the publishing device
updates the ttl using the update schema",
              "type": "integer"
            }
          }
        }
    }
}
```

2. 属性定义

属性定义如表 10-19 所示。

表 10-19 属性定义

属 性 名	属性值类型	访问模式	是否强制	描 述
n	字符串	只读		资源的友好名称
if	数组	只读		资源支持的接口集合
rt	数组	只读		资源类型
id	字符串	只读		指定资源的 ID 示例
ttl	整型			指示 RD 的时间,保留已发布资源的时间。经过这段时间(秒)后,RD 会使链接无效。为了使链接保持活动,发布设备使用更新模式更新 ttl
links	数组			一组简单或单独的 OCF 链接(数组)。除了 OCF 链路所需的属性之外,还需要该集合中链接的标识符
di	字符串			设备唯一标识符
sel	对象		是	一种选择标准,关于设备所要发布的 RD 上,以此选择可发现的资源目录

3. CRUDN 行为

CRUDN 行为如表 10-20 所示。

表 10-20 CRUDN 行为

资 源	创建	检索	更新	删除	通知
/oic/rd		get	post	delete	

10.11 可发现资源

检索可发现资源集合。URI 示例为“/oic/res”,资源类型为“oic. wk. res”。

1. Swagger 2.0 定义

```
{
  "swagger": "2.0",
  "info": {
    "title": "Discoverable Resources Baseline Interface",
    "version": "v1 - 20160622",
    "license": {
      "name": "copyright 2016 - 2017 Open Connectivity Foundation, Inc. All rights reserved.",
      "x - description": "Redistribution and use in source and binary forms, with or without modification, are permitted provided that the following conditions are met:\n 1. Redistributions of source code must retain the above copyright notice, this list of conditions and the following disclaimer. \n 2. Redistributions in binary form must reproduce the above copyright notice, this list of conditions and the following disclaimer in the documentation and/orother materials provided with the distribution. \n\n THIS SOFTWARE IS PROVIDED BY THE Open Connectivity Foundation, INC. \"AS IS\" AND ANY EXPRESS OR IMPLIED WARRANTIES, INCLUDING, BUT NOT LIMITED TO, THE IMPLIED WARRANTIES OF MERCHANTABILITY AND FITNESS FOR A PARTICULAR PURPOSE OR WARRANTIES OF NON - INFRINGEMENT, ARE DISCLAIMED. \ n IN NO EVENT SHALL THE Open Connectivity Foundation, INC. OR
```

```
CONTRIBUTORS BE LIABLE FOR ANY DIRECT, INDIRECT, INCIDENTAL, SPECIAL, EXEMPLARY, OR CONSEQUENTIAL DAMAGES
(INCLUDING, BUT NOT LIMITED TO, PROCUREMENT OF SUBSTITUTE GOODS OR SERVICES; LOSS OF USE, DATA, OR PROFITS;
OR BUSINESS INTERRUPTION)\n HOWEVER CAUSED AND ON ANY THEORY OF LIABILITY, WHETHER IN CONTRACT, STRICT
LIABILITY, OR TORT (INCLUDING NEGLIGENCE OR OTHERWISE) ARISING IN ANY WAY OUT OF THE USE OF THIS SOFTWARE,
EVEN IF ADVISED OF THE POSSIBILITY OF SUCH DAMAGE.\n"
      }
    },
    "schemes": ["http"],
    "consumes": ["application/json"],
    "produces": ["application/json"],
    "paths": {
      "/oic-res-BaselineInterfaceURI" : {
        "get": {
          "description": "Baseline representation of /oic/res; list of discoverable resources\nRetrieve the
discoverable resource set, baseline interface\n",
          "parameters": [
            {"$ref": "#/parameters/interface-baseline"}
          ],
          "responses": {
              "200": {
                "description" : "",
                "x-example":
                  [
                   {
                    "rt": ["oic.wk.res"],
                    "if": ["oic.if.baseline", "oic.if.ll" ],
                    "links":
                      [
                        {
                          "href": "/humidity",
                          "rt":   ["oic.r.humidity"],
                          "if":   ["oic.if.s"],
                          "p":    {"bm": 3},
                          "eps": [
                              {"ep": "coaps://[fe80::b1d6]:1111", "pri": 2},
                              {"ep": "coaps://[fe80::b1d6]:1122"},
                              {"ep": "coap+tcp://[2001:db8:a::123]:2222", "pri": 3}
                          ]
                        },
                        {
                          "href": "/temperature",
                          "rt":   ["oic.r.temperature"],
                          "if":   ["oic.if.s"],
                          "p":    {"bm": 3},
                          "eps": [
                              {"ep": "coaps://[[2001:db8:a::123]:2222"}
                          ]
                        }
                      ]
                   }
                  ]
                  ,
                "schema": { "$ref": "#/definitions/sbaseline" }
              }
```

```
          }
        }
      },
      "/oic-res-llInterfaceURI" : {
        "get": {
          "description": "Link list representation of /oic/res; list of discoverable
resources\nRetrieve the discoverable resource set, link list interface\n",
          "parameters": [
            {"$ref": "#/parameters/interface-ll"}
          ],
          "responses": {
              "200": {
              "description" : "",
              "x-example":
                [
                  {
                    "href": "/humidity",
                    "rt":   ["oic.r.humidity"],
                    "if":   ["oic.if.s"],
                    "p":    {"bm": 3},
                    "eps": [
                      {"ep": "coaps://[fe80::b1d6]:1111", "pri": 2},
                      {"ep": "coaps://[fe80::b1d6]:1122"},
                      {"ep": "coaps+tcp://[2001:db8:a::123]:2222", "pri": 3}
                    ]
                  },
                  {
                    "href": "/temperature",
                    "rt":   ["oic.r.temperature"],
                    "if":   ["oic.if.s"],
                    "p":    {"bm": 3},
                    "eps": [
                      {"ep": "coaps://[[2001:db8:a::123]:2222"}
                    ]
                  }
                ]
              "schema": { "$ref": "#/definitions/slinklist" }
            }
          }
        }
      }
    },
    "parameters": {
      "interface-ll" : {
        "in" : "query",
        "name" : "if",
        "type" : "string",
        "enum" : ["oic.if.ll"]
      }, 9867
      "interface-baseline" : {
        "in" : "query",
        "name" : "if",
        "type" : "string",
        "enum" : ["oic.if.baseline"]
```

```
      }
    },
    "definitions": {
      "sbaseline" :
          {
        "properties": {
          "if": {
            "description": "The interface set supported by this resource",
            "items": {
              "enum": [
                "oic.if.baseline",
                "oic.if.ll"
              ],
              "type": "string"
            },
            "minItems": 1,
            "readOnly": true,
            "type": "array"
          },
          "links": {
            "items": {
              "properties": {
                "anchor": {
                  "description": "This is used to override the context URI e.g. override the URI of the
containing collection",
                  "format": "uri",
                  "maxLength": 256,
                  "type": "string"
                },
                "di": {
                  "description": "Unique identifier for device (UUID)",
                  "pattern": "^[a-fA-F0-9]{8}-[a-fA-F0-9]{4}-[a-fA-F0-9]{4}-[a-fA-
F0-9]{4}-[a-fA- F0-9]{12}$",
                  "type": "string"
                },
                "eps": {
                  "description": "the Endpoint information of the target Resource",
                  "items": {
                    "properties": {
                      "ep": {
                        "description": "URI with Transport Protocol Suites + Endpoint Locator as specified",
                        "format": "uri",
                        "type": "string"
                      },
                      "pri": {
                        "description":"The priority among multiple Endpoints as specified ",
                        "minimum": 1,
                        "type": "integer"
                      }
                    },
                    "type": "object"
                  },
                  "type": "array"
                },
```

```
                "href": {
                  "description": "This is the target URI, it can be specified as a Relative Reference or
fully-qualified URI. Relative Reference should be used along with the di parameter to make it unique.",
                  "format": "uri",
                  "maxLength": 256,
                  "type": "string"
                },
                "if": {
                  "description": "The interface set supported by this resource",
                  "items": {
                    "enum": [
                      "oic.if.baseline",
                      "oic.if.ll",
                      "oic.if.b",
                      "oic.if.rw",
                      "oic.if.r",
                      "oic.if.a",
                      "oic.if.s"
                    ],
                    "type": "string"
                  },
                  "minItems": 1,
                  "type": "array"
                },
                "ins": {
                  "description": "The instance identifier for this web link in an array of web links -
used in collections",
                  "oneOf": [
                    {
                      "description": "An ordinal number that is not repeated - must be unique in the
collection context",
                      "type": "integer"
                    },
                    {
                      "description": "Any unique string including a URI",
                      "format": "uri",
                      "maxLength": 256,
                      "type": "string"
                    },
                    {
                      "description": "Unique identifier (UUID)",
                      "pattern": "^[a-fA-F0-9]{8}-[a-fA-F0-9]{4}-[a-fA-F0-9]{4}-[a-fA
-F0-9]{4}-[a-fA-F0-9]{12}$",
                      "type": "string"
                    }
                  ]
                },
                "p": {
                  "description": "Specifies the framework policies on the Resource referenced by the target
URI",
                  "properties": {
                    "bm": {
                      "description": "Specifies the framework policies on the Resource referenced by the
target URI for e.g. observable and discoverable",
```

```
            "type": "integer"
          }
        },
        "required": [
          "bm"
        ],
        "type": "object"
      }, 9991
      "rel": {
        "description": "The relation of the target URI referenced by the link to the context URI",
        "oneOf": [
          {
            "default": [
              "hosts"
            ],
            "items": {
              "maxLength": 64,
              "type": "string"
            },
            "minItems": 1,
            "type": "array"
          },
          {
            "default": "hosts",
            "maxLength": 64,
            "type": "string"
          }
        ]
      },
      "rt": {
        "description": "Resource Type",
        "items": {
          "maxLength": 64,
          "type": "string"
        },
        "minItems": 1,
        "type": "array"
      },
      "title": {
        "description": "A title for the link relation. Can be used by the UI to provide a context",
        "maxLength": 64,
        "type": "string"
      },
      "type": {
        "default": "application/cbor", 1
        "description": "A hint at the representation of the resource referenced by the target
URI. This represents the media types that are used for both accepting and emitting",
        "items": {
          "maxLength": 64,
          "type": "string"
        },
        "minItems": 1,
        "type": "array"
      }
```

```
                },
                "required": [
                  "href",
                  "rt",
                  "if"
                ],
                "type": "object"
              },
              "type": "array"
            },
            "mpro": {
              "description": "Supported messaging protocols",
              "maxLength": 64,
              "readOnly": true,
              "type": "string"
            },
            "n": {
              "description": "Human friendly name",
              "maxLength": 64,
              "readOnly": true,
              "type": "string"
            },
            "rt": {
              "description": "Resource Type",
              "items": {
                "maxLength": 64,
                "type": "string"
              },
              "minItems": 1,
              "readOnly": true,
              "type": "array"
            }
          },
          "required": [
            "rt",
            "if",
            "links"
          ],
          "type": "object"
        }
      ,
      "slinklist" :
            {
          "properties": {
            "anchor": {
              "description": "This is used to override the context URI e.g. override the URI of the containing
collection",
              "format": "uri",
              "maxLength": 256,
              "type": "string"
            },
            "di": {
              "description": "Unique identifier for device (UUID)",
              "pattern": "^[a-fA-F0-9]{8}-[a-fA-F0-9]{4}-[a-fA-F0-9]{4}-[a-fA-F0-9]
```

```
{4} - [a - fA - F0 - 19]{12} $ ", 1
                "type": "string"
            },
            "eps": {
              "description": "the Endpoint information of the target Resource",
              "items": {
                "properties": {
                  "ep": {
                    "description": "URI with Transport Protocol Suites + Endpoint Locator as specified ",
                    "format": "uri",
                    "type": "string"
                  },
                  "pri": {
                    "description": "The priority among multiple Endpoints as specified",
                    "minimum": 1,
                    "type": "integer"
                  }
                },
                "type": "object"
              },
              "type": "array"
            },
            "href": {
              "description": "This is the target URI, it can be specified as a Relative Reference or fully-
qualified URI. Relative Reference should be used along with the di parameter to make it unique.",
              "format": "uri",
              "maxLength": 256,
              "type": "string"
            },
            "if": {
              "description": "The interface set supported by this resource",
              "items": {
                "enum": [
                  "oic.if.baseline",
                  "oic.if.ll",
                  "oic.if.b",
                  "oic.if.rw",
                  "oic.if.r",
                  "oic.if.a",
                  "oic.if.s"
                ],
                "type": "string"
              },
              "minItems": 1,
              "type": "array"
            },
            "ins": {
              "description": "The instance identifier for this web link in an array of web links - used in
collections",
              "oneOf": [
                {
                  "description": "An ordinal number that is not repeated - must be unique in the collection
context",
                  "type": "integer"
```

```
            },
            {
              "description": "Any unique string including a URI",
              "format": "uri",
              "maxLength": 256,
              "type": "string"
            },
            {
              "description": "Unique identifier (UUID)",
              "pattern": "^[a-fA-F0-9]{8}-[a-fA-F0-9]{4}-[a-fA-F0-9]{4}-[a-fA-F0-
9]{4}-[a-fA-F0-9]{12}$",
              "type": "string"
            }
          ]
        },
        "p": {
          "description": "Specifies the framework policies on the Resource referenced by the target URI",
          "properties": {
            "bm": {
              "description": "Specifies the framework policies on the Resource referenced by the target
URI for e.g. observable and discoverable",
              "type": "integer"
            }
          },
          "required": [
            "bm"
          ],
          "type": "object"
        },
        "rel": {
          "description": "The relation of the target URI referenced by the link to the context URI",
          "oneOf": [
            {
              "default": [
                "hosts"
              ],
              "items": {
                "maxLength": 64,
                "type": "string"
              },
              "minItems": 1,
              "type": "array"
            },
            {
              "default": "hosts",
              "maxLength": 64,
              "type": "string"
            }
          ]
        },
        "rt": {
          "description": "Resource Type",
          "items": {
            "maxLength": 64,
```

```
                "type": "string"
              },
              "minItems": 1,
              "type": "array"
            },
            "title": {
              "description": "A title for the link relation. Can be used by the UI to provide a context",
              "maxLength": 64,
              "type": "string"
            },
            "type": {
              "default": "application/cbor",
              "description": "A hint at the representation of the resource referenced by the target URI. This
represents the media types that are used for both accepting and emitting",
              "items": {
                "maxLength": 64,
                "type": "string"
              },
              "minItems": 1,
              "type": "array"
            }
          },
          "required": [
            "href",
            "rt",
            "if"
          ],
          "type": "object"
        }
      }
    }
```

2. 属性定义

属性定义如表 10-21 所示。

表 10-21 属性定义

属 性 名	属性值类型	访问模式	是否强制	描 述
mpro	字符串	只读		支持的消息协议
links	数组		是	
if	数组	只读	是	该资源支持的接口集
n	字符串	只读		人性化名称
rt	数组	只读		资源类型
di	多类型			设备唯一标识符
title	字符串			链接关系标题,可用来提供界面内容
eps	数组			目标资源的终端信息
p	项目			规定从目标 URI 引用资源的框架策略
ins	多类型			集合中使用的链接数组中 Web 实例标识符
href	字符串		是	目标 URI,它可以被指定为相对引用或完全限定的 URI。应该与“di”参数一起使用,使其能够唯一

续表

属性名	属性值类型	访问模式	是否强制	描述
type	数组			提示由目标URI引用的资源表示，用于接收和发送的媒体类型
anchor	字符串			用于覆盖内容URI，例如，覆盖包含集合的URI
rel	多类型			内容URI链接引用目标URI的关系

3. CRUDN 行为

CRUDN 行为如表 10-22 所示。

表 10-22 CRUDN 行为

资源	创建	检索	更新	删除	通知
/oic/res		get			

10.12 场景

该资源是通用集合资源，“rts”值应包含“oic. sceneCollection”资源类型，提供指向场景的当前网络连接列表。URI 示例为“/SceneListResURI”，资源类型为“oic. wk. scenelist”“oic. wk. scenemember”“oic. wk. scenecollection”。

1. Swagger 2.0 定义

```
{
  "swagger": "2.0",
  "info": {
    "title": "Scenes (Top level)",
    "version": "v1 - 20160622",
    "license": {
      "name": "copyright 2016 - 2017 Open Connectivity Foundation, Inc. All rights reserved.",
      "x - description": "Redistribution and use in source and binary forms, with or without modification, are permitted provided that the following conditions are met:\n 1. Redistributions of source code must retain the above copyright notice, this list of conditions and the following disclaimer. \n 2. Redistributions in binary form must reproduce the above copyright notice, this list of conditions and the following disclaimer in the documentation and/or other materials provided with the distribution. \n\n THIS SOFTWARE IS PROVIDED BY THE Open Connectivity Foundation, INC. \"AS IS\" AND ANY EXPRESS OR IMPLIED WARRANTIES, INCLUDING, BUT NOT LIMITED TO, THE IMPLIED WARRANTIES OF MERCHANTABILITY AND FITNESS FOR A PARTICULAR PURPOSE OR WARRANTIES OF NON - INFRINGEMENT, ARE DISCLAIMED. \n IN NO EVENT SHALL THE Open Connectivity Foundation, INC. OR CONTRIBUTORS BE LIABLE FOR ANY DIRECT, INDIRECT, INCIDENTAL, SPECIAL, EXEMPLARY, OR CONSEQUENTIAL DAMAGES (INCLUDING, BUT NOT LIMITED TO, PROCUREMENT OF SUBSTITUTE GOODS OR SERVICES; LOSS OF USE, DATA, OR PROFITS; OR BUSINESS INTERRUPTION) \n HOWEVER CAUSED AND ON ANY THEORY OF LIABILITY, WHETHER IN CONTRACT, STRICT LIABILITY, OR TORT (INCLUDING NEGLIGENCE OR OTHERWISE) ARISING IN ANY WAY OUT OF THE USE OF THIS SOFTWARE, EVEN IF ADVISED OF THE POSSIBILITY OF SUCH DAMAGE. \n"
    }
  },
  "schemes": ["http"],
  "consumes": ["application/json"],
  "produces": ["application/json"],
  "paths": {
```

```
"/SceneListResURI" : {
  "get": {
    "description": "Toplevel Scene resource.\nThis resource is a generic collection resource.\nThe rts value shall contain oic.wk.scenecollection resource types.\nProvides the current list of web links pointing to scenes\n",
     "parameters": [
     ],
     "responses": {
         "200": {
          "description" : "",
          "x-example":
            {
                "rt": ["oic.wk.scenelist"],
                "n": "list of scene Collections",
                "rts": ["oic.wk.scenecollection"],
                "links": [
                 ]
            }
          ,
         "schema": { "$ref": "#/definitions/Collection" }
        }
     }
   }
 },
 "/SceneMemberResURI" : {
   "get": {
     "description": "Collection that models a scene member.\nProvides the scene member\n",
     "parameters": [
     ],
     "responses": {
         "200": {
          "description" : "",
          "x-example":
            {
              "rt": ["oic.wk.scenemember"],
              "id": "0685B960-FFFF-46F7-BEC0-9E6234671ADC1",
              "n": "my binary switch (for light bulb) mappings",
              "link": {
                "href": "binarySwitch",
                "rt": ["oic.r.switch.binary"],
                "if": ["oic.if.a", "oic.if.baseline"],
                "eps": [
                    {"ep": "coap://[fe80::b1d6]:1111", "pri": 2},
                    {"ep": "coaps://[fe80::b1d6]:1122"},
                    {"ep": "coap+tcp://[2001:db8:a::123]:2222", "pri": 3}
                ]
              },
              "sceneMappings": [
               {
                 "scene":          "off",
                 "memberProperty":  "value",
                 "memberValue":     true
               },
               {
```

```
                "scene":            "Reading",
                "memberProperty":  "value",
                "memberValue":     false
              },
              {
                "scene":            "TVWatching",
                "memberProperty":  "value",
                "memberValue":     true
              }
            ]
          }
          ,
        "schema": { "$ref": "#/definitions/SceneMember" }
      }
    }
  }
},
"/SceneCollectionResURI" : {
  "get": {
    "description": "Collection that models a set of Scenes. \nThis resource is a generic collection resource with additional parameters. \nThe rts value shall contain oic. scenemember resource types. \nThe additional parameters are\n lastScene, this is the scene value last set by any OCF Client\n sceneValues, this is the list of available scenes\n lastScene shall be listed in sceneValues. \nProvides the current list of web links pointing to scenes\n",
    "parameters": [
    ],
    "responses": {
      "200": {
        "description" : "",
        "x-example":
          {
            "lastScene": "off",
            "sceneValues": "off,Reading,TVWatching",
            "rt":          ["oic.wk.scenecollection"],
            "n":           "My Scenes for my living room",
            "id":          "0685B960-736F-46F7-BEC0-9E6CBD671ADC1",
            "rts":         ["oic.wk.scenemember"],
            "links": [
            ]
          }
          ,
        "schema": { "$ref": "#/definitions/SceneCollection" }
      }
    }
  },
  "post": {
    "description": "Provides the action to change the last set scene selection. \nCalling this method shall update all scene members to the prescribed membervalue. \nWhen this method is called with the same value as the current lastScene value\nthen all scene members shall be updated. \n",
    "parameters": [
      {
        "name": "body",
        "in": "body",
        "required": true,
```

```
            "schema": { "$ref": "#/definitions/SceneCollectionUpdate" },
            "x-example":
              {
                "lastScene": "Reading"
              }
          }
        ],
        "responses": {
          "200": {
            "description" : "Indicates that the value is changed.\nThe changed properties are provided
in the response.\n",
            "x-example":
              {
                "lastScene": "Reading"
              }
              ,
            "schema": { "$ref": "#/definitions/SceneCollectionUpdate" }
          }
        }
      }
    }
  },
  "parameters": {
    "interface" : {
      "in" : "query",
      "name" : "if",
      "type" : "string",
      "enum" : ["oic.if.a", "oic.if.ll", "oic.if.baseline"]
    }
  },
  "definitions": {
    "Collection" :
          {
        "description": "A set (array) of simple or individual OIC Links. In addition to properties required
for an OIC Link, the identifier for that link in this set is also required",
        "items": {
          "properties": {
            "anchor": {
              "description": "This is used to override the context URI e.g. override the URI of the
containing collection",
              "format": "uri",
              "maxLength": 256,
              "type": "string"
            },
            "di": {
              "description": "Unique identifier for device (UUID)",
              "pattern": "^[a-fA-F0-9]{8}-[a-fA-F0-9]{4}-[a-fA-F0-9]{4}-[a-fA-F0-9]
{4}-[a-fA-F0-9]{12}$",
              "type": "string"
            },
            "drel": {
              "description": "When specified this is the default relationship to use when an OIC Link does
not specify an explicit relationship with *rel* parameter",
              "type": "string"
```

```
},
"eps": {
  "description": "the Endpoint information of the target Resource",
  "items": {
    "properties": {
      "ep": {
        "description": "URI with Transport Protocol Suites + Endpoint Locator as specified",
        "format": "uri",
        "type": "string"
      },
      "pri": {
        "description": "The priority among multiple Endpoints as specified",
        "minimum": 1,
        "type": "integer"
      }
    },
    "type": "object"
  },
  "type": "array"
},
"href": {
  "description": "This is the target URI, it can be specified as a Relative Reference or fully-
qualified URI. Relative Reference should be used along with the di parameter to make it unique.",
  "format": "uri",
  "maxLength": 256,
  "type": "string"
},
"id": {
  "description": "Instance ID of this specific resource",
  "maxLength": 64,
  "readOnly": true,
  "type": "string"
},
"if": {
  "description": "The interface set supported by this resource",
  "items": {
    "enum": [
      "oic.if.baseline",
      "oic.if.ll",
      "oic.if.b",
      "oic.if.rw",
      "oic.if.r",
      "oic.if.a",
      "oic.if.s"
    ],
    "type": "string"
  },
  "minItems": 1,
  "type": "array"
},
"ins": {
  "description": "The instance identifier for this web link in an array of web links - used in
collections",
  "oneOf": [
```

```
                {
                  "description": " An ordinal number that is not repeated - must be unique in the
collection context",
                  "type": "integer"
                },
                {
                  "description": "Any unique string including a URI",
                  "format": "uri",
                  "maxLength": 256,
                  "type": "string"
                },
                {
                  "description": "Unique identifier (UUID)",
                  "pattern": "^[a-fA-F0-9]{8}-[a-fA-F0-9]{4}-[a-fA-F0-9]{4}-[a-fA-
F0-9]{4}-[a-fA- F0-9]{12}$",
                  "type": "string"
                }
              ]
            },
            "links": {
              "description": "All forms of links in a collection",
              "oneOf": [
                {
                  "description": " A set (array) of simple or individual OIC Links. In addition to
properties required for an OIC Link, the identifier for that link in this set is also required",
                  "items": {
                    "properties": {
                      "anchor": {
                        "description": "This is used to override the context URI e.g. override the URI of
the containing collection",
                        "format": "uri",
                        "maxLength": 256,
                        "type": "string"
                      },
                      "di": {
                        "description": "Unique identifier for device (UUID)",
                        "pattern": "^[a-fA-F0-9]{8}-[a-fA-F0-9]{4}-[a-fA-F0-9]{4}-[a-
fA-F0-9]{4}-[a-fA-F0-9]{12}$",
                        "type": "string"
                      },
                      "eps": {
                        "description": "the Endpoint information of the target Resource",
                        "items": {
                          "properties": {
                            "ep": {
                              "description": "URI with Transport Protocol Suites + Endpoint Locator as
specified",
                              "format": "uri",
                              "type": "string"
                            },
                            "pri": {
                              "description": "The priority among multiple Endpoints as specified",
                              "minimum": 1,
                              "type": "integer"
```

```
                }
              },
              "type": "object"
            },
            "type": "array"
          },
          "href": {
            "description": "This is the target URI, it can be specified as a Relative Reference or
fully-qualified URI. Relative Reference should be used along with the di parameter to make it unique.",
            "format": "uri",
            "maxLength": 256,
            "type": "string"
          },
          "if": {
            "description": "The interface set supported by this resource",
            "items": {
              "enum": [
                "oic.if.baseline",
                "oic.if.ll",
                "oic.if.b",
                "oic.if.rw",
                "oic.if.r",
                "oic.if.a",
                "oic.if.s"
              ],
              "type": "string"
            },
            "minItems": 1,
            "type": "array"
          },
          "ins": {
            "description": "The instance identifier for this web link in an array of web links -
used in collections",
            "oneOf": [
              {
                "description": "An ordinal number that is not repeated - must be unique in
the collection context",
                "type": "integer"
              },
              {
                "description": "Any unique string including a URI",
                "format": "uri",
                "maxLength": 256,
                "type": "string"
              },
              {
                "description": "Unique identifier (UUID)",
                "pattern": "^[a-fA-F0-9]{8}-[a-fA-F0-9]{4}-[a-fA-F0-9]{4}-
[a-fA-F0-1 9]{4}-[a-fA-F0-9]{12}$",
                "type": "string"
              }
            ]
          },
          "p": {
```

```
"description": "Specifies the framework policies on the Resource referenced by
the target URI",
        "properties": {
          "bm": {
            "description": "Specifies the framework policies on the Resource referenced
by the target URI for e.g. observable and discoverable",
            "type": "integer"
          }
        },
        "required": [
          "bm"
        ],
        "type": "object"
      },
      "rel": {
        "description": "The relation of the target URI referenced by the link to the
context URI",
        "oneOf": [
          {
            "default": [
              "hosts"
            ],
            "items": {
              "maxLength": 64,
              "type": "string"
            },
            "minItems": 1,
            "type": "array"
          },
          {
            "default": "hosts",
            "maxLength": 64,
            "type": "string"
          }
        ]
      },
      "rt": {
        "description": "Resource Type",
        "items": {
          "maxLength": 64,
          "type": "string"
        },
        "minItems": 1,
        "type": "array"
      },
      "title": {
        "description": "A title for the link relation. Can be used by the UI to
provide a context",
        "maxLength": 64,
        "type": "string"
      },
      "type": {
        "default": "application/cbor",
        "description": "A hint at the representation of the resource referenced by the
```

```
target URI. This represents the media types that are used for both accepting and emitting",
                        "items": {
                          "maxLength": 64,
                          "type": "string"
                        },
                        "minItems": 1,
                        "type": "array"
                      }
                    },
                    "required": [
                      "href",
                      "rt",
                      "if"
                    ],
                    "type": "object"
                  },
                  "type": "array"
                }
              ]
            },
            "n": {
              "description": "Friendly name of the resource",
              "maxLength": 64,
              "readOnly": true,
              "type": "string"
            },
            "p": {
              "description": "Specifies the framework policies on the Resource referenced by the target URI",
              "properties": {
                "bm": {
                  "description": "Specifies the framework policies on the Resource referenced by the target URI
for e.g. observable and discoverable",
                  "type": "integer"
                }
              },
              "required": [
                "bm"
              ],
              "type": "object"
            },
            "rel": {
              "description": "The relation of the target URI referenced by the link to the context URI",
              "oneOf": [
                {
                  "default": [
                    "hosts"
                  ],
                  "items": {
                    "maxLength": 64,
                    "type": "string"
                  },
                  "minItems": 1,
                  "type": "array"
                },
```

```
            {
              "default": "hosts",
              "maxLength": 64,
              "type": "string"
            }
          ]
        },
        "rt": {
          "description": "Resource Type",
          "items": {
            "maxLength": 64,
            "type": "string"
          },
          "minItems": 1,
          "type": "array"
        },
        "rts": {
          "description": "Defines the list of allowable resource types (for Target and anchors) in
links included in the collection; new links being created can only be from this list",
          "items": {
            "maxLength": 64,
            "type": "string"
          },
          "minItems": 1,
          "readOnly": true,
          "type": "array"
        },
        "title": {
          "description": "A title for the link relation. Can be used by the UI to provide a context",
          "maxLength": 64,
          "type": "string"
        },
        "type": {
          "default": "application/cbor",
          "description": "A hint at the representation of the resource referenced by the target URI.
This represents the media types that are used for both accepting and emitting",
          "items": {
            "maxLength": 64,
            "type": "string"
          },
          "minItems": 1,
          "type": "array"
        }
      },
      "required": [
        "href",
        "rt",
        "if"
      ],
      "type": "object"
    },
    "type": "array"
  }
,
```

```
"SceneMember" :
        {
    "properties": {
      "SceneMappings": {
        "description": "array of mappings per scene, can be 1",
        "items": {
          "properties": {
            "memberProperty": {
              "description": "property name that will be mapped",
              "readOnly": true,
              "type": "string"
            },
            "memberValue": {
              "description": "value of the Member Property",
              "readOnly": true,
              "type": "string"
            },
            "scene": {
              "description": "Specifies a scene value that will acted upon",
              "type": "string"
            }
          },
          "required": [
            "scene",
            "memberProperty",
            "memberValue"
          ],
          "type": "object"
        },
        "type": "array"
      },
      "id": {
        "description": "Can be an value that is unique to the use context or a UUIDv4",
        "type": "string"
      },
      "if": {
        "description": "The interface set supported by this resource",
        "items": {
          "enum": [
            "oic.if.baseline",
            "oic.if.ll",
            "oic.if.b",
            "oic.if.lb",
            "oic.if.rw",
            "oic.if.r",
            "oic.if.a",
            "oic.if.s"
          ],
          "type": "string"
        },
        "minItems": 1,
        "readOnly": true,
        "type": "array"
      },
```

```
            "link": {
              "description": "web link that points at a resource",
              "properties": {
                "anchor": {
                  "description": "This is used to override the context URI e.g. override the URI of the
containing collection",
                  "format": "uri",
                  "maxLength": 256,
                  "type": "string"
                },
                "di": {
                  "description": "Unique identifier for device (UUID)",
                  "pattern": "^[a-fA-F0-9]{8}-[a-fA-F0-9]{4}-[a-fA-F0-9]{4}-[a-fA-F0-
9]{4}-[a-fA-F0-9]{12}$",
                  "type": "string"
                },
                "eps": {
                  "description": "the Endpoint information of the target Resource",
                  "items": {
                    "properties": {
                      "ep": {
                        "description": "URI with Transport Protocol Suites + Endpoint Locator as specified",
                        "format": "uri",
                        "type": "string"
                      },
                      "pri": {
                        "description": "The priority among multiple Endpoints as specified",
                        "minimum": 1,
                        "type": "integer"
                      }
                    },
                    "type": "object"
                  },
                  "type": "array"
                },
                "href": {
                  "description": "This is the target URI, it can be specified as a Relative Reference or
fully-qualified URI. Relative Reference should be used along with the di parameter to make it unique.",
                  "format": "uri",
                  "maxLength": 256,
                  "type": "string"
                },
                "if": {
                  "description": "The interface set supported by this resource",
                  "items": {
                    "enum": [
                      "oic.if.baseline",
                      "oic.if.ll",
                      "oic.if.b",
                      "oic.if.rw",
                      "oic.if.r",
                      "oic.if.a",
                      "oic.if.s"
                    ],
```

```
                "type": "string"
              },
              "minItems": 1,
              "type": "array"
            },
            "ins": {
              "description": "The instance identifier for this web link in an array of web links - used
in collections",
              "oneOf": [
                {
                    "description": "An ordinal number that is not repeated - must be unique in the
collection context",
                  "type": "integer"
                },
                {
                  "description": "Any unique string including a URI",
                  "format": "uri",
                  "maxLength": 256,
                  "type": "string"
                },
                {
                  "description": "Unique identifier (UUID)",
                  "pattern": "^[a-fA-F0-9]{8}-[a-fA-F0-9]{4}-[a-fA-F0-9]{4}-[a-fA-
F0-9]{4}-[a-fA-F0-9]{12}$",
                  "type": "string"
                }
              ]
            },
            "p": {
              "description": "Specifies the framework policies on the Resource referenced by the target
URI",
              "properties": {
                "bm": {
                  "description": "Specifies the framework policies on the Resource referenced by the
target URI for e.g. observable and discoverable",
                  "type": "integer"
                }
              },
              "required": [
                "bm"
              ],
              "type": "object"
            },
            "rel": {
              "description": "The relation of the target URI referenced by the link to the context URI",
              "oneOf": [
                {
                  "default": [
                    "hosts"
                  ],
                  "items": {
                    "maxLength": 64,
                    "type": "string"
                  },
```

```
                "minItems": 1,
                "type": "array"
              },
              {
                "default": "hosts",
                "maxLength": 64,
                "type": "string"
              }
            ]
          },
          "rt": {
            "description": "Resource Type",
            "items": {
              "maxLength": 64,
              "type": "string"
            },
            "minItems": 1,
            "type": "array"
          },
          "title": {
            "description": "A title for the link relation. Can be used by the UI to provide a context",
            "maxLength": 64,
            "type": "string"
          },
          "type": {
            "default": "application/cbor",
            "description": "A hint at the representation of the resource referenced by the target URI.
This represents the media types that are used for both accepting and emitting",
            "items": {
              "maxLength": 64,
              "type": "string"
            },
            "minItems": 1,
            "type": "array"
          }
        },
        "required": [
          "href",
          "rt",
          "if"
        ],
        "type": "string"
      },
      "n": {
        "description": "Used to name the Scene collection",
        "type": "string"
      },
      "rt": {
        "description": "Resource Type",
        "items": {
          "maxLength": 64,
          "type": "string"
        },
        "minItems": 1,
```

```
        "readOnly": true,
        "type": "array"
      }
    },
    "required": [
      "link"
    ],
    "type": "object"
  }
,
"SceneCollection" :
      {
    "properties": {
      "id": {
        "description": "A unique string that could be a hash or similarly unique",
        "type": "string"
      },
      "if": {
        "description": "The interface set supported by this resource",
        "items": {
          "enum": [
            "oic.if.baseline",
            "oic.if.ll",
            "oic.if.b",
            "oic.if.lb",
            "oic.if.rw",
            "oic.if.r",
            "oic.if.a",
            "oic.if.s"
          ],
          "type": "string"
        },
        "minItems": 1,
        "readOnly": true,
        "type": "array"
      },
      "lastScene": {
        "description": "Last selected Scene, shall be part of sceneValues",
        "type": "string"
      },
      "links": {
        "description": "Array of OIC web links that are reference from this collection",
        "items": {
          "allOf": [
            {
              "properties": {
                "anchor": {
                  "description": "This is used to override the context URI e.g. override the URI of
the containing collection",
                  "format": "uri",
                  "maxLength": 256,
                  "type": "string"
                },
                "di": {
```

```
                    "description": "Unique identifier for device (UUID)",
                    "pattern": "^[a-fA-F0-9]{8}-[a-fA-F0-9]{4}-[a-fA-F0-9]{4}-[a-
fA-F0-9]{4}-[a- fA-F0-9]{12}$",
                    "type": "string"
                  },
                  "eps": {
                    "description": "the Endpoint information of the target Resource",
                    "items": {
                      "properties": {
                        "ep": {
                          "description": "URI with Transport Protocol Suites + Endpoint Locator as
specified in 10.2.1",
                          "format": "uri",
                          "type": "string"
                        },
                        "pri": {
                          "description": "The priority among multiple Endpoints as specified",
                          "minimum": 1,
                          "type": "integer"
                        }
                      },
                      "type": "object"
                    },
                    "type": "array"
                  },
                  "href": {
                    "description": "This is the target URI, it can be specified as a Relative Reference
or fully-qualified URI. Relative Reference should be used along with the di parameter to make it unique.",
                    "format": "uri",
                    "maxLength": 256,
                    "type": "string"
                  },
                  "if": {
                    "description": "The interface set supported by this resource",
                    "items": {
                      "enum": [
                        "oic.if.baseline",
                        "oic.if.ll",
                        "oic.if.b",
                        "oic.if.rw",
                        "oic.if.r",
                        "oic.if.a",
                        "oic.if.s"
                      ],
                      "type": "string"
                    },
                    "minItems": 1,
                    "type": "array"
                  },
                  "ins": {
                    "description": "The instance identifier for this web link in an array of web links -
used in collections",
                    "oneOf": [
                      {
```

```
                    "description": "An ordinal number that is not repeated - must be unique in the
collection context",
                    "type": "integer"
                  },
                  {
                    "description": "Any unique string including a URI",
                    "format": "uri",
                    "maxLength": 256,
                    "type": "string"
                  },
                  {
                    "description": "Unique identifier (UUID)",
                    "pattern": "^[a-fA-F0-9]{8}-[a-fA-F0-9]{4}-[a-fA-F0-9]{4}-
[a-fA-F0-9]{4}-[a-fA-F0-9]{12}$",
                    "type": "string"
                  }
                ]
              },
              "p": {
                "description": "Specifies the framework policies on the Resource referenced by the
target URI",
                "properties": {
                  "bm": {
                    "description": "Specifies the framework policies on the Resource
referenced by the target URI for e.g. observable and discoverable",
                    "type": "integer"
                  }
                },
                "required": [
                  "bm"
                ],
                "type": "object"
              },
              "rel": {
                "description": "The relation of the target URI referenced by the link to the context
URI",
                "oneOf": [
                  {
                    "default": [
                      "hosts"
                    ],
                    "items": {
                      "maxLength": 64,
                      "type": "string"
                    },
                    "minItems": 1,
                    "type": "array"
                  },
                  {
                    "default": "hosts",
                    "maxLength": 64,
                    "type": "string"
                  }
                ]
```

```
                },
                "rt": {
                  "description": "Resource Type",
                  "items": {
                    "maxLength": 64,
                    "type": "string"
                  },
                  "minItems": 1,
                  "type": "array"
                },
                "title": {
                  "description": "A title for the link relation. Can be used by the UI to provide a
context",
                  "maxLength": 64,
                  "type": "string"
                },
                "type": {
                  "default": "application/cbor",
                  "description": "A hint at the representation of the resource referenced by the
target URI. This represents the media types that are used for both accepting and emitting",
                  "items": {
                    "maxLength": 64,
                    "type": "string"
                  },
                  "minItems": 1,
                  "type": "array"
                }
              },
              "required": [
                "href",
                "rt",
                "if"
              ],
              "type": "object"
            },
            {
              "required": [
                "ins"
              ]
            }
          ]
        },
        "type": "array"
      },
      "n": {
        "description": "Used to name the Scene collection",
        "type": "string"
      },
      "rt": {
        "description": "Resource Type",
        "items": {
          "maxLength": 64,
          "type": "string"
        },
```

```
              "minItems": 1,
              "readOnly": true,
              "type": "array"
            },
            "rts": {
              "description": " Defines the list of allowable resource types in links included in the
collection; new links being created can only be from this list",
              "items": {
                "maxLength": 64,
                "type": "string"
              },
              "minItems": 1,
              "readOnly": true,
              "type": "array"
            },
            "sceneValues": {
              "description": "All available scene values",
              "readOnly": true,
              "type": "string"
            }
          },
          "required": [
            "lastScene",
            "sceneValues",
            "rts",
            "id"
          ],
          "type": "object"
        }
      ,
      "SceneCollectionUpdate" :
              {
          "properties": {
            "id": {
              "description": "A unique string that could be a hash or similarly unique",
              "type": "string"
            },
            "if": {
              "description": "The interface set supported by this resource",
              "items": {
                "enum": [
                  "oic.if.baseline",
                  "oic.if.ll",
                  "oic.if.b",
                  "oic.if.lb",
                  "oic.if.rw",
                  "oic.if.r",
                  "oic.if.a",
                  "oic.if.s"
                ],
                "type": "string"
              },
              "minItems": 1,
              "readOnly": true,
```

```
            "type": "array"
          },
          "lastScene": {
            "description": "Last selected Scene, shall be part of sceneValues",
            "type": "string"
          },
          "links": {
           "description": "Array of OIC web links that are reference from this collection",
            "items": {
              "allOf": [
                {
                  "properties": {
                    "anchor": {
                      "description": "This is used to override the context URI e.g. override the URI of
the containing collection",
                      "format": "uri",
                      "maxLength": 256,
                      "type": "string"
                    },
                    "di": {
                      "description": "Unique identifier for device (UUID)",
                      "pattern": "^[a-fA-F0-9]{8}-[a-fA-F0-9]{4}-[a-fA-F0-9]{4}-[a-fA
-F0-9]{4}-[a- fA-F0-9]{12}$",
                      "type": "string"
                    },
                    "eps": {
                      "description": "the Endpoint information of the target Resource",
                      "items": {
                        "properties": {
                          "ep": {
                            "description": "URI with Transport Protocol Suites + Endpoint Locator as
specified",
                            "format": "uri",
                            "type": "string"
                          },
                          "pri": {
                            "description":"The priority among multiple Endpoints as specified ",
                            "minimum": 1,
                            "type": "integer"
                          }
                        },
                        "type": "object"
                      },
                      "type": "array"
                    },
                    "href": {
                      "description": "This is the target URI, it can be specified as a Relative Reference
or fully-qualified URI. Relative Reference should be used along with the di parameter to make it unique.",
                      "format": "uri",
                      "maxLength": 256,
                      "type": "string"
                    },
                    "if": {
                      "description": "The interface set supported by this resource",
```

```
"items": {
  "enum": [
    "oic.if.baseline",
    "oic.if.ll",
    "oic.if.b",
    "oic.if.rw",
    "oic.if.r",
    "oic.if.a",
    "oic.if.s"
  ],
  "type": "string"
},
"minItems": 1,
"type": "array"
},
"ins": {
  "description": "The instance identifier for this web link in an array of web links - used in collections",
  "oneOf": [
    {
      "description": "An ordinal number that is not repeated - must be unique in the collection context",
      "type": "integer"
    },
    {
      "description": "Any unique string including a URI",
      "format": "uri",
      "maxLength": 256,
      "type": "string"
    },
    {
      "description": "Unique identifier (UUID)",
      "pattern": "^[a-fA-F0-9]{8}-[a-fA-F0-9]{4}-[a-fA-F0-9]{4}-[a-fA-F0-9]{4}-[a-fA-F0-9]{12}$",
      "type": "string"
    }
  ]
},
"p": {
  "description": "Specifies the framework policies on the Resource referenced by the target URI",
  "properties": {
    "bm": {
      "description": "Specifies the framework policies on the Resource referenced by the target URI for e.g. observable and discoverable",
      "type": "integer"
    }
  },
  "required": [
    "bm"
  ],
  "type": "object"
},
"rel": {
```

```
                "description": "The relation of the target URI referenced by the link to the context
URI",
                "oneOf": [
                  {
                    "default": [
                      "hosts"
                    ],
                    "items": {
                      "maxLength": 64,
                      "type": "string"
                    },
                    "minItems": 1,
                    "type": "array"
                  },
                  {
                    "default": "hosts",
                    "maxLength": 64,
                    "type": "string"
                  }
                ]
              },
              "rt": {
                "description": "Resource Type",
                "items": {
                  "maxLength": 64,
                  "type": "string"
                },
                "minItems": 1,
                "type": "array"
              },
              "title": {
                "description": "A title for the link relation. Can be used by the UI to provide a
context",
                "maxLength": 64,
                "type": "string"
              },
              "type": {
                "default": "application/cbor",
                  "description": "A hint at the representation of the resource referenced by the
target URI. This represents the media types that are used for both accepting and emitting",
                "items": {
                  "maxLength": 64,
                  "type": "string"
                },
                "minItems": 1,
                "type": "array"
              }
            },
            "required": [
              "href",
              "rt",
              "if"
            ],
            "type": "object"
```

```
                },
                {
                  "required": [
                    "ins"
                  ]
                }
              ]
            },
            "type": "array"
          },
          "n": {
            "description": "Used to name the Scene collection",
            "type": "string"
          },
          "rt": {
            "description": "Resource Type",
            "items": {
              "maxLength": 64,
              "type": "string"
            },
            "minItems": 1,
            "readOnly": true,
            "type": "array"
          },
          "rts": {
            "description": "Defines the list of allowable resource types in links included in the
collection; new links being created can only be from this list",
            "items": {
              "maxLength": 64,
              "type": "string"
            },
            "minItems": 1,
            "readOnly": true,
            "type": "array"
          },
          "sceneValues": {
            "description": "All available scene values",
            "readOnly": true,
            "type": "string"
          }
        },
        "required": [
          "lastScene"
        ],
        "type": "object"
      }
    }
}
```


2. 属性定义

属性定义如表 10-23 所示。

表 10-23 属性定义

属 性 名	属性值类型	访问模式	是否强制	描 述
rts	多类型			定义了集合中包含链接中允许的资源类型(对于目标和锚点)的列表,正在创建的新链接只能来自此列表
drel	字符串			在 OCF 链接未通过"rel"参数指明一个显式关系时,指定使用该默认关系
lastScene	字符串	读写	是	最后选择的场景,应该是"sceneValues"的一部分
sceneValues	字符串	只读	是	所有可用的场景值
n	字符串	读写		用于命名场景集合
id	对象	读写	是	一个唯一的字符串
links	数组	读写		从该集合中引用的 OCF Web 链接数组
id	字符串	读写		可以是对使用上下文或 UUIDv4 唯一的值
SceneMappings	数组	读写		每个场景的映射数组,可以是 1
link	字符串	读写	是	指向一个资源的 Web 链接
rt	数组	只读		资源类型
if	数组	只读		资源支持的接口集合
type	数组			提示由目标 URI 引用的资源表示,用于接收和发送的媒体类型
n	字符串	只读		人性化名称
id	字符串	只读		指定资源的 ID 示例
di	多类型			设备唯一标识符
anchor	字符串			用于覆盖内容 URI,例如,覆盖包含集合的 URI
rel	多类型			内容 URI 链接引用目标 URI 的关系
title	字符串			链接关系标题,可被 URI 用来提供内容
eps	数组			目标资源的终端信息
p	对象			规定从目标 URI 引用资源的框架策略

3. CRUDN 行为

CRUDN 行为如表 10-24 所示。

表 10-24 CRUDN 行为

资 源	创建	检索	更新	删除	通知
/SceneListResURI		get			

第 11 章 应用资源类型规范

CHAPTER 11

OCF 核心规范指定了可以实现 IoT 用法和生态系统的 OCF 框架。该 OCF 核心框架可扩展支持那些简单设备(受限性设备)和更多功能性设备(智能设备)。

本章基于 OCF 核心规范所定义资源类型,详细说明那些已经被 OCF 定义或者有可能被 OCF 设备使用到的资源,列出 OCF 当前指定的所有资源类型。资源由应用程序配置文件设备定义使用,本章提及的资源类型,可以由任何在 OCF 集合里或设备代表中的 OCF 认证设备所使用。

应用程序配置文件设备规范(例如,为智慧家居和医疗保健所创建的)指定了适合于该配置文件的设备类型,这些规范使用了本章定义的资源类型。

11.1 基准模型构造

11.1.1 概述

本部分对资源类型的构建基础做简单的描述。

1. URI

本章提到的 URI 是非规范化的,可能是供应商定义的。一个资源的实例由 URI 指示。当一个 OCF 设备中的同一个资源类型多于一个实例时,需要使用不同 URI 来为不同的实例做指示。其实现应遵循 OCF 核心规范中关于 URI 群体定义的要求,详细说明请参考本书的第 2 章。

2. 接口

OCF 核心规范指定所有资源类型至少有一个接口与它们关联,这个接口在资源发现阶段会被广播。此外,OCF 核心规范定义了许多可以被资源类型实例所应用的接口。

本部分定义的所有资源类型相关联的默认接口,应为资源类型定义中枚举列出的支持接口,除非资源类型定义具有传感器或执行器接口作为默认接口。在这种情况下,应选择其中一个作为默认接口。因此,托管这样资源类型的服务器端,应该允许“oic. if. s”(传感器接口)或“oic. if. a”(执行器接口)作为经由“/oic/res”暴露的接口,以及被授权的基准接口(“oic. if. baseline”)。除了指定为默认的接口之外,服务器端还可以支持其他接口。

注意,被设备所暴露的那些有关资源实例可见的功能,应该遵循当地(每个国家或立法地区)的监管要求或其他限制(例如,在一些管辖地区,用二进制开关去远程上电一些连接设备是受限制的;锁定状态可以是只读的,取决于上下文)。在这种情况下,设备不应暴露资源的执行器接口(“oic. if. a”),设备应使用“/oic/res”的传感器接口(“oic. if. s”)作为默认接口以及任何强制接口。

3. RAML 定义

本部分使用的 RAML 定义是规范性的,扩展所有定义的 JSON 有效载荷应符合指定的 JSON 模式,定义的模式包括所有 OCF 核心规范定义(强制)属性的扩展。

RAML 定义用于描述指定资源类型上 CRUDN 操作的有效载荷。CRUDN 操作在 OCF 核心规范中定义。OCF 核心规范还指定 CRUDN 操作的有效内容中的附加属性。本章文档中的 RAML 定义本身不足以创建实现,需要添加在核心规范中定义的附加属性来创建兼容实现。本章使用 RAML 特定支持的响应子集,对这些响应的使用,在表 5-4 中定义了 RAML 中的返回码。请注意,成功和错误条件的实际值是在 OCF 核心规范中定义的。

RAML 定义将 OCF CRUDN 行为映射到表 11-1 中定义的 RAML。

表 11-1　CRUDN 与 RAML 定义之间的转换

资　源	创建	检索	更新	删除	通知
/example	put 或 post	get	put 或 post	delete	

通知不是 RAML 定义的一部分,而是在核心规范中定义的。通知的语义与 CRUDN 检索值相同。在本部分定义的所有资源类型,都支持在本书的第 6 章 OCF 的功能交互中定义的通知。

11.1.2　属性定义

本节介绍公共属性、资源属性和基本资源模式。

1. 公共属性

OCF 核心规范指定了大量可能被 OCF 资源所定义的属性。公共属性"if"和"rt"应该被所有在本部分定义的资源类型指定;它们通过发现 OCF 服务器端和它的可用资源,由 OCF 核心规范定义的"/oic/res"资源类型来公开。公共属性"p"和"n"可以被本部分定义的所有资源类型所指定,公共属性如表 11-2 所示。

表 11-2　OCF 资源的公共属性

属性名	属性标题	属　性　值	属性值类型	访问模式	描　　述
if	接口	参见第 3 章	字符串	只读	核心规范定义;资源支持的接口
rt	资源类型	参见第 3 章	字符串	只读	核心规范定义;资源类型。资源类型在本部分定义
n	名称	参见第 3 章	字符串	只读	核心规范定义;人类可以理解的资源名称
id	ID	参见第 3 章	字符串	只读	核心规范定义;资源的唯一标识符

如果 OCF 客户端要求这些属性包含在响应检索操作提供的资源表示中,则客户端应通过在查询参数中指定此属性来选择 OCF 核心规范定义的基准接口("oic. if. baseline")。

建议托管支持批处理接口的智能家居设备集合用"oic. if. baseline"填充"bifLink"参数,用于集合中的所有资源。这是为了确保对集合的检索操作在响应中提供公共属性"rt"和"if"。

2. 资源属性

指定 CRUDN 操作的属性是使用 JSON 模式定义的(请参阅 JSON Schema)。基本资源类型是围绕表示物理属性的单个值来表达的。此类资源类型使用表 11-3 定义的属性指定。表中的"强制性"意味着该属性应被定义为整体资源类型模式的一部分;实际包含属性作为返回或生成有效载荷的一部分,取决于被调用的操作模式。

表 11-3　JSON 模式中资源类型的属性定义

属性名	属性标题	属　性　值	属性值类型	属性值规则	访问模式	是否强制	描　　述
Value	取决于资源	取决于资源	取决于资源	取决于资源	取决于资源	是	资源的当前值
Range	范围	[最小值,最大值]	数组	线性范围	只读	否	输入值的范围,指定为两个元素数组
Step	步长	取决于资源	整数或数字	取决于资源	只读	否	指定范围的步长值
Precision	精度	取决于资源	数字	取决于资源	只读	否	值的精度

对于本质上具有多个物理参数的资源,可以用指定不同物理参数的多个属性替换。值的类型应在资源类型的 RAML 定义中指示,并且应适用于所传递的值。所有属性名称和属性值区分大小写。

3. 基本资源模式

本节定义的所有资源类型表示由 JSON 模式指出。资源类型 RAML 定义嵌入到资源类型特定的模式元素中。可以根据资源类型定义和允许的 CRUDN 操作创建或更新资源。该操作具有不同含义的响应代码,如表 11-4 所示。

表 11-4　响应代码及含义

响应代码	含　　义
200	响应的有效载荷将确认更改,RAML 定义将包含一个模式来定义有效负载
201	有效负载是由于创建操作而由服务器端创建资源的 URL。RAML 定义将包含定义有效负载的模式
204	一切顺利,没有提供有效载荷,RAML 定义不包含模式;RAML 定义可以省略此值,因为它被认为是 OCF 服务器端的默认行为
403	情况 1:在使用查询参数选择特定属性值资源上检索的情况下;如果服务器端不支持提供的值,则应返回此响应。响应有效内容应包含查询参数的允许值 情况 2:由于提供的有效负载问题,服务器端无法创建或更新资源。对于更新,除非在资源类型定义中另有说明,响应有效载荷应包括为 200 定义的相同模式;指示当前资源属性值

11.1.3　示例资源定义

本节包括代表执行器资源的 RAML 示例和执行传感器资源的 RAML 示例。

1. 代表执行器资源的 RAML 示例

```
#%RAML 0.8 title:
OICExampleActuator
version: v1.0
 /ActuatorExample:
description: |
  ResourceActuatorExample description.
  If the ActuatorExample is implemented as the example in the RAML the next values apply:  The name of the
Resource is "ResourceExample Name"
  The Resource Type is "oic.r.actuatorexample"
  The Interface (if) is denoting an Actuator by having the value oic.if.a.
  The unique identification is "actuator_example_id"
  The value of the ActuatorExample is modeled as integer
  The range of the value of ActuatorExample is between 0 and 100
```

```
get:
 description: |
     retrieves the example Resource.
   responses:
         200:
           body:
             application/json:
               schema: |
                 {
                  "id": "http://openinterconnect.org/schemas/oic.r.actuatorexample.json", "$schema":
"http://json-schema.org/draft-04/schema#",
                  "title": "AcutatorExample",
                  "definitions": {
                    "oic.r.actuatorexample: {
                      "type": "object",
                      "properties": {
                        "value": { "type": "string" },
                        "range": {
                           "type": "array",
                          "items": {
                             "type":
   "integer"
                          }
                        }
                      }
                    }
                 . },
                  "type": "object",
                  "allOf": [
                    {"$ref": "oic.core.json#/definitions/oic.core"},
                    {"$ref": oic.baseResource.json#/definitions/oic.r.baseresource"},
                    {"$ref": "#/definitions/oic.r.actuatorexample"}
                  ],
                  "required": ["value"]
               }
              example: |
                {
                  "n":      "ActuatorExample Name",
                  "id":     "actuator_example_id",
                  "rt":     ["oic.r.actuatorexample"],
                  "value":  "0",
                  "range":  ["0,100"]
                }
    post:
      description: |
        sets the Actuator value
        example only updates the value of the Resource
        it does not change the Resource name, although it is allowed to do so.
      body:
          applicaton/json:
          schema: |
              {
                "id": "http://openinterconnect.org/schemas/oic.r.actuatorexample.json",
                "$schema": "http://json-schema.org/draft-04/schema#",
```

```
                "title": "AcutatorExample",
                "definitions": {
                  "oic.r.actuatorexample": {
                    "type": "object",
                    "properties": {
                      "value": { "type": "string" },
                      "range": {
                         "type": "array",
                        "items": {
                          "type": "integer"
                        }
                      }
                    }
                  }
                },
                "type": "object",
                "allOf": [
                  {"$ref": "oic.core.json#/definitions/oic.core"},
                  {"$ref": "oic.baseResource.json#/definitions/oic.r.baseresource"},
                  {"$ref": "#/definitions/oic.r.actuatorexample"}
                ],
                "required": ["value"]
              }
          example: |
              {
                "id":  "actuator_example_id",
               "value" : 5
              }
      responses:
        200:
          body:
            application/json:
              schema: |
                {
                 "id": "http://openinterconnect.org/schemas/oic.r.actuatorexample.json",
                 "$schema": "http://json-schema.org/draft-04/schema#",
                 "title": "AcutatorExample",
                 "definitions": {
                   "oic.r.actuatorexample": {
                     "type": "object",
                     "properties": {
                       "value": { "type": "string" },
                       "range": {
                           "type": "array",
                         "items": {
                        "type":"integer"
                       }
                     }
                   }
                 }
               },
               "type": "object",
               "allOf": [
                 {"$ref": "oic.core.json#/definitions/oic.core"},
```

```
        {"$ref": "oic.baseResource.json#/definitions/oic.r.baseresource"},
        {"$ref": "#/definitions/oic.r.actuatorexample"}
       ],
      "required": ["value"]
     }
    example: |
     {
       "id":  "actuator_example_id",
       "value": 5
     }
  204:
```

2. 指定传感器资源的 RAML 示例

```
#%RAML 0.8
title: OICExampleSensor
version: v1.0
/SensorExample:
  description: |
    SensorExample description.
      If the SensorExample is implemented as the example in the RAML the next values apply:
      The name of the Resource is "ResourceExample_Name"
      The Resource Type is "oic.r.sensorexample"
      The Interface (if) is denoting a Sensor by having the value oic.if.s.
      The unique identification is "sensor_example_id"
      The value of the ResourceSensorExample is modeled as integer
    get:
     description: |
       retrieves the example Resource.
     responses:
       200:
         body:
           application/json:
             schema: |
               {
                "id": "http://openinterconnect.org/schemas/oic.r.sensorexample.json",
                "$schema": "http://json-schema.org/draft-04/schema#",
                "title": "SensorExample",
                "definitions": {
                  "oic.r.sensorexample": {
                    "type": "object",
                    "properties": {
                      "value": { "type": "string" },
                      "range": {
                        "type": "array",
                        "items": {
                          "type": "integer"
                        }
                      }
                    }
                  }
                },
               "type": "object",
               "allOf": [
```

```
        {"$ref": "oic.core.json#/definitions/oic.core"},
        {"$ref": "oic.baseResource.json#/definitions/oic.r.baseresource"},
        {"$ref": "#/definitions/oic.r.sensorexample"},
      ],
      "required": ["value"]
    }
  example: |
    {
      "n":     "SensorExample_Name",
      "rt":    ["oic.r.sensorexample"],
      "id":    "sensor_example_id",
      "value": "3"      }
```

11.1.4 可观察的资源类型

OCF核心规范定义了一种机制，通过该机制，资源可以将自己作为"可观察"通告给OCF客户端。在本部分中定义的所有资源类型可被观察到。通过使用策略链路参数，使资源类型可否观察，是完全依赖于实现的，也就是有条件通知。

所有可观察资源都可以申请一种条件，由于某种观察动作而产生通知，这些条件可以是基于时间的、基于值的或者基于值和时间的。这是由条件通知("oic.r.value.conditional")资源类型和一个可观察资源实例构成的，是由拥有["oic.r.<resource>,oic.r.value.conditional"]的"rt"服务器端暴露出的资源。

1. 条件通知属性概要

表11-5所示的条件通知属性概括了条件通知资源类型的多个属性，资源类型实例中至少要有一个来源于该表格的属性。

表11-5 条件通知属性

名 称	类型	操作	是否强制	描 述
阈值	数值	读写	否	在通知产生之前观察值变化量
最小通知间隔	整型	读写	否	通知发送之前经过的最短时间
最大通知间隔	整型	读写	否	通知发送之前经过的最长时间

所有属性一旦被暴露，则必须被设置初始值。所有属性都可以0值暴露，这表明该属性的功能没有被激活。任意客户端可以将被暴露的值更新至任意访问控制列表，这种改变是全局的，而且会产生通知发送给所有观察者。通知者可以拒绝对属性值的更新，在这种情况下，诊断有效载荷应该包含在拒绝响应中，去指示该值的有效范围。

2. 属性定义：阈值

阈值指两个通知之间的最小值更改。当自上次通知以来的更改大于或等于此值时，应发送通知(在最小通知间隔的限制内)。测量是针对发送最后一个通知中的值进行，因此，所有通知(在可能存在的任何最大通知间隔约束内)将携带至少阈值的值。属性值为0，表示不应用阈值。

3. 属性定义：最小通知间隔

最小通知间隔指在通知之间发生的最短时间(以毫秒为单位)。如果满足值变更条件(阈值等于、超过、不存在则值变化)，到期前不得发送通知，直到期限届满为止。如果属性存在并设置为0，则不运行最小通知间隔计时器；如果属性存在并且值大于0，那么最小通知周期定时器应该等于该值。属性值本身最初由通知程序填充。如果属性不存在,则最小通知期间由通知者决定。每次发送通知时,计时器都将被复位。

4. 属性定义:最大通知间隔

最大通知间隔指在通知之间通知器不得超过的最长时间(以毫秒为单位)。当计时器到期时,应发送通知。如果属性存在并设置为 0,则不运行最大通知间隔计时器;如果属性存在且值大于 0,则最大通知间隔计时器应该等于该值。属性值本身最初由通知器填充。当最小通知间隔和最大通知间隔都存在,并且都不为零时,最大通知间隔的值应大于最小通知间隔。如果属性不存在,该值应由通知器设置。每次发送通知时,计时器都将被复位。

5. 管理状态机

最小通知间隔和最大通知间隔计时器在通知被发送后都会复位。当条件(阈值)和最小通知间隔都满足时,通知将会被发送。如果观察到的属性值随后在最小通知间隔到期之前落在阈值之下,则通知人可能不采取任何行动,或者在包含当前观察到的属性值(在通知时)的最小通知间隔到期时,可以发送通知。如果没有计时器限制,那么只要观察到的属性值改变了大于或等于阈值的量,就会发送通知。

总体条件通知逻辑定义如图 11-1 所示,条件通知示例流程提供了一个说明性的顺序,代码表示如下:

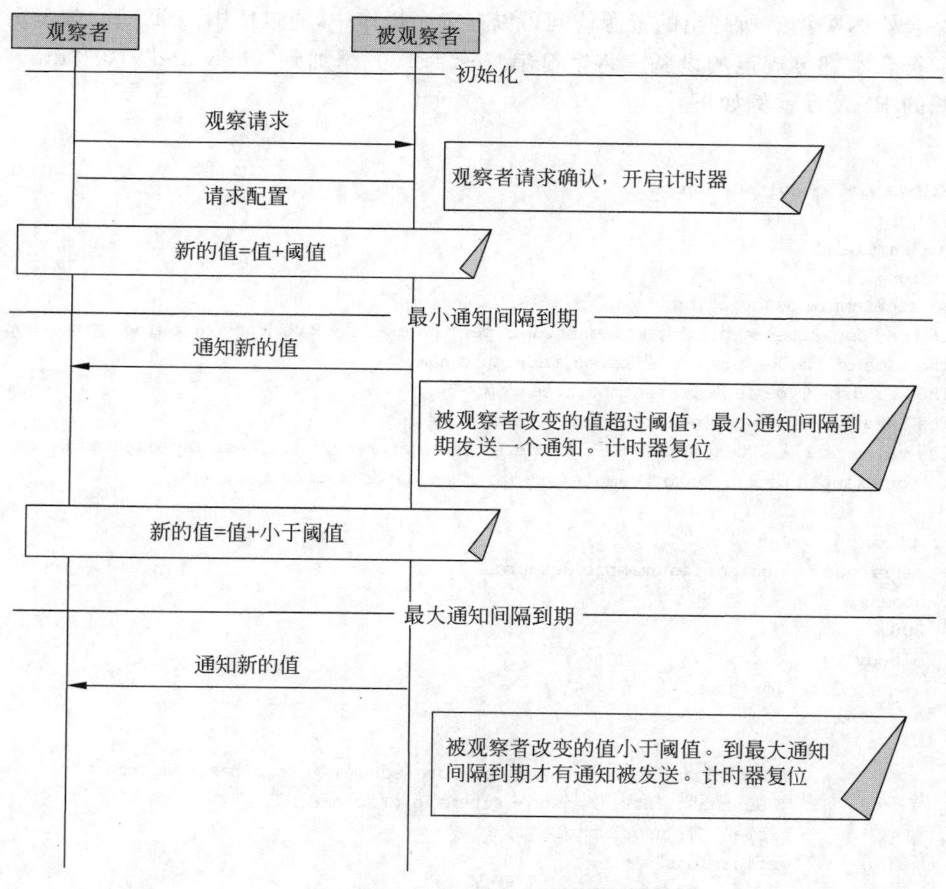

图 11-1 条件通知示例流程

```
If minnotifyperiod expired:
  If observed value changed:
    If change amount >= threshold:
      Send notification with current value
      Reset minnotifyperiod , maxnotifyperiod
If maxnotifyperiod expired:
  Get current value
  Send notification with current value
  Reset minnotifyperiod, maxnotifyperiod
```

11.1.5 复合资源类型

复合资源类型是由一个或多个单一或其他复合资源类型组成的资源，如下文的 RAML 定义所示。可以将复合资源类型视为新的单个资源类型。复合资源型机制是一个强大的概念，因为它使用现有的资源类型形成一个新的组合，表达更多的上下文资源，而不指定新的单一资源类型。

通过引用现有资源值链接到集合的方法，定义复合资源类型。链接是通过使用“Links”的数组来完成的；有关更多详细信息，请参阅核心规范（见第 3 章资源表示）。请注意，下面列出的示例，包含定义部分的模式，仅用于描述目的。数组的属性名称是链接。关系型应为包含，表示该复合资源含有其他资源类型构成综合资源类型。所列出的资源访问可以在单个操作中，通过使用 OCF 核心规范定义“oic.if.ll”接口来实现。符合本部分规范的设备应将字符串"res.1.1.0"添加到“oic.wk.d”中的“dmv”属性中。

复合资源的 RAML 示例如下：

```
#%RAML 0.8
title: OICExampleCompositeResource
version: v1.0
/CompositeExample:
  description: |
    CompositeExample description.
      If the CompositeExample is implemented as per the example RAML the following values apply:
      The name of the Resource is "CompositeExample Name"
      The Resource Type is "oic.r.compositeexample"
      The Interface (if) can denote Sensor or Actuator
      The value of the ActuatorExample is modeled as 2 references to other implemented Resources
      In the example oic.r.SensorExample and oic.r.ActuatorExample are used.
  get:
    description: |
        retrieves the composite example Resource.
      responses:
        200:
          body:
            application/json:
              schema: |
                {
                  "id": "http://openinterconnect.org/schemas/oic.r.baseResource#",
                  "$schema": "http://json-schema.org/schema#",
                  "title": "SensorExample",
                  "definitions": {
                    "oic.r.compositeexample": {
                      "type": "object",
                      "properties": {
```

```
                "links": {
                    "type": "array",
                    "items": {
                        "$ref": "oic.oic-link-schema.json#"
                    }
                }
            }
        }
    },
    "type": "object",
    "allOf": [
        {"$ref": "oic.core.json#/definitions/oic.core"},
        {"$ref": "oic.baseResource.json#/definitions/oic.r.baseresource"},
        {"$ref": "#/definitions/oic.r.compositeexample"}
    ],
    "required": ["n","id","links"]
}
example: |
    {
        "n": "CompositeExample Name",
        "id": "composite_example_id",
        "links": [
            {
                "href": "/my_1st_reference",
                "rel": "contains",
                "rt": ["oic.r.actuatorexample"],
                "if": ["oic.if.a"]
            },
            {
                "href": "/my_2nd_reference",
                "rel": "contains",
                "rt": ["oic.r.sensorexample"],
                "if": ["oic.if.s"]
            }
        ]
    }
```

11.1.6 基础资源

基础资源模式是所有其他资源构建的基础。URI 示例为“/BaseResourceSchemaResURI”，资源类型为“oic. baseresource”。

1. RAML 定义

```
#%RAML 0.8
title: OICBaseResourceSchema
version: v1.1.0-20160519
traits:
  - interface-a :
      queryParameters:
        if:
          enum: ["oic.if.a"]
  - interface-baseline :
      queryParameters:
```

```
        if:
          enum: ["oic.if.baseline"]
  /BaseResourceSchemaResURI:
    description: |
    This is the base resource schema in which all other resources defined in this specification build
    get:
      description: |
        retrieves the state of the resource.
      responses :
        200:
          body:
            application/json:
              schema: |
                {
                  "id":
"http://openinterconnect.org/iotdatamodels/schemas/oic.basecorecomposite.json#", "$schema": "http://
json-schema.org/draft-04/schema#",
                  "description" : "Copyright (c) 2016 Open Connectivity Foundation, Inc. All rights reserved.",
                  "title": "Base and Core Composite Resource",
                  "definitions": {
                    "oic.core": {
                      "properties": {
                        "rt": {
                          "type": "array",
                          "items" : [
                            {
                              "type" : "string",
                              "maxLength": 64
                            }
                          ],
                          "minItems" : 1,
                          "description": "ReadOnly, Resource Type"
                        },
                        "if": {
                          "type" : "array",
                          "description": "ReadOnly, The interface set supported by this resource",
                          "items": {
                            "type": "string",
                            "enum" : ["oic.if.baseline", "oic.if.ll", "oic.if.b", "oic.if.lb",
"oic.if.rw", "oic.if.r", "oic.if.a", "oic.if.s"]
                          }
                        },
                        "n": {
                          "type": "string",
                          "description": "ReadOnly, Friendly name of the resource"
                        },
                        "id": {
                          "type": "string",
                          "description": "ReadOnly, Instance ID of this specific resource"
                        }
                      }
                    },
                    "oic.r.baseresource": {
                      "properties": {
```

```
                    "value": {
                      "anyOf": [
                        {"type": "array"},
                        {"type": "string"},
                        {"type": "boolean"},
                        {"type": "integer"},
                        {"type": "number"},
                        {"type": "object"}
                      ],
                      "description": "The value sensed or actuated by this Resource"
                    },
                    "range": {
                      "type": "array",
                      "description": "The valid range for the value Property",
                      "minItems": 2,
                      "maxItems": 2,
                      "items": {
                        "anyOf": [
                          {"type": "number"},
                          {"type": "integer"}
                        ]
                      }
                    }
                  }
                }
              },
              "type": "object",
              "allOf": [
                {"$ref": "#definitions/oic.core"},
                {"$ref": "#/definitions/oic.r.baseresource"}
              ]
            }
          example: |
            {
              "rt" :      ["oic.baseresource"],
              "if":       ["oic.if.baseline"],
              "id":       "unique_example_id",
              "value":    "someValue",
              "range":    [0,100]
            }
  post:
    description: |
      sets the read - write resource properties
    body:
      application/json:
        schema: |
          {
            "id": " http://openinterconnect. org/iotdatamodels/schemas/oic. baseResource. json # ", "
  $ schema": "http://json - schema. org/draft - 04/schema # ",
            "description" : "Copyright (c) 2016 Open Connectivity Foundation, Inc. All rights
 reserved. ",
            "title": "Base Resource",
            "definitions": {
              "oic.r.baseresource": {
```

```
                    "type": "object",
                    "properties": {
                      "value": {
                        "anyOf": [
                          {"type": "array"},
                          {"type": "string"},
                          {"type": "boolean"},
                          {"type": "integer"},
                          {"type": "number"},
                          {"type": "object"}
                        ],
                        "description": "The value sensed or actuated by this Resource"
                      },
                      "range": {
                        "type": "array",
                        "description": "The valid range for the value Property",
                        "minItems": 2,
                        "maxItems": 2,
                        "items": {
                          "anyOf": [
                            {"type": "number"},
                            {"type": "integer"}
                          ]
                        }
                      }
                    }
                  }
                },
                "type": "object",
                "allOf": [
                  {"$ref": "oic.core.json#/definitions/oic.core"},
                  {"$ref": "#/definitions/oic.r.baseresource"}
                ]
              }
          example: |
            {
              "value": "newValue"
            }
      responses :
        200:
          body:
            application/json:
              schema: |
                {
                  "id": "http://openinterconnect.org/iotdatamodels/schemas/oic.baseResource.json#", "
$schema": "http://json-schema.org/draft-04/schema#",
                  "description" : "Copyright (c) 2016 Open Connectivity Foundation, Inc. All rights reserved.",
                  "title": "Base Resource",
                  "definitions": {
                    "oic.r.baseresource": {
                      "type": "object",
                      "properties": {
                        "value": {
                          "anyOf": [
```

```
              {"type": "array"},
              {"type": "string"},
              {"type": "boolean"},
              {"type": "integer"},
              {"type": "number"},
              {"type": "object"}
            ],
            "description": "The value sensed or actuated by this Resource"
          },
          "range": {
            "type": "array",
            "description": "The valid range for the value Property",
            "minItems": 2,
            "maxItems": 2,
            "items": {
              "anyOf": [
                {"type": "number"},
                {"type": "integer"}
              ]
            }
          }
        }
      }
    },
    "type": "object",
    "allOf": [
      {"$ref": "oic.core.json#/definitions/oic.core"},
      {"$ref": "#/definitions/oic.r.baseresource"}
    ]
  }
example: |
  {
    "value": "newValue"
  }
```

2. 属性定义

属性定义如表 11-6 所示。

表 11-6　属性定义

属　性　名	属性值类型	访问模式	说　明
rt	数组		资源类型
maxLength	整型		最大长度
if	数组	只读	该资源支持的接口
n	字符串	只读	资源的友好名称
id	字符串	只读	此特定资源的实例 ID
value	多类型		
range	数组	读写	value 属性的有效范围

3. CRUDN 行为

CRUDN 行为如表 11-7 所示。

表 11-7 CRUDN 行为

资　源	创建	检索	更新	删除	通知
/BaseResourceSchemaResURI		get	post		get

11.2 资源类型定义概述

本节介绍目前 OCF 所有资源类型的定义，如表 11-8 所示，按字母顺序排列的资源类型列出了完整集。

表 11-8 按字母顺序的资源类型列表

友好名称(非正式)	资源类型	注　释
Acceleration Sensor	oic. r. sensor. acceleration	加速传感器
Activity Count	oic. r. sensor. activity. count	行为计数
Air Quality	oic. r. airquality	空气质量
Air Quality Collection	oic. r. airqualitycollection	空气质量集合
Altimeter	oic. r. altimeter	高度计
Atmospheric Pressure	oic. r. sensor. atmosphericpressure	气压
Air Flow	oic. r. airflow	气流
Air Flow Control	oic. r. airflowcontrol	气流控制
Audio Controls	oic. r. audio	音频控制
Auto Focus	oic. r. autofocus	自动对焦
Automatic Document Feeder	oic. r. automaticdocumentfeeder	自动文档输送
Auto White Balance	oic. r. colour. autowhitebalance	自动白平衡
Basic Resource Schema	Not Applicable	基础资源模式
Battery	oic. r. energy. battery	电池
Binary Switch	oic. r. switch. binary	二进制开关
Brightness	oic. r. light. brightness	亮度
Button Switch	oic. r. button	按钮开关
Carbon Dioxide Sensor	oic. r. sensor. carbondioxide	二氧化碳传感器
Carbon Monoxide Sensor	oic. r. sensor. carbonmonoxide	一氧化碳传感器
Clock	oic. r. clock	时钟
Colour Chroma	oic. r. colour. chroma	颜色的色度
Colour RGB	oic. r. colour. rgb	颜色的 RGB
Colour Saturation	oic. r. colour. saturation	颜色的饱和度
Consumable	oic. r. consumable	耗材
Consumable Collection	oic. r. consumablecollection	耗材集合
Contact Sensor	oic. r. sensor. contact	接触传感器
Delay Defrost	oic. r. delaydefrost	除霜
Demand Response Load Control (DRLC)	oic. r. energy. drlc	需求响应负荷控制
Dimming	oic. r. light. dimming	调整亮度
Door	oic. r. door	门
Ecomode	oic. r. ecomode	节能模式

续表

友好名称(非正式)	资源类型	注　释
Energy Consumption	oic. r. energy. consumption	能耗
Energy Overload/Circuit Breaker	oic. r. energy. overload	能耗过载制动
Energy Usage	oic. r. energy. usage	能耗使用
Generic Sensor	oic. r. sensor	通用传感器
Geolocation Sensor	oic. r. sensor. geolocation	位置传感器
Glass Break Sensor	oic. r. sensor. glassbreak	玻璃破碎传感器
Heart Rate Zone Sensor	oic. r. sensor. heart. zone	心率传感器
Heating Zone	oic. r. heatingzone	加热区
Heating Zone Collection	oic. r. heatingzonecollection	加热区集合
Height	oic. r. height	高度
Humidity	oic. r. humidity	湿度
Icemaker	oic. r. icemaker	制冰
Illuminance Sensor	oic. r. sensor. illuminance	照度传感器
Lock	oic. r. lock. status	锁
Lock Code	oic. r. lock. code	锁密码
Magnetic Field Direction	oic. r. sensor. magneticfielddirection	磁场方向
Media	oic. r. media	媒体
Media Source	oic. r. media. source	媒体源
Media Source List	oic. r. mediasourcelist	媒体源列表
Media Source Input	oic. r. media. input	媒体源输入
Media Source Output	oic. r. media. output	媒体源输出
Mode	oic. r. mode	模式
Movement	oic. r. movement. linear	移动
Motion Sensor	oic. r. sensor. motion	运动传感器
Night Mode	oic. r. nightmode	夜间模式
Open Level	oic. r. openlevel	打开程度
Operational State	oic. r. operational. state	操作状态
Pan Tilt Zoom Movement	oic. r. ptz	平移变焦运动
Presence Sensor	oic. r. sensor. presence	出现传感器
Ramp Time	oic. r. light. ramptime	斜升时间
Refrigeration	oic. r. refrigeration	冷冻
Selectable Levels	oic. r. selectablelevels	可选等级
Signal Strength	oic. r. signalstrength	信号强度
Sleep Sensor	oic. r. sensor. sleep	睡眠传感器
Smoke Sensor	oic. r. sensor. smoke	烟雾传感器
Speech Synthesis	oic. r. speech. tts	语音合成
Temperature	oic. r. temperature	温度
Three Axis Sensor	oic. r. sensor. threeaxis	三轴传感器
Time Period	oic. r. time. period	时段
Touch Sensor	oic. r. sensor. touch	接触传感器
UV Radiation	oic. r. sensor. radiation. uv	紫外辐射
Value Conditional	oic. r. value. conditional	条件值
Water Sensor	oic. r. sensor. water	水传感器
Weight	oic. r. weight	重量

所有资源类型应根据 OCF 核心规范，在本书第 3 章创建，所有与资源类型的比较都不区分大小写。所有资源类型都以“oic. r”作为前缀，表示它是 OCF 定义的资源类型，本章将依次介绍每个资源的定义。

11.3 应用资源类型举例

本节以气流为例，对应用资源的定义、操作等进行描述，其他应用资源请参阅 OCF 规范文档，在此不再一一赘述。

气流资源描述与空气流相关的属性。方向是气流的方向性(如果适用)。方向值取决于该单元的功能。速度是表示本机当前速度水平的整数。范围是代表速度值的最小值、最大值的数组。如果没有出现，则默认值为[0,100]。可以存在自动模式，打开的时候，速度由设备自动控制。URI 示例为“/AirFlowResURI”，资源类型为“oic. r. airflow”。

1. RAML 定义

```
#%RAML 0.8
title: OICAirFlow
version: v1.1.0-20160519
traits:
  - interface :
     queryParameters:
       if:
         enum: ["oic.if.a", "oic.if.baseline"]
/AirFlowResURI:
  description: |
    This resource describes the properties associated with air flow.
    The direction is the directionality of the air flow if applicable.
    Direction values are dependent on the capabilities of the unit.
    The speed is an integer representing the current speed level for the unit.
    The range is an array of the min,max values for the speed level.
  is : ['interface']
  get:
    description: |
      Retrieves the current air flow values.
    responses :
      200:
        body:
          application/json:
            schema: |
              {
                "id": "http://openinterconnect.org/iotdatamodels/schemas/oic.r.airFlow.json#",
                "$schema": "http://json-schema.org/draft-04/schema#",
                "description" : ""Copyright (c) 2016 Open Connectivity Foundation, Inc. All rights reserved.",
                "title": "Air Flow",
                "definitions": {
                  "oic.r.airflow": {
                    "type": "object",
                    "properties": {
                      "direction": {
                        "type": "string",
```

```
                        "description": "Directionality of the air flow"
                      },
                      "speed": {
                        "type": "integer",
                        "description": "Current speed level"
                      },
                      "range": {
                        "type": "array",
                        "description": "ReadOnly, Min,max values for the speed level",
                        "items": {
                          "type": "integer"
                        }
                      }
                    }
                  }
                },
                "type": "object",
                "allOf": [
                  {"$ref": "oic.core.json#/definitions/oic.core"},
                  {"$ref": "oic.baseResource.json#/definitions/oic.r.baseresource"},
                  {"$ref": "#/definitions/oic.r.airflow"}
                ],
                "required": ["speed"]
              }
            example: |
              {
                "rt":        ["oic.r.airflow"],
                "id":        "unique_example_id",
                "direction": "left",
                "speed":     5,
                "range":     [1,7]
              }
  post:
    description: |
      Sets the current air flow values.
      Only direction and speed may be set by an update operation.
    body:
      application/json:
        schema: |
          {
            "id": "http://openinterconnect.org/iotdatamodels/schemas/oic.r.airFlow.json#", "$schema": "
http://json-schema.org/draft-04/schema#",
            "description" : "Copyright (c) 2016 Open Connectivity Foundation, Inc. All rights reserved.",
            "title": "Air Flow",
            "definitions": {
              "oic.r.airflow": {
                "type": "object",
                "properties": {
                  "direction": {
                    "type": "string",
                    "description": "Directionality of the air flow"
                  },
                  "speed": {
                    "type": "integer",
```

```
                    "description": "Current speed level"
                  },
                  "range": {
                    "type": "array",
                    "description": "ReadOnly, Min,max values for the speed level",
                    "items": {
                      "type": "integer"
                    }
                  }
                }
              }
            },
            "type": "object",
            "allOf": [
              {"$ref": "oic.core.json#/definitions/oic.core"},
              {"$ref": "oic.baseResource.json#/definitions/oic.r.baseresource"},
              {"$ref": "#/definitions/oic.r.airflow"}
            ],
            "required": ["speed"]
          }
        example: |
          {
            "id":         "unique_example_id",
            "direction": "right",
            "speed":      3
          }
    responses :
      200:
        body:
          application/json:
            schema: |
              {
                "id": "http://openinterconnect.org/iotdatamodels/schemas/oic.r.airFlow.json#", "
$schema": "http://json-schema.org/draft-04/schema#",
                "description" : "Copyright (c) 2016 Open Connectivity Foundation, Inc. All rights reserved.",
                "title": "Air Flow",
                "definitions": {
                  "oic.r.airflow": {
                    "type": "object",
                    "properties": {
                      "direction": {
                        "type": "string",
                        "description": "Directionality of the air flow"
                      },
                      "speed": {
                        "type": "integer",
                        "description": "Current speed level"
                      },
                      "range": {
                        "type": "array",
                        "description": "ReadOnly, Min,max values for the speed level",
                        "items": {
                          "type": "integer"
                        }
```

```
                    }
                  }
                }
              },
              "type": "object",
              "allOf": [
                {"$ref": "oic.core.json#/definitions/oic.core"},
                {"$ref": "oic.baseResource.json#/definitions/oic.r.baseresource"},
                {"$ref": "#/definitions/oic.r.airflow"}
              ],
              "required": ["speed"]
            }
          example: |
            {
              "id":         "unique_example_id",
              "direction":  "right",
              "speed":      3
            }
      403:
        description: |
          This response is generated by the OIC Server when the client sends:
          An update with an invalid property value for direction.
          An update with an out of range property value for speed.
          The server responds with the current resource representation.
        body:
          application/json:
            schema: |
              {
                "id": "http://openinterconnect.org/iotdatamodels/schemas/oic.r.airFlow.json#",
                "$schema": "http://json-schema.org/draft-04/schema#",
                "description" : "Copyright (c) 2016 Open Connectivity Foundation, Inc. All rights
reserved.",
                "title": "Air Flow",
                "definitions": {
                  "oic.r.airflow": {
                    "type": "object",
                    "properties": {
                      "direction": {
                        "type": "string",
                        "description": "Directionality of the air flow"
                      },
                      "speed": {
                        "type": "integer",
                        "description": "Current speed level"
                      },
                      "range": {
                        "type": "array",
                        "description": "ReadOnly, Min,max values for the speed level",
                        "items": {
                          "type": "integer"
                        }
                      }
                    }
                  }
```

```
        },
        "type": "object",
        "allOf": [
          {"$ref": "oic.core.json#/definitions/oic.core"},
          {"$ref": "oic.baseResource.json#/definitions/oic.r.baseresource"},
          {"$ref": "#/definitions/oic.r.airflow"}
        ],
        "required": ["speed"]
      }
    example: |
      {
        "id":         "unique_example_id",
        "direction": "right",
        "speed":      3
      }
```

2. 属性定义

属性定义如表 11-9 所示。

表 11-9 属性定义

属 性 名	属性值类型	是否强制	访问模式	说 明
direction	字符串		读写	空气流的方向性
speed	整型	是	读写	目前的速度等级
range	数组		读写	速度等级的最大、最小值

3. CRUDN 行为

CRUDN 行为如表 11-10 所示。

表 11-10 CRUDN 行为

资 源	创建	检索	更新	删除	通知
/AirFlowResURI		get	post		get

第 12 章 OCF 开发方法及案例

CHAPTER 12

本章主要介绍 OCF 的具体开发方法，包括基于 Mac、Windows、Linux、Android 和 Arduino 开发的方法，从软件工具、编译方法、实例代码到综合案例进行描述。

12.1 基于 Mac 的开发方法

本节主要包括 Mac OSX 环境下的编译方法、APP 实例和实例代码。

12.1.1 Mac OSX 环境下的编译方法

(1) 为了能够在 Mac OSX 环境下编译代码，需要安装 Xcode，其下载地址为 https://developer.apple.com/xcode/downloads/。

(2) 在 Mac 命令行下执行：

```
$ cd <top directory of the project>
$ scons SYS_VERSION = yyy
```

其中，yyy 是 OSX 的版本号，例如 10.10。

(3) 运行编译成功的示例程序，首先要在命令行执行以下指令：

```
export DYLD_LIBRARY_PATH = <iotivity root>/out/darwin/x86_64/release
```

12.1.2 APP 实例

该实例描述基本的服务器端和客户端，展示基本的资源发现过程和资源操作。

(1) 编译程序成功后，分别生成执行文件 ocservercoll 和 occlientcoll，如图 12-1 和图 12-2 所示。

(2) 从图 12-1 中可以看到，服务器端启动后创建了 fan、light 和 room 三个资源。

(3) 客户端启动后，发现服务器端创建的资源和已经存在的资源，并对资源进行操作，如图 12-3 和图 12-4 所示。

12.1.3 实例代码

本节包括两部分代码：服务器端(ocservercoll.cpp)和客户端(occlientcoll.cpp)。

```
SimpleClientServer — ocservercoll -t 1 — 80x24
occlient.o              occlientslow.o          ocservercoll.o
occlientbasicops        ocserver                ocserverslow
occlientbasicops.o      ocserver.o              ocserverslow.o
occlientcoll            ocserverbasicops
liyonghuadeMacBook-Pro:SimpleClientServer liyonghua$ ./ocservercoll
16:22.942358000 INFO: ocservercontainer: Usage : ocservercoll -t <Test Case>
16:22.942392000 INFO: ocservercontainer: Test Case 1 : Create room resource with
 default collection entity handler.
16:22.942396000 INFO: ocservercontainer: Test Case 2 : Create room resource with
 application collection entity handler.
liyonghuadeMacBook-Pro:SimpleClientServer liyonghua$ ./ocservercoll -t 1
18:36.180022000 DEBUG: ocservercontainer: OCServer is starting...
18:36.181041000 INFO: ocservercontainer: Created fan resource with result: OC_ST
ACK_OK
18:36.181051000 INFO: ocservercontainer: Created light resource with result: OC_
STACK_OK
18:36.181057000 INFO: ocservercontainer: Created room resource with result: OC_S
TACK_OK
18:36.181062000 INFO: ocservercontainer: OC Bind Contained Resource to resource:
 OC_STACK_OK
18:36.181065000 INFO: ocservercontainer: OC Bind Contained Resource to resource:
 OC_STACK_OK
18:36.181085000 INFO: ocservercontainer: Entering ocserver main loop...
```

图 12-1　启动 ocservercoll

```
SimpleClientServer — occlientcoll -t 1 -c 0 — 80x24
liyonghuadeMacBook-Pro:SimpleClientServer liyonghua$ ./occlientcoll
17:28.26580000 INFO: occlient: Usage : occlientcoll -t <Test Case> -c <CA connec
tivity Type>
17:28.26615000 INFO: occlient: -c 0 : Default auto-selection
17:28.26618000 INFO: occlient: -c 1 : IP Connectivity Type
17:28.26621000 INFO: occlient: Test Case 1 : Discover Resources && Initiate GET
Request on an available resource using default interface.
17:28.26624000 INFO: occlient: Test Case 2 : Discover Resources && Initiate GET
Request on an available resource using batch interface.
17:28.26626000 INFO: occlient: Test Case 3 : Discover Resources && Initiate GET
Request on an available resource using link list interface.
17:28.26629000 INFO: occlient: Test Case 4 : Discover Resources && Initiate GET
& PUT Request on an available resource using default interface.
17:28.26635000 INFO: occlient: Test Case 5 : Discover Resources && Initiate GET
& PUT Request on an available resource using batch interface.
17:28.26638000 INFO: occlient: Test Case 6 : Discover Resources && Initiate GET
& PUT Request on an available resource using link list interface.
17:28.26657000 INFO: occlient: Test Case 7 : Discover Resources && Initiate GET
Request on an unavailable resource using default interface.
17:28.26662000 INFO: occlient: Test Case 8 : Discover Resources && Initiate GET
Request on an unavailable resource using batch interface.
17:28.26666000 INFO: occlient: Test Case 9 : Discover Resources && Initiate GET
Request on an unavailable resource using link list interface.
liyonghuadeMacBook-Pro:SimpleClientServer liyonghua$ ./occlientcoll -t 1 -c 0
```

图 12-2　启动 occlientcoll

```
SimpleClientServer — occlientcoll -t 1 -c 0 — 80×24
liyonghuadeMacBook-Pro:SimpleClientServer liyonghua$ ./occlientcoll -t 1 -c 0
18:39.69057000 INFO: occlient: Entering occlient main loop...
18:41.74565000 INFO: occlient: Entering discoveryReqCB (Application Layer CB)
18:41.74598000 INFO: occlient: StackResult: OC_STACK_OK
18:41.74603000 INFO: occlient: Callback Context recvd successfully
18:41.74608000 INFO: occlient: Device =============> Discovered @ fe80::3e15:c2f
f:feda:41d8:50701
18:41.74617000 INFO: PayloadLog: Payload Type: Discovery
18:41.74622000 INFO: PayloadLog:        SID: 101dd639-1976-4ba1-a9d4-98e5afc7a7a
9
18:41.74625000 INFO: PayloadLog:        Interface:
18:41.74630000 INFO: PayloadLog:                oic.if.ll
18:41.74635000 INFO: PayloadLog:        Resource #1
18:41.74641000 INFO: PayloadLog:        URI:/oic/sec/doxm
18:41.74644000 INFO: PayloadLog:        Resource Types:
18:41.74647000 INFO: PayloadLog:                oic.r.doxm
18:41.74649000 INFO: PayloadLog:        Interfaces:
18:41.74652000 INFO: PayloadLog:                oic.if.baseline
18:41.74655000 INFO: PayloadLog:        Bitmap: 1
18:41.74658000 INFO: PayloadLog:        Secure?: false
18:41.74661000 INFO: PayloadLog:        Port: 0
18:41.74663000 INFO: PayloadLog:
18:41.74666000 INFO: PayloadLog:        Resource #2
18:41.74669000 INFO: PayloadLog:        URI:/oic/sec/pstat
```

图 12-3 客户端发现资源

```
SimpleClientServer — occlientcoll -t 1 -c 0 — 80×24
Get Query: /a/room?if=oic.if.baseline
18:43.80262000 INFO: occlient: StackResult: OC_STACK_OK
18:43.80306000 INFO: occlient: SEQUENCE NUMBER: 16777216
18:43.80316000 INFO: occlient: Callback Context for Get recvd successfully 1
18:43.80322000 INFO: PayloadLog: Payload Type: Representation
18:43.80333000 INFO: PayloadLog:        Resource #1
18:43.80341000 INFO: PayloadLog:        URI:/a/room
18:43.80367000 INFO: PayloadLog:        Resource Types:
18:43.80371000 INFO: PayloadLog:                core.room
18:43.80375000 INFO: PayloadLog:        Interfaces:
18:43.80379000 INFO: PayloadLog:                oic.if.baseline
18:43.80384000 INFO: PayloadLog:                oic.if.b
18:43.80388000 INFO: PayloadLog:                oic.if.ll
18:43.80391000 INFO: PayloadLog:        Values:
18:43.80395000 INFO: PayloadLog:        Resource #2
18:43.80399000 INFO: PayloadLog:        URI:/a/light
18:43.80401000 INFO: PayloadLog:        Resource Types:
18:43.80404000 INFO: PayloadLog:                core.light
18:43.80407000 INFO: PayloadLog:        Interfaces:
18:43.80409000 INFO: PayloadLog:                oic.if.baseline
18:43.80412000 INFO: PayloadLog:        Values:
18:43.80415000 INFO: PayloadLog:        Resource #3
18:43.80418000 INFO: PayloadLog:        URI:/a/fan
18:43.80420000 INFO: PayloadLog:        Resource Types:
```

图 12-4 客户端通过 GET 操作获取资源信息

1. ocservercoll.cpp

```
#include <stdio.h>
#include <string.h>
#include <string>
#include <stdlib.h>
#include <unistd.h>
#include <signal.h>
#include <pthread.h>
#include <ocstack.h>
#include <logger.h>
#include "ocpayload.h"
const char * getResult(OCStackResult result);
#define TAG PCF("ocservercontainer")
volatile sig_atomic_t gQuitFlag = 0;
int gLightUnderObservation = 0;
void createResources();
typedef struct LIGHTRESOURCE{
    OCResourceHandle handle;
    bool state;
    int power;
} LightResource;
static LightResource light;
char * gLightResourceUri = (char * )"/a/light";
char * gRoomResourceUri = (char * )"/a/room";
char * gFanResourceUri = (char * )"/a/fan";
typedef enum
{
    TEST_INVALID = 0,
    TEST_DEFAULT_COLL_EH,
    TEST_APP_COLL_EH,
    MAX_TESTS
} SERVER_TEST;
void PrintUsage()
{
    OIC_LOG(INFO, TAG, "Usage : ocservercoll -t <Test Case>");
    OIC_LOG(INFO, TAG,
            "Test Case 1 : Create room resource with default collection entity handler.");
    OIC_LOG(INFO, TAG,
            "Test Case 2 : Create room resource with application collection entity handler.");
}
unsigned static int TEST = TEST_INVALID;
static void
PrintReceivedMsgInfo(OCEntityHandlerFlag flag, OCEntityHandlerRequest * ehRequest)
{
    const char * typeOfMessage;
    const char * typeOfMethod;
    switch (flag)
    {
        case OC_REQUEST_FLAG:
            typeOfMessage = "OC_REQUEST_FLAG";
            break;
        case OC_OBSERVE_FLAG:
            typeOfMessage = "OC_OBSERVE_FLAG";
```

```
            break;
        default:
            typeOfMessage = "UNKNOWN";
    }
    if (ehRequest == NULL)
    {
        typeOfMethod = "UNKNOWN";
    }
    else if (ehRequest->method == OC_REST_GET)
    {
        typeOfMethod = "OC_REST_GET";
    }
    else
    {
        typeOfMethod = "OC_REST_PUT";
    }
    OIC_LOG_V(INFO, TAG, "Receiving message type: %s, method %s", typeOfMessage,
            typeOfMethod);
}
//实体处理句柄,只用于不存在的资源
OCEntityHandlerResult
OCDeviceEntityHandlerCb (OCEntityHandlerFlag flag,
        OCEntityHandlerRequest *entityHandlerRequest, char* uri, void* /*callbackParam*/)
{
    OIC_LOG_V(INFO, TAG, "Inside device default entity handler - flags: 0x%x, uri: %s", flag, uri);
    OCEntityHandlerResult ehResult = OC_EH_OK;
    OCEntityHandlerResponse response;
    if (!entityHandlerRequest)
    {
        OIC_LOG(ERROR, TAG, "Invalid request pointer");
        return OC_EH_ERROR;
    }
    if (entityHandlerRequest->resource == NULL)
    {
        OIC_LOG(INFO, TAG, "Received request from client to a non-existing resource");
        ehResult = OC_EH_RESOURCE_NOT_FOUND;
    }
    else
    {
        OIC_LOG_V(INFO, TAG, "Device Handler: Received unsupported request from client %d",
                    entityHandlerRequest->method);
        ehResult = OC_EH_ERROR;
    }
    if (!((ehResult == OC_EH_ERROR) || (ehResult == OC_EH_FORBIDDEN)))
    {
        //生成响应消息.注意,这个需要一些与请求消息有关的信息
        response.requestHandle = entityHandlerRequest->requestHandle;
        response.resourceHandle = entityHandlerRequest->resource;
        response.ehResult = ehResult;
        response.payload = nullptr;
        response.numSendVendorSpecificHeaderOptions = 0;
        memset(response.sendVendorSpecificHeaderOptions,
              0, sizeof response.sendVendorSpecificHeaderOptions)
        //表明该响应消息不在持久的缓存中
```

```
        response.persistentBufferFlag = 0;
        //发送响应消息
        if (OCDoResponse(&response) != OC_STACK_OK)
        {
            OIC_LOG(ERROR, TAG, "Error sending response");
            ehResult = OC_EH_ERROR;
        }
    }
    return ehResult;
}
//Room资源的实体处理句柄
OCEntityHandlerResult OCEntityHandlerRoomCb(OCEntityHandlerFlag flag,
                                OCEntityHandlerRequest * ehRequest,
                                void * /* callback */)
{
    OCEntityHandlerResult ret = OC_EH_OK;
    OCEntityHandlerResponse response;
    OIC_LOG_V(INFO, TAG, "Callback for Room");
    PrintReceivedMsgInfo(flag, ehRequest );
    if(ehRequest && flag == OC_REQUEST_FLAG )
    {
        std::string query = (const char *)ehRequest->query;
        OCRepPayload* payload = OCRepPayloadCreate();
        //请求操作类型是GET
        if(OC_REST_GET == ehRequest->method)
        {
            if(query.find(OC_RSRVD_INTERFACE_DEFAULT) != std::string::npos)
            {
                OCRepPayloadSetUri(payload, gRoomResourceUri);
                OCRepPayloadSetPropString(payload, "name", "John's Room");
                OCRepPayload *tempPayload = OCRepPayloadCreate();
                OCRepPayloadSetUri(tempPayload, gLightResourceUri);
                OCRepPayloadAppend(payload, tempPayload);
                OCRepPayload *tempPayload2 = OCRepPayloadCreate();
                OCRepPayloadSetUri(tempPayload2, gFanResourceUri);
                OCRepPayloadAppend(payload, tempPayload2);
            }
            else if(query.find(OC_RSRVD_INTERFACE_LL) != std::string::npos)
            {
                OCRepPayloadSetUri(payload, gRoomResourceUri);
                OCRepPayload *tempPayload = OCRepPayloadCreate();
                OCRepPayloadSetUri(tempPayload, gLightResourceUri);
                OCRepPayloadAppend(payload, tempPayload);
                OCRepPayload *tempPayload2 = OCRepPayloadCreate();
                OCRepPayloadSetUri(tempPayload2, gFanResourceUri);
                OCRepPayloadAppend(payload, tempPayload2);
            }
            else if(query.find(OC_RSRVD_INTERFACE_BATCH) != std::string::npos)
            {
                OCRepPayloadSetUri(payload, gRoomResourceUri);
                OCRepPayload *tempPayload = OCRepPayloadCreate();
                OCRepPayloadSetUri(tempPayload, gLightResourceUri);
                OCRepPayloadSetPropBool(tempPayload, "state", false);
                OCRepPayloadSetPropInt(tempPayload, "power", 0);
```

```
        OCRepPayloadAppend(payload, tempPayload);
        OCRepPayload * tempPayload2 = OCRepPayloadCreate();
        OCRepPayloadSetUri(tempPayload2, gFanResourceUri);
        OCRepPayloadSetPropBool(tempPayload2, "state", true);
        OCRepPayloadSetPropInt(tempPayload2, "speed", 10);
        OCRepPayloadAppend(payload, tempPayload2);
    }
    if (ret == OC_EH_OK)
    {
        //生成响应消息.注意,这个需要一些与请求消息有关的信息
        response.requestHandle = ehRequest->requestHandle;
        response.resourceHandle = ehRequest->resource;
        response.ehResult = ret;
        response.payload = reinterpret_cast<OCPayload*>(payload);
        response.numSendVendorSpecificHeaderOptions = 0;
        memset(response.sendVendorSpecificHeaderOptions,
            0, sizeof response.sendVendorSpecificHeaderOptions);
        memset(response.resourceUri, 0, sizeof response.resourceUri);
        //表明该响应消息不在持久的缓存中
        response.persistentBufferFlag = 0;
        //发送响应消息
        if (OCDoResponse(&response) != OC_STACK_OK)
        {
            OIC_LOG(ERROR, TAG, "Error sending response");
            ret = OC_EH_ERROR;
        }
    }
}
//请求操作类型是 PUT
else if(OC_REST_PUT == ehRequest->method)
{
    if(query.find(OC_RSRVD_INTERFACE_DEFAULT) != std::string::npos)
    {
        if(ret != OC_EH_ERROR)
        {
            OCRepPayloadSetUri(payload, gRoomResourceUri);
            OCRepPayloadSetPropString(payload, "name", "John's Room");
        }
    }
    if(query.find(OC_RSRVD_INTERFACE_LL) != std::string::npos)
    {
        if(ret != OC_EH_ERROR)
        {
            OCRepPayloadSetUri(payload, gRoomResourceUri);
        }
        if(ret != OC_EH_ERROR)
        {
            OCRepPayload * tempPayload = OCRepPayloadCreate();
            OCRepPayloadSetUri(tempPayload, gLightResourceUri);
            OCRepPayloadAppend(payload, tempPayload);
        }
        if(ret != OC_EH_ERROR)
        {
            OCRepPayload * tempPayload = OCRepPayloadCreate();
```

```
                OCRepPayloadSetUri(tempPayload, gFanResourceUri);
                OCRepPayloadAppend(payload, tempPayload);
            }
        }
        if(query.find(OC_RSRVD_INTERFACE_BATCH ) != std::string::npos)
        {
            if(ret != OC_EH_ERROR)
            {
                OCRepPayloadSetUri(payload, gRoomResourceUri);
            }
            if(ret != OC_EH_ERROR)
            {
                OCRepPayload * tempPayload = OCRepPayloadCreate();
                OCRepPayloadSetUri(tempPayload, gLightResourceUri);
                OCRepPayloadSetPropBool(tempPayload, "state", true);
                OCRepPayloadSetPropInt(tempPayload, "power", 0);
                OCRepPayloadAppend(payload, tempPayload);
            }
            if(ret != OC_EH_ERROR)
            {
                OCRepPayload * tempPayload = OCRepPayloadCreate();
                OCRepPayloadSetUri(tempPayload, gFanResourceUri);
                OCRepPayloadSetPropBool(tempPayload, "state", false);
                OCRepPayloadSetPropInt(tempPayload, "speed", 0);
                OCRepPayloadAppend(payload, tempPayload);
            }
        }
        if (ret == OC_EH_OK)
        {
            //生成响应消息.注意,这个需要一些与请求消息有关的信息
            response.requestHandle = ehRequest->requestHandle;
            response.resourceHandle = ehRequest->resource;
            response.ehResult = ret;
            response.payload = reinterpret_cast<OCPayload *>(payload);
            response.numSendVendorSpecificHeaderOptions = 0;
            memset(response.sendVendorSpecificHeaderOptions,
                  0, sizeof response.sendVendorSpecificHeaderOptions);
            memset(response.resourceUri, 0, sizeof response.resourceUri);
            //表明该响应消息不在持久的缓存中
            response.persistentBufferFlag = 0;
            //发送响应消息
            if (OCDoResponse(&response) != OC_STACK_OK)
            {
                OIC_LOG(ERROR, TAG, "Error sending response");
                ret = OC_EH_ERROR;
            }
        }
    }
    else
    {
        OIC_LOG_V (INFO, TAG, "Received unsupported method %d from client",
              ehRequest->method);
        OCRepPayloadDestroy(payload);
        ret = OC_EH_ERROR;
```

```
        }
    }
    else if (ehRequest && flag == OC_OBSERVE_FLAG)
    {
        gLightUnderObservation = 1;
    }
    return ret;
}
//Light资源的实体处理句柄
OCEntityHandlerResult OCEntityHandlerLightCb(OCEntityHandlerFlag flag,
    OCEntityHandlerRequest * ehRequest,void * /* callbackParam */)
{
    OCEntityHandlerResult ret = OC_EH_OK;
    OCEntityHandlerResponse response;
    OIC_LOG_V(INFO, TAG, "Callback for Light");
    PrintReceivedMsgInfo(flag, ehRequest );
    if(ehRequest && flag == OC_REQUEST_FLAG)
    {
        OCRepPayload * payload = OCRepPayloadCreate();
        if(OC_REST_GET == ehRequest->method)
        {
            OCRepPayloadSetUri(payload, gLightResourceUri);
            OCRepPayloadSetPropBool(payload, "state", false);
            OCRepPayloadSetPropInt(payload, "power", 0);
        }
        else if(OC_REST_PUT == ehRequest->method)
        {
            OCRepPayloadSetUri(payload, gLightResourceUri);
            OCRepPayloadSetPropBool(payload, "state", true);
            OCRepPayloadSetPropInt(payload, "power", 0);
        }
        else
        {
            OIC_LOG_V (INFO, TAG, "Received unsupported method %d from client",
                ehRequest->method);
            ret = OC_EH_ERROR;
        }
        if (ret == OC_EH_OK)
        {
            //生成响应消息.注意,这个需要一些与请求消息有关的信息
            response.requestHandle = ehRequest->requestHandle;
            response.resourceHandle = ehRequest->resource;
            response.ehResult = ret;
            response.payload = reinterpret_cast<OCPayload *>(payload);
            response.numSendVendorSpecificHeaderOptions = 0;
            memset(response.sendVendorSpecificHeaderOptions,
                0, sizeof response.sendVendorSpecificHeaderOptions);
            memset(response.resourceUri, 0, sizeof response.resourceUri);
            //表明该响应消息不在持久的缓存中
            response.persistentBufferFlag = 0;
            //发送响应消息
            if (OCDoResponse(&response) != OC_STACK_OK)
            {
                OIC_LOG(ERROR, TAG, "Error sending response");
```

```
                ret = OC_EH_ERROR;
            }
        }
        else
        {
            OCRepPayloadDestroy(payload);
        }
    }
    else if (ehRequest && flag == OC_OBSERVE_FLAG)
    {
        gLightUnderObservation = 1;
    }
    return ret;
}
//Fan 资源的实体处理句柄
OCEntityHandlerResult OCEntityHandlerFanCb(OCEntityHandlerFlag flag,
        OCEntityHandlerRequest * ehRequest, void * /* callback */)
{
    OCEntityHandlerResult ret = OC_EH_OK;
    OCEntityHandlerResponse response;
    OIC_LOG_V(INFO, TAG, "Callback for Fan");
    PrintReceivedMsgInfo(flag, ehRequest );
    if(ehRequest && flag == OC_REQUEST_FLAG)
    {
        OCRepPayload * payload = OCRepPayloadCreate();
        if(OC_REST_GET == ehRequest -> method)
        {
            OCRepPayloadSetUri(payload, gFanResourceUri);
            OCRepPayloadSetPropBool(payload, "state", true);
            OCRepPayloadSetPropInt(payload, "speed", 10);
        }
        else if(OC_REST_PUT == ehRequest -> method)
        {
            OCRepPayloadSetUri(payload, gFanResourceUri);
            OCRepPayloadSetPropBool(payload, "state", false);
            OCRepPayloadSetPropInt(payload, "speed", 0);
        }
        else
        {
            OIC_LOG_V (INFO, TAG, "Received unsupported method %d from client",
                    ehRequest -> method);
            ret = OC_EH_ERROR;
        }
        if (ret == OC_EH_OK)
        {
            //生成响应消息.注意,这个需要一些与请求消息有关的信息
            response.requestHandle = ehRequest -> requestHandle;
            response.resourceHandle = ehRequest -> resource;
            response.ehResult = ret;
            response.payload = reinterpret_cast < OCPayload * >(payload);
            response.numSendVendorSpecificHeaderOptions = 0;
            memset(response.sendVendorSpecificHeaderOptions,
                    0, sizeof response.sendVendorSpecificHeaderOptions);
            memset(response.resourceUri, 0, sizeof response.resourceUri);
```

```
            //表明该响应消息不在持久的缓存中
            response.persistentBufferFlag = 0;
            //发送响应消息
            if (OCDoResponse(&response) != OC_STACK_OK)
            {
                OIC_LOG(ERROR, TAG, "Error sending response");
                ret = OC_EH_ERROR;
            }
        }
        OCRepPayloadDestroy(payload);
    }
    else if (ehRequest && flag == OC_OBSERVE_FLAG)
    {
        gLightUnderObservation = 1;
    }
    return ret;
}
/*SIGINT处理程序：将gQuitFlag设置为1,以便正常终止*/
void handleSigInt(int signum)
{
    if (signum == SIGINT)
    {
        gQuitFlag = 1;
    }
}
void *ChangeLightRepresentation (void *param)
{
    (void)param;
    OCStackResult result = OC_STACK_ERROR;
    while (!gQuitFlag)
    {
        sleep(10);
        light.power += 5;
        if (gLightUnderObservation)
        {
            OIC_LOG_V(INFO, TAG,
                " =====>Notifying stack of new power level %d\n", light.power);
            result = OCNotifyAllObservers (light.handle, OC_NA_QOS);
            if (OC_STACK_NO_OBSERVERS == result)
            {
                gLightUnderObservation = 0;
            }
        }
    }
    return NULL;
}
int main(int argc, char* argv[])
{
    pthread_t threadId;
    int opt;
    while ((opt = getopt(argc, argv, "t:")) != -1)
    {
        switch(opt)
        {
```

```
        case 't':
            TEST = atoi(optarg);
            break;
        default:
            PrintUsage();
            return -1;
        }
    }
    if(TEST <= TEST_INVALID || TEST >= MAX_TESTS)
    {
        PrintUsage();
        return -1;
    }
    OIC_LOG(DEBUG, TAG, "OCServer is starting...");
    if (OCInit(NULL, 0, OC_SERVER) != OC_STACK_OK)
    {
        OIC_LOG(ERROR, TAG, "OCStack init error");
        return 0;
    }
    OCSetDefaultDeviceEntityHandler(OCDeviceEntityHandlerCb, NULL);
    /*声明并创建示例资源: light */
    createResources();
    /*创建一个用于更改 light 资源表示的线程 */
    pthread_create (&threadId, NULL, ChangeLightRepresentation, (void *)NULL);
    //使用 Ctrl + C 退出循环
    OIC_LOG(INFO, TAG, "Entering ocserver main loop...");
    signal(SIGINT, handleSigInt);
    while (!gQuitFlag)
    {
        if (OCProcess() != OC_STACK_OK)
        {
            OIC_LOG(ERROR, TAG, "OCStack process error");
            return 0;
        }
        sleep(2);
    }
    /*取消 light 资源的线程并等待它终止 */
    pthread_cancel(threadId);
    pthread_join(threadId, NULL);
    OIC_LOG(INFO, TAG, "Exiting ocserver main loop...");
    if (OCStop() != OC_STACK_OK)
    {
        OIC_LOG(ERROR, TAG, "OCStack process error");
    }
    return 0;
}
void createResources()
{
    light.state = false;
    OCResourceHandle fan;
    OCStackResult res = OCCreateResource(&fan,
        "core.fan",
        OC_RSRVD_INTERFACE_DEFAULT,
        "/a/fan",
```

```
            OCEntityHandlerFanCb,
            NULL,
            OC_DISCOVERABLE|OC_OBSERVABLE);
    OIC_LOG_V(INFO, TAG, "Created fan resource with result: %s", getResult(res));
    OCResourceHandle light;
    res = OCCreateResource(&light,
            "core.light",
            OC_RSRVD_INTERFACE_DEFAULT,
            "/a/light",
            OCEntityHandlerLightCb,
            NULL,
            OC_DISCOVERABLE|OC_OBSERVABLE);
    OIC_LOG_V(INFO, TAG, "Created light resource with result: %s", getResult(res));
    OCResourceHandle room;
    if(TEST == TEST_APP_COLL_EH)
    {
        res = OCCreateResource(&room,
                "core.room",
                OC_RSRVD_INTERFACE_BATCH,
                "/a/room",
                OCEntityHandlerRoomCb,
                NULL,
                OC_DISCOVERABLE);
    }
    else
    {
        res = OCCreateResource(&room,
                "core.room",
                OC_RSRVD_INTERFACE_BATCH,
                "/a/room",
                NULL,
                NULL,
                OC_DISCOVERABLE);
    }
    OIC_LOG_V(INFO, TAG, "Created room resource with result: %s", getResult(res));
    OCBindResourceInterfaceToResource(room, OC_RSRVD_INTERFACE_LL);
    OCBindResourceInterfaceToResource(room, OC_RSRVD_INTERFACE_DEFAULT);
    res = OCBindResource(room, light);
    OIC_LOG_V(INFO, TAG, "OC Bind Contained Resource to resource: %s", getResult(res));
    res = OCBindResource(room, fan);
    OIC_LOG_V(INFO, TAG, "OC Bind Contained Resource to resource: %s", getResult(res));
}
```

2. occlientcoll.cpp

```
#include <stdio.h>
#include <stdlib.h>
#include <string.h>
#include <signal.h>
#include <unistd.h>
#include <ocstack.h>
#include <iostream>
#include <sstream>
#include "ocpayload.h"
```

```
#include "payload_logging.h"
#include "logger.h"
const char * getResult(OCStackResult result);
std::string getQueryStrForGetPut();
#define TAG ("occlient")
#define DEFAULT_CONTEXT_VALUE 0x99
#ifndef MAX_LENGTH_IPv4_ADDR
#define MAX_LENGTH_IPv4_ADDR 16
#endif
typedef enum
{
    TEST_INVALID = 0,
    TEST_GET_DEFAULT,
    TEST_GET_BATCH,
    TEST_GET_LINK_LIST,
    TEST_PUT_DEFAULT,
    TEST_PUT_BATCH,
    TEST_PUT_LINK_LIST,
    TEST_UNKNOWN_RESOURCE_GET_DEFAULT,
    TEST_UNKNOWN_RESOURCE_GET_BATCH,
    TEST_UNKNOWN_RESOURCE_GET_LINK_LIST,
    MAX_TESTS
} CLIENT_TEST;
/* 可以在客户端初始化的连接类型列表,用户输入验证时需要 */
typedef enum {
    CT_ADAPTER_DEFAULT = 0,
    CT_IP,
    MAX_CT
} CLIENT_ConnectivityType_TYPE;
unsigned static int TestType = TEST_INVALID;
unsigned static int ConnectivityType = 0;
typedef struct
{
    char text[30];
    CLIENT_TEST test;
} testToTextMap;
testToTextMap queryInterface[] = {
        {"invalid", TEST_INVALID},
        {"?if=oic.if.baseline", TEST_GET_DEFAULT},
        {"?if=oic.if.b", TEST_GET_BATCH},
        {"?if=oic.if.ll", TEST_GET_LINK_LIST},
        {"?if=oic.if.baseline", TEST_UNKNOWN_RESOURCE_GET_DEFAULT},
        {"?if=oic.if.b", TEST_UNKNOWN_RESOURCE_GET_BATCH},
        {"?if=oic.if.ll", TEST_UNKNOWN_RESOURCE_GET_LINK_LIST},
        {"?if=oic.if.baseline", TEST_PUT_DEFAULT},
        {"?if=oic.if.b", TEST_PUT_BATCH},
        {"?if=oic.if.ll", TEST_PUT_LINK_LIST},
};
//以下变量决定了用于发送单播消息的接口协议(IP协议等).默认设为IP协议
static OCConnectivityType ConnType = CT_ADAPTER_IP;
static const char * RESOURCE_DISCOVERY_QUERY = "/oic/res";
//observe 注册句柄
OCDoHandle gObserveDoHandle;
//超过阈值后,客户端将注销以进一步观察
```

```
int gNumObserveNotifies = 1;
int gQuitFlag = 0;
/* SIGINT 处理器:设置 gQuitFlag 为 1 以便正常终止 */
void handleSigInt(int signum)
{
    if (signum == SIGINT)
    {
        gQuitFlag = 1;
    }
}
//函数声明
OCStackApplicationResult getReqCB(void* ctx, OCDoHandle handle, OCClientResponse * clientResponse);
int InitGetRequestToUnavailableResource(OCClientResponse * clientResponse);
int InitObserveRequest(OCClientResponse * clientResponse);
int InitPutRequest(OCClientResponse * clientResponse);
int InitGetRequest(OCClientResponse * clientResponse);
int InitDiscovery();
OCPayload* putPayload()
{
    OCRepPayload* payload = OCRepPayloadCreate();
    if(!payload)
    {
        std::cout << "Failed to create put payload object"<< std::endl;
        std::exit(1);
    }
    OCRepPayloadSetPropInt(payload, "power", 15);
    OCRepPayloadSetPropBool(payload, "state", true);
    return (OCPayload*) payload;
}
void PrintUsage()
{
    OIC_LOG(INFO, TAG, "Usage : occlientcoll -t <Test Case> -c <CA connectivity Type>");
    OIC_LOG(INFO, TAG, "-c 0 : Default auto-selection");
    OIC_LOG(INFO, TAG, "-c 1 : IP Connectivity Type");
    OIC_LOG(INFO, TAG, "Test Case 1 : Discover Resources && Initiate GET Request on an "\
        "available resource using default interface.");
    OIC_LOG(INFO, TAG, "Test Case 2 : Discover Resources && Initiate GET Request on an "\
            "available resource using batch interface.");
    OIC_LOG(INFO, TAG, "Test Case 3 : Discover Resources && Initiate GET Request on an "\
            "available resource using link list interface.");
    OIC_LOG(INFO, TAG, "Test Case 4 : Discover Resources && Initiate GET & PUT Request on an "\
            "available resource using default interface.");
    OIC_LOG(INFO, TAG, "Test Case 5 : Discover Resources && Initiate GET & PUT Request on an "\
            "available resource using batch interface.");
    OIC_LOG(INFO, TAG, "Test Case 6 : Discover Resources && Initiate GET & PUT Request on an "\
            "available resource using link list interface.");
    OIC_LOG(INFO, TAG, "Test Case 7 : Discover Resources && Initiate GET Request on an "\
            "unavailable resource using default interface.");
    OIC_LOG(INFO, TAG, "Test Case 8 : Discover Resources && Initiate GET Request on an "\
            "unavailable resource using batch interface.");
    OIC_LOG(INFO, TAG, "Test Case 9 : Discover Resources && Initiate GET Request on an "\
            "unavailable resource using link list interface.");
}
OCStackApplicationResult putReqCB(void* ctx, OCDoHandle /* handle */,
```

```
                                OCClientResponse * clientResponse)
{
    if(clientResponse == NULL)
    {
        OIC_LOG(INFO, TAG, "The clientResponse is NULL");
        return OC_STACK_DELETE_TRANSACTION;
    }
    if(ctx == (void * )DEFAULT_CONTEXT_VALUE)
    {
        OIC_LOG_V(INFO, TAG, "Callback Context for PUT query recvd successfully");
        OIC_LOG_PAYLOAD(INFO, clientResponse - > payload);
    }
    return OC_STACK_KEEP_TRANSACTION;
}
OCStackApplicationResult getReqCB(void * ctx, OCDoHandle / * handle * /,
                                OCClientResponse * clientResponse)
{
    OIC_LOG_V(INFO, TAG, "StackResult: % s",
        getResult(clientResponse - > result));
    if(ctx == (void * )DEFAULT_CONTEXT_VALUE)
    {
        OIC_LOG_V(INFO, TAG, "SEQUENCE NUMBER: % d", clientResponse - > sequenceNumber);
        if(clientResponse - > sequenceNumber == 0)
        {
            OIC_LOG_V(INFO, TAG, "Callback Context for GET query recvd successfully");
            OIC_LOG_PAYLOAD(INFO, clientResponse - > payload);
        }
        else
        {
            OIC_LOG_V(INFO, TAG, "Callback Context for Get recvd successfully % d",
                    gNumObserveNotifies);
            OIC_LOG_PAYLOAD(INFO, clientResponse - > payload);;
            gNumObserveNotifies++;
            if (gNumObserveNotifies == 3)
            {
                if (OCCancel (gObserveDoHandle, OC_LOW_QOS, NULL, 0) != OC_STACK_OK)
                {
                    OIC_LOG(ERROR, TAG, "Observe cancel error");
                }
            }
        }
    }
    if(TestType == TEST_PUT_DEFAULT || TestType == TEST_PUT_BATCH || TestType == TEST_PUT_LINK_LIST)
    {
        InitPutRequest(clientResponse);
    }
    return OC_STACK_KEEP_TRANSACTION;
}
//当设备被发现时此函数被调用
OCStackApplicationResult discoveryReqCB(void * ctx, OCDoHandle / * handle * /,OCClientResponse * clientResponse)
{
    OIC_LOG(INFO, TAG,
        "Entering discoveryReqCB (Application Layer CB)");
    OIC_LOG_V(INFO, TAG, "StackResult: % s",
```

```
        getResult(clientResponse->result));
    if (ctx == (void*) DEFAULT_CONTEXT_VALUE)
    {
        OIC_LOG_V(INFO, TAG, "Callback Context recvd successfully");
    }
    OIC_LOG_V(INFO, TAG,
            "Device =============> Discovered @ %s:%d",
            clientResponse->devAddr.addr,
            clientResponse->devAddr.port);
    OIC_LOG_PAYLOAD(INFO, clientResponse->payload);
    ConnType = clientResponse->connType;
    if(TestType == TEST_UNKNOWN_RESOURCE_GET_DEFAULT || TestType == TEST_UNKNOWN_RESOURCE_GET_BATCH ||
TestType == TEST_UNKNOWN_RESOURCE_GET_LINK_LIST)
    {
        InitGetRequestToUnavailableResource(clientResponse);
    }
    else
    {
        InitGetRequest(clientResponse);
    }
    return OC_STACK_KEEP_TRANSACTION;
}
int InitGetRequestToUnavailableResource(OCClientResponse * clientResponse)
{
    OCStackResult ret;
    OCCallbackData cbData;
    std::ostringstream getQuery;
    getQuery << "/SomeUnknownResource";
    cbData.cb = getReqCB;
    cbData.context = (void*)DEFAULT_CONTEXT_VALUE;
    cbData.cd = NULL;
    ret = OCDoResource(NULL, OC_REST_GET, getQuery.str().c_str(),
&clientResponse->devAddr, 0, ConnType, OC_LOW_QOS,
&cbData, NULL, 0);
    if (ret != OC_STACK_OK)
    {
        OIC_LOG(ERROR, TAG, "OCStack error");
    }resource
    return ret;
}
int InitObserveRequest(OCClientResponse * clientResponse)
{
    OCStackResult ret;
    OCCallbackData cbData;
    OCDoHandle handle;
    std::ostringstream obsReg;
    obsReg << getQueryStrForGetPut();
    cbData.cb = getReqCB;
    cbData.context = (void*)DEFAULT_CONTEXT_VALUE;
    cbData.cd = NULL;
    OIC_LOG_V(INFO, TAG, "OBSERVE payload from client =");
    OCPayload* payload = putPayload();
    OIC_LOG_PAYLOAD(INFO, payload);
```

```
    OCPayloadDestroy(payload);
    ret = OCDoResource(&handle, OC_REST_OBSERVE, obsReg.str().c_str(),
&clientResponse->devAddr, 0, ConnType,
                    OC_LOW_QOS, &cbData, NULL, 0);
    if (ret != OC_STACK_OK)
    {
        OIC_LOG(ERROR, TAG, "OCStack resource error");
    }
    else
    {
        gObserveDoHandle = handle;
    }
    return ret;
}
int InitPutRequest(OCClientResponse * clientResponse)
{
    OCStackResult ret;
    OCCallbackData cbData;
    //进行 PUT 查询
    std::ostringstream getQuery;
    getQuery << "coap://" << clientResponse->devAddr.addr << ":" <<
            clientResponse->devAddr.port <<
            "/a/room" << queryInterface[TestType].text;
    cbData.cb = putReqCB;
    cbData.context = (void *)DEFAULT_CONTEXT_VALUE;
    cbData.cd = NULL;
    OIC_LOG_V(INFO, TAG, "PUT payload from client = ");
    OCPayload * payload = putPayload();
    OIC_LOG_PAYLOAD(INFO, payload);
    OCPayloadDestroy(payload);
    ret = OCDoResource(NULL, OC_REST_PUT, getQuery.str().c_str(),
&clientResponse->devAddr, putPayload(), ConnType,
                    OC_LOW_QOS, &cbData, NULL, 0);
    if (ret != OC_STACK_OK)
    {
        OIC_LOG(ERROR, TAG, "OCStack resource error");
    }
    return ret;
}
int InitGetRequest(OCClientResponse * clientResponse)
{
    OCStackResult ret;
    OCCallbackData cbData;
    //进行 GET 查询
    std::ostringstream getQuery;
    getQuery << "/a/room" << queryInterface[TestType].text;
    std::cout << "Get Query: " << getQuery.str() << std::endl;
    cbData.cb = getReqCB;
    cbData.context = (void *)DEFAULT_CONTEXT_VALUE;
    cbData.cd = NULL;
    ret = OCDoResource(NULL, OC_REST_GET, getQuery.str().c_str(),
```

```
&clientResponse->devAddr, 0, ConnType, OC_LOW_QOS,
&cbData, NULL, 0);
    if (ret != OC_STACK_OK)
    {
        OIC_LOG(ERROR, TAG, "OCStack resource error");
    }
    return ret;
}
int InitDiscovery()
{
    OCStackResult ret;
    OCCallbackData cbData;
    /*开始发现查询*/
    char szQueryUri[MAX_QUERY_LENGTH] = { 0 };
    strcpy(szQueryUri, RESOURCE_DISCOVERY_QUERY);
    cbData.cb = discoveryReqCB;
    cbData.context = (void*)DEFAULT_CONTEXT_VALUE;
    cbData.cd = NULL;
    ret = OCDoResource(NULL, OC_REST_DISCOVER, szQueryUri, NULL, 0, ConnType,
                       OC_LOW_QOS,
&cbData, NULL, 0);
    if (ret != OC_STACK_OK)
    {
        OIC_LOG(ERROR, TAG, "OCStack resource error");
    }
    return ret;
}
int main(int argc, char* argv[])
{
    int opt;
    while ((opt = getopt(argc, argv, "t:c:")) != -1)
    {
        switch (opt)
        {
            case 't':
                TestType = atoi(optarg);
                break;
            case 'c':
                ConnectivityType = atoi(optarg);
                break;
            default:
                PrintUsage();
                return -1;
        }
    }
    if ((TestType <= TEST_INVALID || TestType >= MAX_TESTS) ||
        ConnectivityType >= MAX_CT)
    {
        PrintUsage();
        return -1;
    }
```

```
    /* 初始化 OCStack */
    if (OCInit(NULL, 0, OC_CLIENT) != OC_STACK_OK)
    {
        OIC_LOG(ERROR, TAG, "OCStack init error");
        return 0;
    }
    if(ConnectivityType == CT_ADAPTER_DEFAULT || ConnectivityType == CT_IP)
    {
        ConnType = CT_ADAPTER_IP;
    }
    else
    {
        OIC_LOG(INFO, TAG, "Default Connectivity type selected...");
        ConnType = CT_ADAPTER_IP;
    }
    InitDiscovery();
    //使用 Ctrl + C 退出循环
    OIC_LOG(INFO, TAG, "Entering occlient main loop...");
    signal(SIGINT, handleSigInt);
    while (!gQuitFlag)
    {
        if (OCProcess() != OC_STACK_OK)
        {
            OIC_LOG(ERROR, TAG, "OCStack process error");
            return 0;
        }
        sleep(2);
    } OIC_LOG(INFO, TAG, "Exiting occlient main loop...");
    if (OCStop() != OC_STACK_OK)
    {
        OIC_LOG(ERROR, TAG, "OCStack stop error");
    }
    return 0;
}
std::string getQueryStrForGetPut()
{
    return "/a/room";
}
```

12.2 基于 Windows 的开发方法

本节包括软件工具的安装、Windows 环境下的编译方法和 APP 实例。

12.2.1 软件工具的安装

(1). 安装 Visual Studio 2015 或 Visual Studio 2013，支持所有版本，包括免费社区版。注意，确保在安装时勾选“C++Common Tools”选项。该选项在安装界面的 Programming languages 目录下的 Visual C++中能找到。

(2) 从 Web 安装依赖项并添加到 PATH。

① Python2.7（选择"Install just for me"）

② SCons（如果 SCons 不能找到 Python，尝试安装 32 位版本的 Python）

③ 下载 7-Zip。对于"64-bit x64"使用".exe"版本。启动 setup.exe 并确保 7-zip 安装路径存在于环境变量中。

④ 下载 cmake-3.6.0-rc1-win64-x64.msi。确保它的安装路径存在于环境变量中。

以下是为了编译 IoTivity 而添加环境变量的示范指令，假设所有依赖项已经安装到 64 位系统中默认位置。注意，此命令必须从管理员命令提示符运行。

```
setx PATH %PATH%;C:\Python27\;C:\Python27\Scripts;C:\Program Files (x86)\CMake\bin;C:\Program Files\7
-Zip
```

(3) 下载安装 Iotivity 工程 1.2.1 及以上版本。

12.2.2 Windows 环境下的编译方法

(1) 打开命令提示符，进入 Iotivity 工程根目录，并使用"run.bat"方便脚本调用 SCons。

(2) 或者直接调用 SCons。

```
scons TARGET_OS = windows TARGET_ARCH = amd64 \RELEASE = 0 WITH_RA = 0 TARGET_TRANSPORT = IP SECURED = 1 WITH_
TCP = 0 \BUILD_SAMPLE = ON LOGGING = OFF TEST = 1
```

12.2.3 APP 实例

该实例描述基本的服务器端和客户端，展示基本的资源发现过程和资源操作。实例实现同 Linux 系统，详见 Linux 实例代码。

(1) 编译程序成功后，分别执行生成的执行文件 simpleserver.exe 和 simpleclient.exe。可以直接在 Iotivity 的根目录下执行指令：

```
C:\path\to\iotivity\> run server
C:\path\to\iotivity\> run client
```

(2) server 启动后，将创建资源并等待，如图 12-5 所示。

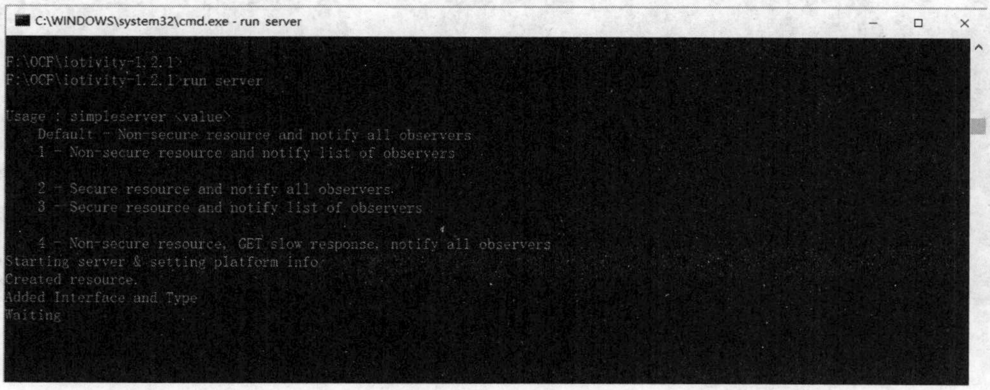

图 12-5 启动 simpleserver

(3) 启动 client 后，将发现 server 创建的资源以及已经存在的资源，如图 12-6 所示。

```
选择C:\WINDOWS\system32\cmd.exe
F:\OCF\iotivity-1.2.1>run client

---------------------------------------------------------------------
Usage : simpleclient <ObserveType> <TransportType>
   ObserveType : 1 - Observe
   ObserveType : 2 - ObserveAll
   TransportType : 1 - IP
   TransportType : 2 - TCP
---------------------------------------------------------------------

<===Setting ObserveType to Observe and ConnectivityType to IP===>

Finding Resource...
Finding Resource for second time...
In foundResource
In foundResource
In foundResource
In foundResource
In foundResource
Found resource 31313131-3131-3131-3131-313131313131/oic/sec/doxm for the first time on server with ID: 31313131-3131-313
1-3131-313131313131
DISCOVERED Resource:
Found resource 31313131-3131-3131-3131-313131313131/a/light for the first time on server with ID: 31313131-3131-3131-313
1-313131313131
        URI of the resource: /oic/sec/doxm
DISCOVERED Resource:
        Host address of the resource: coaps://[fe80::6583:3eed:8a26:e1ff%252]:63154
        URI of the resource: /a/lightFound resource 31313131-3131-3131-3131-313131313131/oic/d
```

图 12-6　启动 simpleclient 并发现资源

(4) client 发现资源后，对 server 的资源进行 GET 操作，如图 12-7 和图 12-8 所示。

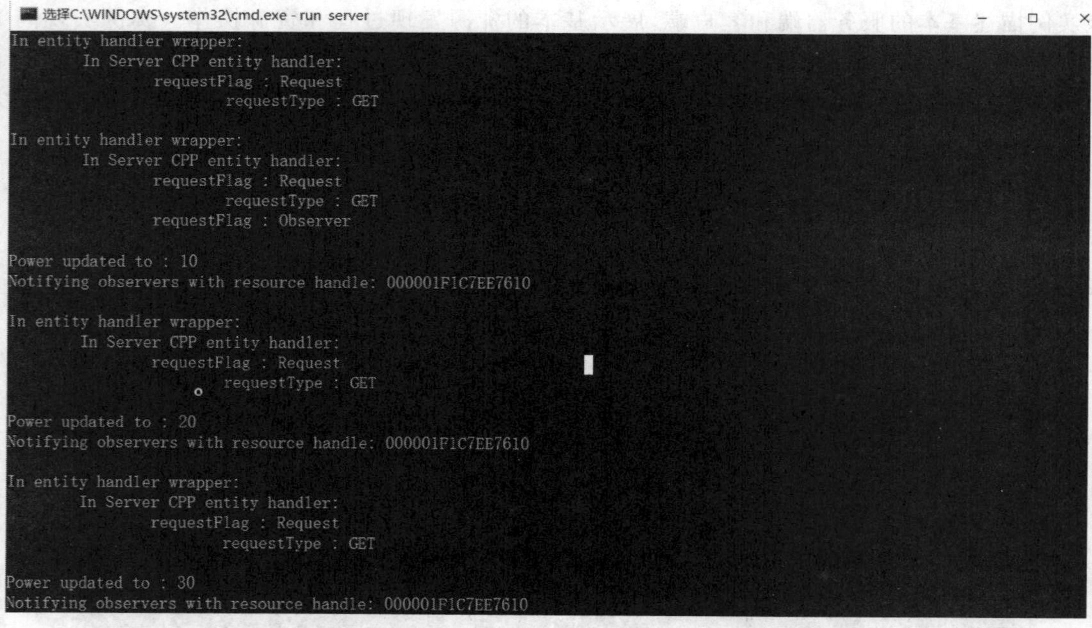

图 12-7　server 对 GET 操作做出响应

选择C:\WINDOWS\system32\cmd.exe

```
OBSERVE RESULT:
        SequenceNumber: 1
        state: false
        power: 10
        name: John's light
OBSERVE RESULT:
        SequenceNumber: 2
        state: false
        power: 20
        name: John's light
OBSERVE RESULT:
        SequenceNumber: 3
        state: false
        power: 30
        name: John's light
OBSERVE RESULT:
        SequenceNumber: 4
        state: false
        power: 40
        name: John's light
OBSERVE RESULT:
        SequenceNumber: 5
        state: false
        power: 50
        name: John's light
OBSERVE RESULT:
        SequenceNumber: 6
        state: false
        power: 60
```

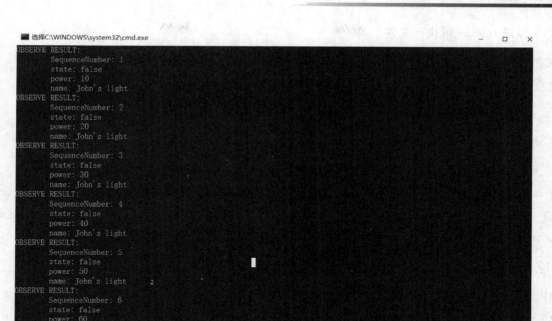

图 12-8　client 进行 GET 操作

12.3　基于 Linux 的开发方法

本节包括软件工具的安装、Linux 环境下的编译方法、APP 实例和代码。

12.3.1　软件工具的安装

软件工具安装步骤如下。

(1) Linux 系统版本：Ubuntu 12.04，64 位(推荐使用 12.04.1 以上的版本，其余版本会存在兼容性问题)。

(2) 源码下载。官方网站下载地址为 https://www.iotivity.org/downloads。

git/gerrit 下载地址为：https://gerrit.iotivity.org。

(3) 搭建工具和库文件。

在“iotivity”根目录下运行以下指令安装：

```
$ sudo apt - get install git - core scons ssh build - essential g++doxygen valgrind
$ sudo apt - get install libboost - dev libboost - program - options - dev libboost - thread - dev uuid - dev
libssl - dev libtool libglib2.0 - dev
sudo apt - get install libboost - all - dev
```

注：Csdk 需要安装 tinycbor，在根目录下运行 git clone https://github.com/01org/tinycbor.git extlibs/tinycbor/tinycbor -b v0.2.1 即可。

12.3.2　Linux 环境下的编译方法

编译方法如下。

(1) 编译 release 可执行文件。

```
$ scons
```

(2) 编译 debug 可执行文件。

```
$ scons RELEASE = false
```

(3) 运行 samples。

```
$ export LD_LIBRARY_PATH = < iotivity >/out/linux/x86_64/release
```

C++ sample 目录：< iotivity >/out/linux/x86_64/release/resource/examples。

C sample 目录：< iotivity >/out/linux/x86_64/release/resource/csdk/stack/samples/linux/SimpleClientServer

12.3.3 APP 实例

该实例描述基本的服务器端和客户端，展示基本的资源发现过程和资源操作。

(1) 程序编译成功后，命令行进入 C++ sample 目录，运行生成的 simpleserver 和 simpleclient，如图 12-9 所示。

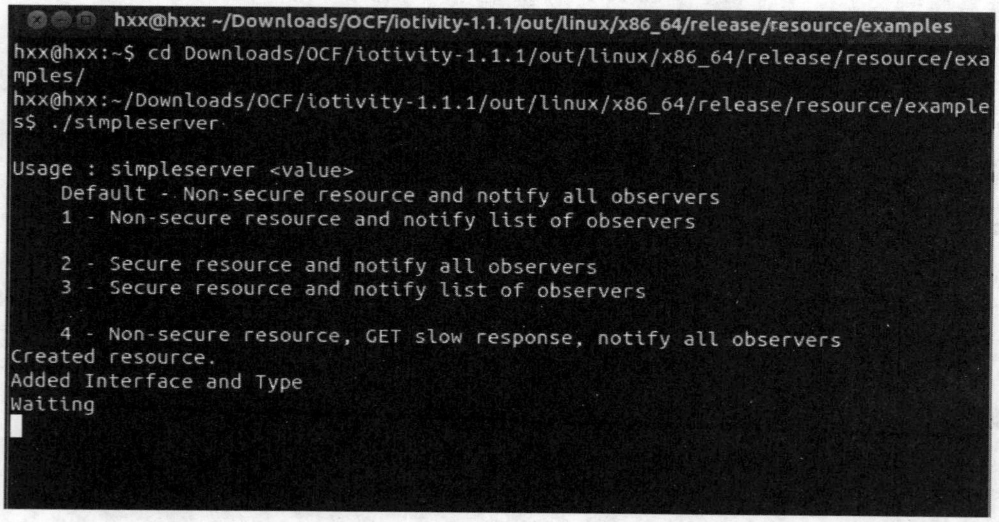

```
hxx@hxx: ~/Downloads/OCF/iotivity-1.1.1/out/linux/x86_64/release/resource/examples
hxx@hxx:~$ cd Downloads/OCF/iotivity-1.1.1/out/linux/x86_64/release/resource/exa
mples/
hxx@hxx:~/Downloads/OCF/iotivity-1.1.1/out/linux/x86_64/release/resource/example
s$ ./simpleserver

Usage : simpleserver <value>
    Default - Non-secure resource and notify all observers
    1 - Non-secure resource and notify list of observers

    2 - Secure resource and notify all observers
    3 - Secure resource and notify list of observers

    4 - Non-secure resource, GET slow response, notify all observers
Created resource.
Added Interface and Type
Waiting
```

图 12-9 启动 simpleserver

(2) 启动 simpleclient 后，将发现 simpleserver 创建的资源以及已经存在的资源，如图 12-10 所示。

(3) simpleclient 发现资源后，对 simpleserver 的资源进行 GET、PUT 和 POST 等操作，如图 12-11 和图 12-12 所示。

12.3.4 实例代码

本节包括 simpleserver.cpp 和 simpleclient.cpp 的代码。

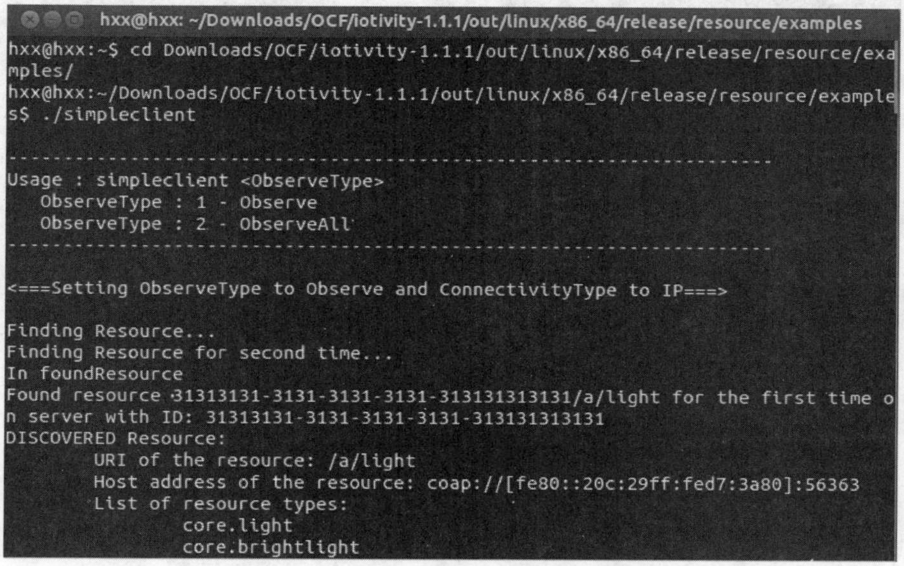

```
hxx@hxx: ~/Downloads/OCF/iotivity-1.1.1/out/linux/x86_64/release/resource/examples
hxx@hxx:~$ cd Downloads/OCF/iotivity-1.1.1/out/linux/x86_64/release/resource/exa
mples/
hxx@hxx:~/Downloads/OCF/iotivity-1.1.1/out/linux/x86_64/release/resource/example
s$ ./simpleclient

---------------------------------------------------------------------
Usage : simpleclient <ObserveType>
   ObserveType : 1 - Observe
   ObserveType : 2 - ObserveAll
---------------------------------------------------------------------

<===Setting ObserveType to Observe and ConnectivityType to IP===>

Finding Resource...
Finding Resource for second time...
In foundResource
Found resource 31313131-3131-3131-3131-313131313131/a/light for the first time o
n server with ID: 31313131-3131-3131-3131-313131313131
DISCOVERED Resource:
        URI of the resource: /a/light
        Host address of the resource: coap://[fe80::20c:29ff:fed7:3a80]:56363
        List of resource types:
                core.light
                core.brightlight
```

图 12-10　启动 simpleclient 并发现资源

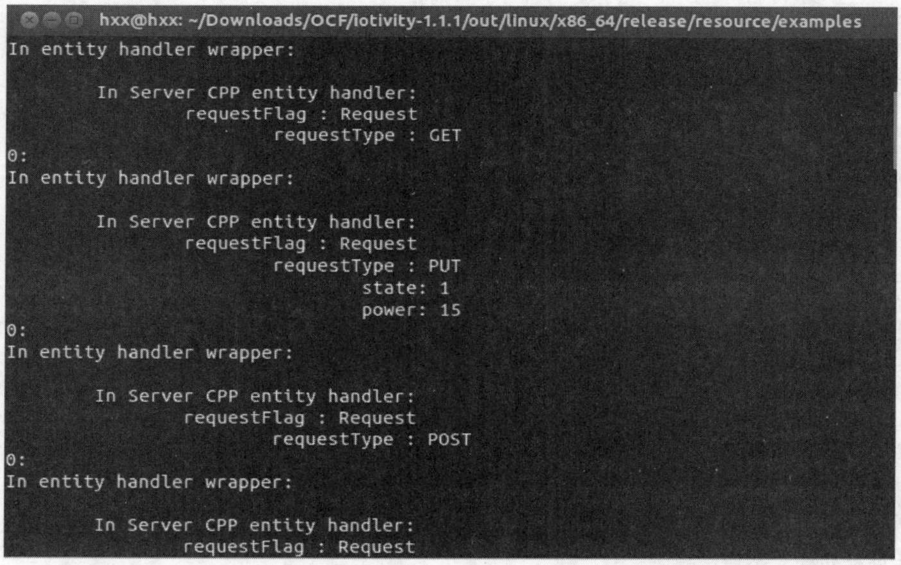

```
hxx@hxx: ~/Downloads/OCF/iotivity-1.1.1/out/linux/x86_64/release/resource/examples
In entity handler wrapper:

        In Server CPP entity handler:
                requestFlag : Request
                        requestType : GET
0:
In entity handler wrapper:

        In Server CPP entity handler:
                requestFlag : Request
                        requestType : PUT
                                state: 1
                                power: 15
0:
In entity handler wrapper:

        In Server CPP entity handler:
                requestFlag : Request
                        requestType : POST
0:
In entity handler wrapper:

        In Server CPP entity handler:
                requestFlag : Request
```

图 12-11　simpleserver 对操作的响应

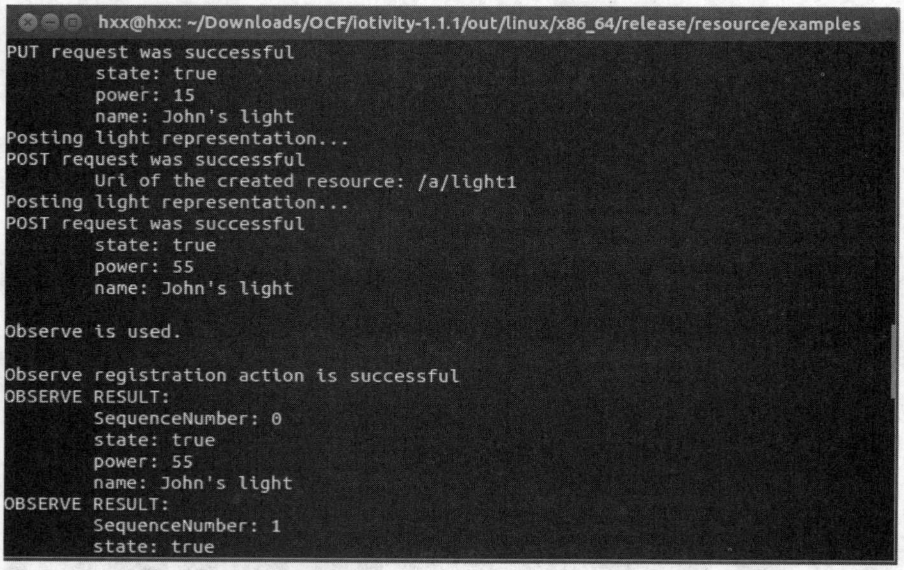

```
hxx@hxx: ~/Downloads/OCF/iotivity-1.1.1/out/linux/x86_64/release/resource/examples
PUT request was successful
        state: true
        power: 15
        name: John's light
Posting light representation...
POST request was successful
        Uri of the created resource: /a/light1
Posting light representation...
POST request was successful
        state: true
        power: 55
        name: John's light

Observe is used.

Observe registration action is successful
OBSERVE RESULT:
        SequenceNumber: 0
        state: true
        power: 55
        name: John's light
OBSERVE RESULT:
        SequenceNumber: 1
        state: true
```

图 12-12　simpleclient 对资源执行操作

1. simpleserver.cpp

```
#include <functional>
#include <pthread.h>
#include <mutex>
#include <condition_variable>
#include "OCPlatform.h"
#include "OCApi.h"
using namespace OC;
using namespace std;
namespace PH = std::placeholders;
static const char * SVR_DB_FILE_NAME = "./oic_svr_db_server.dat";
int gObservation = 0;
void * ChangeLightRepresentation (void * param);
void * handleSlowResponse (void * param, std::shared_ptr<OCResourceRequest> pRequest);
//指定通知所有的 observers 或者部分 observers
//false:通知所有 observers
//true:通知部分 observers
bool isListOfObservers = false;
//指定安全或非安全
//false:非安全资源
//true:安全资源
bool isSecure = false;
//指定实体处理句柄是否做慢响应
bool isSlowResponse = false;
//这个类表示一个名为 lightResource 的资源,该资源有两个名为 state 和 power 的简单属性
class LightResource
{
public:
    //从 TB 客户端访问这些属性
    std::string m_name;
    bool m_state;
```

```
    int m_power;
    std::string m_lightUri;
    OCResourceHandle m_resourceHandle;
    OCRepresentation m_lightRep;
    ObservationIds m_interestedObservers;
public:
    //构造函数
    LightResource()
        :m_name("John's light"), m_state(false), m_power(0), m_lightUri("/a/light"),
              m_resourceHandle(nullptr) {
        m_lightRep.setUri(m_lightUri);
        m_lightRep.setValue("state", m_state);
        m_lightRep.setValue("power", m_power);
        m_lightRep.setValue("name", m_name);
    }
    //注意,这不需要是一个成员函数,对于没有访问的类,可以使用一个自由函数来实现
    //这个函数在内部调用 API 函数 registerResource
    void createResource()
    {
        //资源的 URI
        std::string resourceURI = m_lightUri;
        //资源类型名称.在此是指 light
        std::string resourceTypeName = "core.light";
        //资源接口
        std::string resourceInterface = DEFAULT_INTERFACE;
        //资源属性定义在 ocstack.h
        uint8_t resourceProperty;
        if(isSecure)
        {
            resourceProperty = OC_DISCOVERABLE | OC_OBSERVABLE | OC_SECURE;
        }
        else
        {
            resourceProperty = OC_DISCOVERABLE | OC_OBSERVABLE;
        }
        EntityHandler cb = std::bind(&LightResource::entityHandler, this,PH::_1);
        //此 API 会在内部创建和注册资源
        OCStackResult result = OCPlatform::registerResource(
                                m_resourceHandle, resourceURI, resourceTypeName,
                                resourceInterface, cb, resourceProperty);
        if (OC_STACK_OK != result)
        {
            cout << "Resource creation was unsuccessful\n";
        }
    }
    OCStackResult createResource1()
    {
        //资源的 URI
        std::string resourceURI = "/a/light1";
        //资源类型名称.在此是指 light
        std::string resourceTypeName = "core.light";
        //资源接口
        std::string resourceInterface = DEFAULT_INTERFACE;
        //资源属性定义在 ocstack.h
```

```
    uint8_t resourceProperty;
    if(isSecure)
    {
        resourceProperty = OC_DISCOVERABLE | OC_OBSERVABLE | OC_SECURE;
    }
    else
    {
        resourceProperty = OC_DISCOVERABLE | OC_OBSERVABLE;
    }
    EntityHandler cb = std::bind(&LightResource::entityHandler, this,PH::_1);
    OCResourceHandle resHandle;
    //此 API 会在内部创建和注册资源
    OCStackResult result = OCPlatform::registerResource(
                                resHandle, resourceURI, resourceTypeName,
                                resourceInterface, cb, resourceProperty);
    if (OC_STACK_OK != result)
    {
        cout << "Resource creation was unsuccessful\n";
    }
    return result;
}
OCResourceHandle getHandle()
{
    return m_resourceHandle;
}
//put 资源表示
//从资源表示中获取数据
//以及更新内部状态
void put(OCRepresentation& rep)
{
    try {
        if (rep.getValue("state", m_state))
        {
            cout << "\t\t\t\t" << "state: " << m_state << endl;
        }
        else
        {
            cout << "\t\t\t\t" << "state not found in the representation" << endl;
        }
        if (rep.getValue("power", m_power))
        {
            cout << "\t\t\t\t" << "power: " << m_power << endl;
        }
        else
        {
            cout << "\t\t\t\t" << "power not found in the representation" << endl;
        }
    }
    catch (exception& e)
    {
        cout << e.what() << endl;
    }
}
//post 资源表示
```

```
    //post 可以创建新的资源或者简单地扮演 put 的角色
    //从资源表示中获取数据
    //以及更新内部状态
    OCRepresentation post(OCRepresentation& rep)
    {
        static int first = 1;
        //第一次尝试创建资源时
        if(first)
        {
            first = 0;
            if(OC_STACK_OK == createResource1())
            {
                OCRepresentation rep1;
                rep1.setValue("createduri", std::string("/a/light1"));
                return rep1;
            }
        }
        //从第二次起只执行 put 操作
        put(rep);
        return get();
    }
    //获取更新的资源表示
    //在发送之前更新最新的内部状态的表示
    OCRepresentation get()
    {
        m_lightRep.setValue("state", m_state);
        m_lightRep.setValue("power", m_power);
        return m_lightRep;
    }

    void addType(const std::string& type) const
    {
        OCStackResult result = OCPlatform::bindTypeToResource(m_resourceHandle, type);
        if (OC_STACK_OK != result)
        {
            cout << "Binding TypeName to Resource was unsuccessful\n";
        }
    }
    void addInterface(const std::string& interface) const
    {
        OCStackResult result = OCPlatform::bindInterfaceToResource(m_resourceHandle, interface);
        if (OC_STACK_OK != result)
        {
            cout << "Binding TypeName to Resource was unsuccessful\n";
        }
    }
private:
//这是简单处理句柄的实现
//处理句柄可以人为地以多种方式实现
OCEntityHandlerResult entityHandler(std::shared_ptr<OCResourceRequest> request)
{
    cout << "\tIn Server CPP entity handler:\n";
    OCEntityHandlerResult ehResult = OC_EH_ERROR;
    if(request)
```

```
{
    //获取资源类型和请求标志
    std::string requestType = request->getRequestType();
    int requestFlag = request->getRequestHandlerFlag();
    if(requestFlag & RequestHandlerFlag::RequestFlag)
    {
        cout << "\t\trequestFlag : Request\n";
        auto pResponse = std::make_shared<OC::OCResourceResponse>();
        pResponse->setRequestHandle(request->getRequestHandle());
        pResponse->setResourceHandle(request->getResourceHandle());
        //检查查询参数(如果有的话)
        QueryParamsMap queries = request->getQueryParameters();
        if (!queries.empty())
        {
            std::cout << "\nQuery processing upto entityHandler" << std::endl;
        }
        for (auto it : queries)
        {
            std::cout << "Query key: " << it.first << " value : " << it.second << std:: endl;
        }
        //请求类型是 GET
        if(requestType == "GET")
        {
            cout << "\t\t\trequestType : GET\n";
            if(isSlowResponse)                  //慢响应的情况
            {
                static int startedThread = 0;
                if(!startedThread)
                {
                    std::thread t(handleSlowResponse, (void *)this, request);
                    startedThread = 1;
                    t.detach();
                }
                ehResult = OC_EH_SLOW;
            }
            else                                //一般响应的情况
            {
                pResponse->setErrorCode(200);
                pResponse->setResponseResult(OC_EH_OK);
                pResponse->setResourceRepresentation(get());
                if(OC_STACK_OK == OCPlatform::sendResponse(pResponse))
                {
                    ehResult = OC_EH_OK;
                }
            }
        }
        else if(requestType == "PUT")
        {
            cout << "\t\t\trequestType : PUT\n";
            OCRepresentation rep = request->getResourceRepresentation();
            //执行与 put 请求相关的操作
            //更新 light 资源
            put(rep);
            pResponse->setErrorCode(200);
```

```
            pResponse->setResponseResult(OC_EH_OK);
            pResponse->setResourceRepresentation(get());
            if(OC_STACK_OK == OCPlatform::sendResponse(pResponse))
            {
                ehResult = OC_EH_OK;
            }
        }
        else if(requestType == "POST")
        {
            cout << "\t\t\trequestType : POST\n";
            OCRepresentation rep = request->getResourceRepresentation();
            //执行与 POST 请求相关的操作
            OCRepresentation rep_post = post(rep);
            pResponse->setResourceRepresentation(rep_post);
            pResponse->setErrorCode(200);
            if(rep_post.hasAttribute("createduri"))
            {
                pResponse->setResponseResult(OC_EH_RESOURCE_CREATED); pResponse->setNewResourceUri
(rep_post.getValue<std::string>("createduri"));
            }
            else
            {
                pResponse->setResponseResult(OC_EH_OK);
            }
            if(OC_STACK_OK == OCPlatform::sendResponse(pResponse))
            {
                ehResult = OC_EH_OK;
            }
        }
        else if(requestType == "DELETE")
        {
            cout << "Delete request received" << endl;
        }
    }
    if(requestFlag & RequestHandlerFlag::ObserverFlag)
    {
        ObservationInfo observationInfo = request->getObservationInfo();
        if(ObserveAction::ObserveRegister == observationInfo.action)
        {
            m_interestedObservers.push_back(observationInfo.obsId);
        }
        else if(ObserveAction::ObserveUnregister == observationInfo.action)
        {
            m_interestedObservers.erase(std::remove(m_interestedObservers.begin(),m_
interestedObservers.end(),observationInfo.obsId),m_interestedObservers.end());
        }
        pthread_t threadId;
        cout << "\t\trequestFlag : Observer\n";
        gObservation = 1;
        static int startedThread = 0;
        //在 ChangeLightRepresentation 函数中观察操作发生在不同的线程中
        //如果我们还没创建线程,将在这创建一个
        if(!startedThread)
        {
```

```
                pthread_create (&threadId, NULL, ChangeLightRepresentation, (void * )this);
                startedThread = 1;
            }
            ehResult = OC_EH_OK;
        }
    }
    else
    {
        std::cout << "Request invalid" << std::endl;
    }
    return ehResult;
}
};
//ChangeLightRepresentaion 是观察函数
//用于通过 notifyObservers 向协议栈通知资源的任何变化
void * ChangeLightRepresentation (void * param)
{
    LightResource * lightPtr = (LightResource * ) param;
    //此函数持续监视变化
    while (1)
    {
        sleep (3);
        if (gObservation)
        {
            //如果在观察操作中,light 资源发生任何变化
            //将调用 notifyObservors
            //为了演示,改变功率值并通知变化
            lightPtr->m_power += 10;
            cout << "\nPower updated to : " << lightPtr->m_power << endl;
            cout << "Notifying observers with resource handle: " << lightPtr->getHandle() << endl;
            OCStackResult result = OC_STACK_OK;
            if(isListOfObservers)
            {
                std::shared_ptr<OCResourceResponse> resourceResponse = {std::make_shared
<OCResourceResponse>()};
                resourceResponse->setErrorCode(200);
                resourceResponse->setResourceRepresentation(lightPtr->get(), DEFAULT_INTERFACE);
                result = OCPlatform:: notifyListOfObservers ( lightPtr -> getHandle ( ), lightPtr -> m_
interestedObservers, resourceResponse);
            }
            else
            {
                result = OCPlatform::notifyAllObservers(lightPtr->getHandle());
            }
            if(OC_STACK_NO_OBSERVERS == result)
            {
                cout << "No More observers, stopping notifications" << endl;
                gObservation = 0;
            }
        }
    }
    return NULL;
}
void * handleSlowResponse (void * param, std::shared_ptr<OCResourceRequest> pRequest)
```

```
{
    //此函数处理慢响应的情况
    LightResource * lightPtr = (LightResource * ) param;
    //通过使用 sleep 诱导慢响应的情况
    std::cout << "SLOW response" << std::endl;
    sleep (10);
    auto pResponse = std::make_shared<OC::OCResourceResponse>();
    pResponse->setRequestHandle(pRequest->getRequestHandle());
    pResponse->setResourceHandle(pRequest->getResourceHandle());
    pResponse->setResourceRepresentation(lightPtr->get());
    pResponse->setErrorCode(200);
    pResponse->setResponseResult(OC_EH_OK);
    //设置慢响应标志变回 false
    isSlowResponse = false;
    OCPlatform::sendResponse(pResponse);
    return NULL;
}
void PrintUsage()
{
    std::cout << std::endl;
    std::cout << "Usage : simpleserver <value>\n";
    std::cout << " Default - Non-secure resource and notify all observers\n";
    std::cout << " 1 - Non-secure resource and notify list of observers\n\n";
    std::cout << " 2 - Secure resource and notify all observers\n";
    std::cout << " 3 - Secure resource and notify list of observers\n\n";
    std::cout << " 4 - Non-secure resource, GET slow response, notify all observers\n";
}
static FILE * client_open(const char * /* path */, const char * mode)
{
    return fopen(SVR_DB_FILE_NAME, mode);
}
int main(int argc, char * argv[])
{
    PrintUsage();
    OCPersistentStorage ps {client_open, fread, fwrite, fclose, unlink };
    if (argc == 1)
    {
        isListOfObservers = false;
        isSecure = false;
    }
    else if (argc == 2)
    {
        int value = atoi(argv[1]);
        switch (value)
        {
            case 1:
                isListOfObservers = true;
                isSecure = false;
                break;
            case 2:
                isListOfObservers = false;
                isSecure = true;
                break;
            case 3:
```

```
                isListOfObservers = true;
                isSecure = true;
                break;
            case 4:
                isSlowResponse = true;
                break;
            default:
                break;
        }
    }
    else
    {
        return -1;
    }
    //创建 PlatformConfig 对象
    PlatformConfig cfg {
        OC::ServiceType::InProc,
        OC::ModeType::Server,
        "0.0.0.0",                                  //通过设置为"0.0.0.0"来绑定所有可用的接口
        0,                                          //使用随机的可用端口
        OC::QualityOfService::LowQos,
&ps
    };
    OCPlatform::Configure(cfg);
    try
    {
        //创建资源类的实例(在此是'LightResource'类的实例)
        LightResource myLight;
        //调用 light 类的 createResource 函数
        myLight.createResource();
        std::cout << "Created resource." << std::endl;
        myLight.addType(std::string("core.brightlight"));
        myLight.addInterface(std::string(LINK_INTERFACE));
        std::cout << "Added Interface and Type" << std::endl;
        //条件变量将释放给定的互斥量,然后执行非密集型块,直到调用 notify
        //在这种情况下,因为我们没有调用 cv.notify
        //这应该是一个非处理器密集型版本的 while(true)
        std::mutex blocker;
        std::condition_variable cv;
        std::unique_lock < std::mutex > lock(blocker);
        std::cout <<"Waiting" << std::endl;
        cv.wait(lock, []{return false;});
    }
    catch(OCException &e)
    {
        std::cout << "OCException in main : " << e.what() << endl;
    }
    //没有明确的调用停止程序
    //当 OCPlatform::destructor 被调用时,将在内部做程序清理
    return 0;
}
```

2. simpleclient.cpp

```
#include <string>
#include <map>
#include <cstdlib>
#include <pthread.h>
#include <mutex>
#include <condition_variable>
#include "OCPlatform.h"
#include "OCApi.h"
#define maxSequenceNumber 0xFFFFFF
using namespace OC;
static const char * SVR_DB_FILE_NAME = "./oic_svr_db_client.dat";
typedef std::map<OCResourceIdentifier, std::shared_ptr<OCResource>> DiscoveredResourceMap;
DiscoveredResourceMap discoveredResources;
std::shared_ptr<OCResource> curResource;
static ObserveType OBSERVE_TYPE_TO_USE = ObserveType::Observe;
std::mutex curResourceLock;
class Light
{
public:
    bool m_state;
    int m_power;
    std::string m_name;
    Light() : m_state(false), m_power(0), m_name("")
    {
    }
};
Light mylight;
int observe_count()
{
    static int oc = 0;
    return ++oc;
}
void onObserve(const HeaderOptions /* headerOptions */, const OCRepresentation& rep, const int& eCode,
const int& sequenceNumber)
{
    try
    {
        if(eCode == OC_STACK_OK && sequenceNumber != maxSequenceNumber + 1)
        {
            if(sequenceNumber == OC_OBSERVE_REGISTER)
            {
                std::cout << "Observe registration action is successful" << std::endl;
            }
            std::cout << "OBSERVE RESULT:"<< std::endl;
            std::cout << "\tSequenceNumber: "<< sequenceNumber << std::endl;
            rep.getValue("state", mylight.m_state);
            rep.getValue("power", mylight.m_power);
            rep.getValue("name", mylight.m_name);
            std::cout << "\tstate: " << mylight.m_state << std::endl;
            std::cout << "\tpower: " << mylight.m_power << std::endl;
            std::cout << "\tname: " << mylight.m_name << std::endl;
            if(observe_count() == 11)
```

```
            {
                std::cout <<"Cancelling Observe..."<< std::endl;
                OCStackResult result = curResource->cancelObserve();
                std::cout << "Cancel result: "<< result << std::endl;
                sleep(10);
                std::cout << "DONE"<< std::endl;
                std::exit(0);
            }
        }
        else
        {
            if(eCode == OC_STACK_OK)
            {
                std::cout << "Observe registration failed or de-registration action failed/succeeded" <<
std::endl;
            }
            else
            {
                std::cout << "onObserve Response error: " << eCode << std::endl;
                std::exit(-1);
            }
        }
    }
    catch(std::exception& e)
    {
        std::cout << "Exception: " << e.what() << " in onObserve" << std::endl;
    }
}
void onPost2(const HeaderOptions& /*headerOptions*/,
        const OCRepresentation& rep, const int eCode)
{
    try
    {
        if(eCode == OC_STACK_OK || eCode == OC_STACK_RESOURCE_CREATED
                || eCode == OC_STACK_RESOURCE_CHANGED)
        {
            std::cout << "POST request was successful" << std::endl;
            if(rep.hasAttribute("createduri"))
            {
                std::cout << "\tUri of the created resource: "
<< rep.getValue<std::string>("createduri") << std::endl;
            }
            else
            {
                rep.getValue("state", mylight.m_state);
                rep.getValue("power", mylight.m_power);
                rep.getValue("name", mylight.m_name);
                std::cout << "\tstate: " << mylight.m_state << std::endl;
                std::cout << "\tpower: " << mylight.m_power << std::endl;
                std::cout << "\tname: " << mylight.m_name << std::endl;
            }
            if (OBSERVE_TYPE_TO_USE == ObserveType::Observe)
                std::cout << std::endl << "Observe is used." << std::endl << std::endl;
            else if (OBSERVE_TYPE_TO_USE == ObserveType::ObserveAll)
```

```
            std::cout << std::endl << "ObserveAll is used." << std::endl << std::endl;
          curResource->observe(OBSERVE_TYPE_TO_USE, QueryParamsMap(), &onObserve);
        }
        else
        {
          std::cout << "onPost2 Response error: " << eCode << std::endl;
          std::exit(-1);
        }
    }
    catch(std::exception& e)
    {
        std::cout << "Exception: " << e.what() << " in onPost2" << std::endl;
    }
}
void onPost(const HeaderOptions& /*headerOptions*/,const OCRepresentation& rep, const int eCode)
{
    try
    {
        if(eCode == OC_STACK_OK || eCode == OC_STACK_RESOURCE_CREATED
                || eCode == OC_STACK_RESOURCE_CHANGED)
        {
            std::cout << "POST request was successful" << std::endl;
            if(rep.hasAttribute("createduri"))
            {
                std::cout << "\tUri of the created resource: "
<< rep.getValue<std::string>("createduri") << std::endl;
            }
            else
            {
                rep.getValue("state", mylight.m_state);
                rep.getValue("power", mylight.m_power);
                rep.getValue("name", mylight.m_name);
                std::cout << "\tstate: " << mylight.m_state << std::endl;
                std::cout << "\tpower: " << mylight.m_power << std::endl;
                std::cout << "\tname: " << mylight.m_name << std::endl;
            }
            OCRepresentation rep2;
            std::cout << "Posting light representation..."<< std::endl;
            mylight.m_state = true;
            mylight.m_power = 55;
            rep2.setValue("state", mylight.m_state);
            rep2.setValue("power", mylight.m_power);
            curResource->post(rep2, QueryParamsMap(), &onPost2);
        }
        else
        {
            std::cout << "onPost Response error: " << eCode << std::endl;
            std::exit(-1);
        }
    }
    catch(std::exception& e)
    {
        std::cout << "Exception: " << e.what() << " in onPost" << std::endl;
    }
```

```
}
//局部函数,用于给资源 put 不同的状态
void postLightRepresentation(std::shared_ptr<OCResource> resource)
{
    if(resource)
    {
        OCRepresentation rep;
        std::cout << "Posting light representation..."<< std::endl;
        mylight.m_state = false;
        mylight.m_power = 105;
        rep.setValue("state", mylight.m_state);
        rep.setValue("power", mylight.m_power);
        //使用 rep,query map 和回调参数调用资源的 POST API
        resource->post(rep, QueryParamsMap(), &onPost);
    }
}
//对 put 请求的回调句柄
void onPut(const HeaderOptions& /* headerOptions */, const OCRepresentation& rep, const int eCode)
{
    try
    {
        if (eCode == OC_STACK_OK || eCode == OC_STACK_RESOURCE_CHANGED)
        {
            std::cout << "PUT request was successful" << std::endl;
            rep.getValue("state", mylight.m_state);
            rep.getValue("power", mylight.m_power);
            rep.getValue("name", mylight.m_name);
            std::cout << "\tstate: " << mylight.m_state << std::endl;
            std::cout << "\tpower: " << mylight.m_power << std::endl;
            std::cout << "\tname: " << mylight.m_name << std::endl;
            postLightRepresentation(curResource);
        }
        else
        {
            std::cout << "onPut Response error: " << eCode << std::endl;
            std::exit(-1);
        }
    }
    catch(std::exception& e)
    {
        std::cout << "Exception: " << e.what() << " in onPut" << std::endl;
    }
}
//局部函数,用于给资源 put 不同的状态
void putLightRepresentation(std::shared_ptr<OCResource> resource)
{
    if(resource)
    {
        OCRepresentation rep;
        std::cout << "Putting light representation..."<< std::endl;
        mylight.m_state = true;
        mylight.m_power = 15;
        rep.setValue("state", mylight.m_state);
        rep.setValue("power", mylight.m_power);
```

```
        //使用 rep,query map 和回调参数调用资源的 POST API
        resource->put(rep, QueryParamsMap(), &onPut);
    }
}
//对 GET 请求的回调句柄
void onGet(const HeaderOptions& /* headerOptions */, const OCRepresentation& rep, const int eCode)
{
    try
    {
        if(eCode == OC_STACK_OK)
        {
            std::cout << "GET request was successful" << std::endl;
            std::cout << "Resource URI: " << rep.getUri() << std::endl;
            rep.getValue("state", mylight.m_state);
            rep.getValue("power", mylight.m_power);
            rep.getValue("name", mylight.m_name);
            std::cout << "\tstate: " << mylight.m_state << std::endl;
            std::cout << "\tpower: " << mylight.m_power << std::endl;
            std::cout << "\tname: " << mylight.m_name << std::endl;
            putLightRepresentation(curResource);
        }
        else
        {
            std::cout << "onGET Response error: " << eCode << std::endl;
            std::exit(-1);
        }
    }
    catch(std::exception& e)
    {
        std::cout << "Exception: " << e.what() << " in onGet" << std::endl;
    }
}
//局部函数,用于获取 light 资源的表示
void getLightRepresentation(std::shared_ptr<OCResource> resource)
{
    if(resource)
    {
        std::cout << "Getting Light Representation..."<< std::endl;
        //使用回调参数调用资源的 GET API
        QueryParamsMap test;
        resource->get(test, &onGet);
    }
}
//回调发现资源
void foundResource(std::shared_ptr<OCResource> resource)
{
    std::cout << "In foundResource\n";
    std::string resourceURI;
    std::string hostAddress;
    try
    {
        {
            std::lock_guard<std::mutex> lock(curResourceLock);
            if(discoveredResources.find(resource->uniqueIdentifier()) == discoveredResources.end())
```

```
                {
                    std::cout << "Found resource " << resource->uniqueIdentifier() <<
                        " for the first time on server with ID: "<< resource->sid()<< std::endl;
                    discoveredResources[resource->uniqueIdentifier()] = resource;
                }
                else
                {
                    std::cout <<"Found resource "<< resource->uniqueIdentifier() << " again!"<< std::endl;
                }
                if(curResource)
                {
                    std::cout << "Found another resource, ignoring"<< std::endl;
                    return;
                }
            }
            //对资源对象执行操作
            if(resource)
            {
                std::cout <<"DISCOVERED Resource:"<< std::endl;
                //获取资源 URI
                resourceURI = resource->uri();
                std::cout << "\tURI of the resource: " << resourceURI << std::endl;
                //获取资源主机地址
                hostAddress = resource->host();
                std::cout << "\tHost address of the resource: " << hostAddress << std::endl;
                //获取资源类型
                std::cout << "\tList of resource types: " << std::endl;
                for(auto &resourceTypes : resource->getResourceTypes())
                {
                    std::cout << "\t\t" << resourceTypes << std::endl;
                }
                //获取资源接口
                std::cout << "\tList of resource interfaces: " << std::endl;
                for(auto &resourceInterfaces : resource->getResourceInterfaces())
                {
                    std::cout << "\t\t" << resourceInterfaces << std::endl;
                }
                if(resourceURI == "/a/light")
                {
                    curResource = resource;
                    //调用一个局部函数,它将在内部调用对资源指针的 GET API
                    getLightRepresentation(resource);
                }
            }
            else
            {
                //资源是无效的
                std::cout << "Resource is invalid" << std::endl;
            }
        }
        catch(std::exception& e)
        {
            std::cerr << "Exception in foundResource: "<< e.what() << std::endl;
        }
```

```
}
void printUsage()
{
    std::cout << std::endl;
    std::cout << "---------------------------------------------------------------\n";
    std::cout << "Usage : simpleclient <ObserveType>" << std::endl;
    std::cout << " ObserveType : 1 - Observe" << std::endl;
    std::cout << " ObserveType : 2 - ObserveAll" << std::endl;
    std::cout << "---------------------------------------------------------------\n\n";
}
void checkObserverValue(int value)
{
    if (value == 1)
    {
        OBSERVE_TYPE_TO_USE = ObserveType::Observe;
        std::cout << "<=== Setting ObserveType to Observe ===>\n\n";
    }.
    else if (value == 2)
    {
        OBSERVE_TYPE_TO_USE = ObserveType::ObserveAll;
        std::cout << "<=== Setting ObserveType to ObserveAll ===>\n\n";
    }
    else
    {
        std::cout << "<=== Invalid ObserveType selected."
<<" Setting ObserveType to Observe ===>\n\n";
    }
}
static FILE * client_open(const char * /* path */, const char * mode)
{
    return fopen(SVR_DB_FILE_NAME, mode);
}
int main(int argc, char* argv[]) {
    std::ostringstream requestURI;
    OCPersistentStorage ps {client_open, fread, fwrite, fclose, unlink };
    try
    {
        printUsage();
        if (argc == 1)
        {
            std::cout << "<=== Setting ObserveType to Observe and ConnectivityType to IP ===>\n\n";
        }
        else if (argc == 2)
        {
            checkObserverValue(std::stoi(argv[1]));
        }
        else
        {
            std::cout << "<=== Invalid number of command line arguments ===>\n\n";
            return -1;
        }
    }
    catch(std::exception& )
    {
```

```
        std::cout << "<=== Invalid input arguments ===>\n\n";
        return -1;
    }
    //创建 PlatformConfig 对象
    PlatformConfig cfg {
        OC::ServiceType::InProc,
        OC::ModeType::Both,
        "0.0.0.0",
        0,
        OC::QualityOfService::HighQos,
&ps
    };
    OCPlatform::Configure(cfg);
    try
    {
        //使得所有布尔值在此流中打印为 true/false
        std::cout.setf(std::ios::boolalpha);
        //找到所有资源
        requestURI << OC_RSRVD_WELL_KNOWN_URI;   //<< "?rt=core.light";
        OCPlatform::findResource("", requestURI.str(),
                CT_DEFAULT, &foundResource);
        std::cout << "Finding Resource... " << std::endl;
        //查找资源完成两次,以便我们第二次发现原始资源
        //这些资源将具有相同的 uniqueidentifier(仍然是不同的对象)
        //以便我们可以在 foundResource(上面)中验证/显示重复检查代码
        OCPlatform::findResource("", requestURI.str(),
                CT_DEFAULT, &foundResource);
        std::cout << "Finding Resource for second time..." << std::endl;
        //条件变量将释放给定的互斥量,然后执行非密集型块,直到调用 notify
        //在这种情况下,因为没有调用 cv.notify
        //这应该是一个非处理器密集型版本的 while(true)
        std::mutex blocker;
        std::condition_variable cv;
        std::unique_lock<std::mutex> lock(blocker);
        cv.wait(lock);
    }catch(OCException& e)
    {
        oclog() << "Exception in main: "<< e.what();
    }
    return 0;
}
```

12.4 基于 Android 的开发方法

Iotivity 的 Android 代码在 Linux 系统下编译,环境配置参考 12.3 节。

12.4.1 软件工具的安装

软件安装步骤如下。

(1) JDK。

```
$ sudo apt-get update
$ sudo apt-get install openjdk-7-jdk
```

（2）Android SDK、NDK 和 Gradle，下载地址如表 12-1 所示。

表 12-1 软件下载地址

软 件	下 载 地 址
Android NDK	http://developer.android.com/tools/sdk/ndk/index.html
Android SDK	http://developer.android.com/sdk/index.html
Gradle	https://services.gradle.org/distributions/gradle-2.2.1-all.zip

（3）配置系统变量。

```
ANDROID_NDK = <path to android ndk>
ANDROID_HOME = <path to android sdk>
GRADLE_HOME = <path to gradle/bin>
```

配置完系统变量之后如图 12-13 所示。

```
export ANDROID_NDK=/home/tanyang/Downloads/OCF/iotivity-1.1.1/extlibs/
android/ndk/android-ndk-r10d
export PATH=$ANDROID_NDK:$PATH

export ANDROID_HOME=/home/tanyang/Downloads/OCF/iotivity-1.1.1/extlibs/
android/sdk/android-sdk-linux
export PATH=$ANDROID_HOME/tools:$ANDROID_HOME/platform-tools:$PATH

export GRADLE_HOME=/home/tanyang/Downloads/OCF/iotivity-1.1.1/extlibs/
android/gradle/gradle-2.2.1
export PATH=$GRADLE_HOME/bin:$PATH
```

图 12-13 配置变量后

12.4.2 Android 环境下的编译方法

编译方法如下。

（1）编译指令。

```
scons TARGET_OS = android TARGET_ARCH = <target arch> TARGET_TRANSPORT = <target transport> RELEASE = <release mode> SECURED = <secure> ANDROID_HOME = <path to android SDK> ANDROID_NDK = <path to android NDK> GRADLE_HOME = <path to gradle/bin>
```

（2）编译选项。

编译选项如表 12-2 所示。

表 12-2 编译选项

支持的架构	x86（默认）
	x86_64
	armeabi
	armeabi-v7a
支持的传输类型	所有（默认）
	IP
	蓝牙
	低功耗蓝牙

续表

支持的发布模式	1（发布模式，默认）
	0（调试模式）
是否 DTLS（数据传输层安全）	0（否，默认）
	1（是）

IoTivity Base Android API（. aar file）输出目录：< iotivity >/android/android_api/base/build/outputs/aar/iotivity-base-< your arch >-< release mode >. aar。

(3) 运行 samples。

sample APK 目录：< iotivity >/android/examples/< example name >/build/outputs/apk/< example name >-< release mode >. apk。

下载 Android APP 至 Android 终端：$ adb install < iotivity >/android/examples/< example_name >/build/outputs/apk/< apk name >。

12.4.3 APP 实例

APP 的运行步骤如下。

(1) simpleclient 和 simpleserver 的应用程序界面如图 12-14 所示。

(2) 点击 simpleserver 应用程序中的 START 按钮，应用程序会创建一个 light 虚拟资源，如图 12-15 所示。

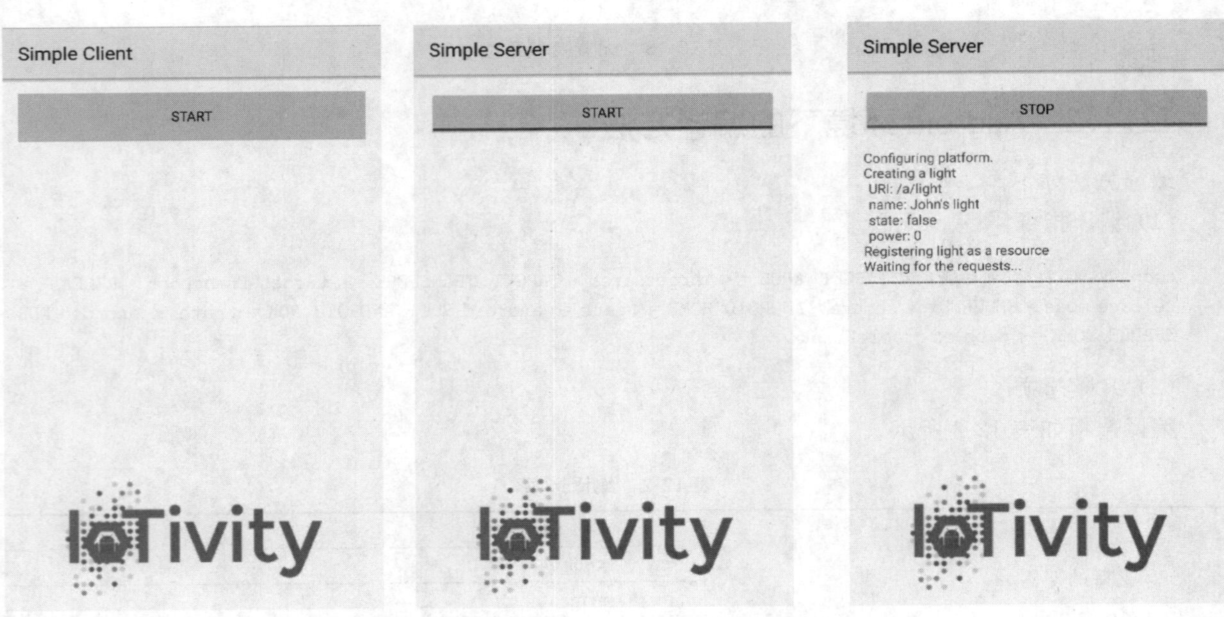

图 12-14 应用程序界面

图 12-15 资源创建

(3) 点击 simpleclient 应用程序中的 START 按钮，可以看到，客户端发现了服务器端创建的资源并将其表示打印了出来，如图 12-16 所示。

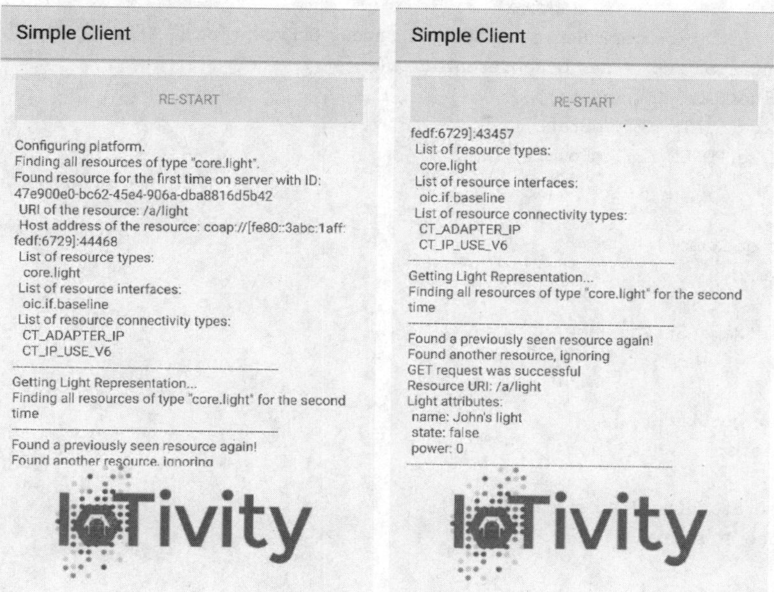

图 12-16　资源创建完成

12.4.4　实例代码

Android 端的示例以 simpleclient 和 simpleserver 为例。这两个示例位于 ioticity 工程的 android/examples 目录下。

1. simpleclient

该工程中主要包含两个 Java 文件，分别为 Light.java 和 SimpleClient.java。

1) Light.java 文件中的代码

```
package org.iotivity.base.examples;
import org.iotivity.base.OcException;
import org.iotivity.base.OcRepresentation;
/* Light,该类创建了一个远程灯资源的对象表示,并根据服务器端的响应更新相应的值 */
public class Light {
   public static final String NAME_KEY = "name";
   public static final String STATE_KEY = "state";
   public static final String POWER_KEY = "power";
   private String mName;
   private boolean mState;
   private int mPower;
   public Light() {
      mName = "";
      mState = false;
      mPower = 0;
   }
   public void setOcRepresentation(OcRepresentation rep) throws OcException {
      mName = rep.getValue(NAME_KEY);
      mState = rep.getValue(Light.STATE_KEY);
      mPower = rep.getValue(Light.POWER_KEY);
```

```
    }
    public OcRepresentation getOcRepresentation() throws OcException {
        OcRepresentation rep = new OcRepresentation();
        rep.setValue(NAME_KEY, mName);
        rep.setValue(STATE_KEY, mState);
        rep.setValue(POWER_KEY, mPower);
        return rep;
    }
    public String getName() {
        return mName;
    }
    public void setName(String name) {
        this.mName = mName;
    }
    public boolean getState() {
        return mState;
    }
    public void setState(boolean state) {
        this.mState = state;
    }
    public int getPower() {
        return mPower;
    }
    public void setPower(int power) {
        this.mPower = power;
    }
    @Override
    public String toString() {
        return "\t" + NAME_KEY + ": " + mName +
                "\n\t" + STATE_KEY + ": " + mState +
                "\n\t" + POWER_KEY + ": " + mPower;
    }
}
```

2）SimpleClient.java 文件中的代码

```
package org.iotivity.base.examples;
import android.app.Activity;
import android.content.Context;
import android.content.Intent;
import android.nfc.NfcAdapter;
import android.os.Bundle;
import android.text.method.ScrollingMovementMethod;
import android.util.Log;
import android.view.View;
import android.widget.Button;
import android.widget.ScrollView;
import android.widget.TextView;
import org.iotivity.base.ErrorCode;
import org.iotivity.base.ModeType;
import org.iotivity.base.ObserveType;
import org.iotivity.base.OcConnectivityType;
import org.iotivity.base.OcException;
import org.iotivity.base.OcHeaderOption;
```

```
import org.iotivity.base.OcPlatform;
import org.iotivity.base.OcRepresentation;
import org.iotivity.base.OcResource;
import org.iotivity.base.OcResourceIdentifier;
import org.iotivity.base.PlatformConfig;
import org.iotivity.base.QualityOfService;
import org.iotivity.base.ServiceType;
import java.util.EnumSet;
import java.util.HashMap;
import java.util.List;
import java.util.Map;
/* SimpleClient 是一个简单的客户端程序,在 simpleServer 开启之后,该程序才会被启动.程序发现有客户端广
播的资源并且对所发现的资源进行不同的操作 (GET, PUT, POST, DELETE 和 OBSERVE) */
public class SimpleClient extends Activity implements
      OcPlatform.OnResourceFoundListener,
      OcResource.OnGetListener,
      OcResource.OnPutListener,
      OcResource.OnPostListener,
      OcResource.OnObserveListener {
   private Map<OcResourceIdentifier, OcResource> mFoundResources = new HashMap<>();
   private OcResource mFoundLightResource = null;
   //local representation of a server's light resource
   private Light mLight = new Light();
   /* 一个本地方法,配置和初始化平台,然后寻找 light 资源 */
   private void startSimpleClient() {
      Context context = this;
      PlatformConfig platformConfig = new PlatformConfig(
            this,
            context,
            ServiceType.IN_PROC,
            ModeType.CLIENT,
            "0.0.0.0",                          //设置"0.0.0.0",绑定可用接口
            0,                                  //使用随机可用端口
            QualityOfService.LOW
      );
      msg("Configuring platform.");
      OcPlatform.Configure(platformConfig);
      try {
         msg("Finding all resources of type \"core.light\".");
         String requestUri = OcPlatform.WELL_KNOWN_QUERY + "?rt=core.light";
         OcPlatform.findResource("",
               requestUri,
               EnumSet.of(OcConnectivityType.CT_DEFAULT),
               this
         );
         sleep(1);
         /* 寻找资源被执行了两次,因此会发现两次资源.这些资源会有相同的唯一标识符,但是不同的对象,
因此,可以在 foundResource 中证实或展现重复的检查代码 */
         msg("Finding all resources of type \"core.light\" for the second time");
         OcPlatform.findResource("",
               requestUri,
               EnumSet.of(OcConnectivityType.CT_DEFAULT),
               this
         );
```

```
    } catch (OcException e) {
        Log.e(TAG, e.toString());
        msg("Failed to invoke find resource API");
    }
    printLine();
}
/*当一个"findResource"请求成功之后,会执行一个事件处理器@param ocResource found resource*/
@Override
public synchronized void onResourceFound(OcResource ocResource) {
    if (null == ocResource) {
        msg("Found resource is invalid");
        return;
    }
    if (mFoundResources.containsKey(ocResource.getUniqueIdentifier())) {
        msg("Found a previously seen resource again!");
    } else {
        msg("Found resource for the first time on server with ID: " + ocResource.getServerId());
        mFoundResources.put(ocResource.getUniqueIdentifier(), ocResource);
    }
    if (null != mFoundLightResource) {
        msg("Found another resource, ignoring");
        return;
    }
    //得到资源的 URI
    String resourceUri = ocResource.getUri();
    //Get the resource host address
    String hostAddress = ocResource.getHost();
    msg("\tURI of the resource: " + resourceUri);
    msg("\tHost address of the resource: " + hostAddress);
    //得到资源的类型
    msg("\tList of resource types: ");
    for (String resourceType : ocResource.getResourceTypes()) {
        msg("\t\t" + resourceType);
    }
    msg("\tList of resource interfaces:");
    for (String resourceInterface : ocResource.getResourceInterfaces()) {
        msg("\t\t" + resourceInterface);
    }
    msg("\tList of resource connectivity types:");
    for (OcConnectivityType connectivityType : ocResource.getConnectivityTypeSet()) {
        msg("\t\t" + connectivityType);
    }
    printLine();
    //在这个例子中只有 light 资源
    if (resourceUri.equals("/a/light")) {
    //将资源引用分配给全局变量使之始终有效
    //超出范围时被 GC 破坏
    mFoundLightResource = ocResource;
    //调用一个本地方法,但是该方法会在内部调用"get"API,用于 foundLightResource
    getLightResourceRepresentation();
    }
}
/*该方法用于获取发现的 light 资源的表示*/
private void getLightResourceRepresentation() {
```

```
    msg("Getting Light Representation...");
    Map<String, String> queryParams = new HashMap<>();
    try {
        //使用 OcResource.OnGetListener 事件调用资源的"get"API。监听器的实现
        sleep(1);
        mFoundLightResource.get(queryParams, this);
    } catch (OcException e) {
        Log.e(TAG, e.toString());
        msg("Error occurred while invoking \"get\" API");
    }
}
/**
 * 当一个"get"请求完成之后,会执行一个事件监听器.
 * @param list              头选项列表
 * @param ocRepresentation 资源的表示
 */
@Override
public synchronized void onGetCompleted(List<OcHeaderOption> list,
                                        OcRepresentation ocRepresentation) {
    msg("GET request was successful");
    msg("Resource URI: " + ocRepresentation.getUri());
    try {
        //将属性值读到一个 light 的本地表示中
        mLight.setOcRepresentation(ocRepresentation);
    } catch (OcException e) {
        Log.e(TAG, e.toString());
        msg("Failed to read the attributes of a light resource");
    }
    msg("Light attributes: ");
    msg(mLight.toString());
    printLine();
    //调用一个本地方法,但是该方法会在内部调用"put"API,用于//foundLightResource.
    putLightRepresentation();
}
/**
 * 当一个"get"请求失败,会执行下面的事件处理器
 * @param throwable exception
 */
@Override
public synchronized void onGetFailed(Throwable throwable) {
    if (throwable instanceof OcException) {
        OcException ocEx = (OcException) throwable;
        Log.e(TAG, ocEx.toString());
        ErrorCode errCode = ocEx.getErrorCode();
        //do something based on errorCode
        msg("Error code: " + errCode);
    }
    msg("Failed to get representation of a found light resource");
}
/*本地方法,为灯的资源推送一个不同状态*/
private void putLightRepresentation() {
    //设置新的值
    mLight.setState(true);
    mLight.setPower(15);
```

```
    msg("Putting light representation...");
    OcRepresentation representation = null;
    try {
        representation = mLight.getOcRepresentation();
    } catch (OcException e) {
        Log.e(TAG, e.toString());
        msg("Failed to get OcRepresentation from a light");
    }
    Map<String, String> queryParams = new HashMap<>();
    try {
        sleep(1);
        mFoundLightResource.put(representation, queryParams, this);
    } catch (OcException e) {
        Log.e(TAG, e.toString());
        msg("Error occurred while invoking \"put\" API");
    }
}
/*
 * 当"put"请求成功,会执行一个事件处理器
 * @param list                头选项列表
 * @param ocRepresentation 一个资源的表示
 */
@Override
public synchronized void onPutCompleted(List<OcHeaderOption> list, OcRepresentation ocRepresentation) {
    msg("PUT request was successful");
    try {
        mLight.setOcRepresentation(ocRepresentation);
    } catch (OcException e) {
        Log.e(TAG, e.toString());
        msg("Failed to create Light representation");
    }
    msg("Light attributes: ");
    msg(mLight.toString());
    printLine();
    //调用一个本地方法,但是该方法会在内部调用"post"API,用于 foundLightResource
    //postLightRepresentation();
}
/* 当"put"请求失败,会执行下面的事件处理器
 * @param throwable exception
 */
@Override
public synchronized void onPutFailed(Throwable throwable) {
    if (throwable instanceof OcException) {
        OcException ocEx = (OcException) throwable;
        Log.e(TAG, ocEx.toString());
        ErrorCode errCode = ocEx.getErrorCode();
        //基于错误代码做一些事情
        msg("Error code: " + errCode);
    }
    msg("Failed to \"put\" a new representation");
}
/* 为该 light 资源发送一个不同的状态 */
private void postLightRepresentation() {
    //设置新值
```

```
    mLight.setState(false);
    mLight.setPower(105);
    msg("Posting light representation...");
    OcRepresentation representation = null;
    try {
        representation = mLight.getOcRepresentation();
    } catch (OcException e) {
        Log.e(TAG, e.toString());
        msg("Failed to get OcRepresentation from a light");
    }
    Map<String, String> queryParams = new HashMap<>();
    try {
        sleep(1);
        mFoundLightResource.post(representation, queryParams, this);
    } catch (OcException e) {
        Log.e(TAG, e.toString());
        msg("Error occurred while invoking \"post\" API");
    }
}
/* 当"post"请求成功,会执行下面的事件处理器
 * @param list     list of the header options
 * @param ocRepresentation 一个资源的表示 */
@Override
public synchronized void onPostCompleted(List<OcHeaderOption> list,
                                         OcRepresentation ocRepresentation) {
    msg("POST request was successful");
    try {
        if (ocRepresentation.hasAttribute(OcResource.CREATED_URI_KEY)) {
            msg("\tUri of the created resource: " +
                    ocRepresentation.getValue(OcResource.CREATED_URI_KEY));
        } else {
            mLight.setOcRepresentation(ocRepresentation);
            msg(mLight.toString());
        }
    } catch (OcException e) {
        Log.e(TAG, e.toString());
    }
    //设置新的值
    mLight.setState(true);
    mLight.setPower(55);
    msg("Posting again light representation...");
    OcRepresentation representation2 = null;
    try {
        representation2 = mLight.getOcRepresentation();
    } catch (OcException e) {
        Log.e(TAG, e.toString());
        msg("Failed to get OcRepresentation from a light");
    }
    Map<String, String> queryParams = new HashMap<>();
    try {
        mFoundLightResource.post(representation2, queryParams, onPostListener2);
    } catch (OcException e) {
        Log.e(TAG, e.toString());
        msg("Error occurred while invoking \"post\" API");
```

```
        }
    }
    /* 当"post"请求失败,会执行下面的事件处理器
     * @param throwable exception */
    @Override
    public synchronized void onPostFailed(Throwable throwable) {
        if (throwable instanceof OcException) {
            OcException ocEx = (OcException) throwable;
            Log.e(TAG, ocEx.toString());
            ErrorCode errCode = ocEx.getErrorCode();
            //do something based on errorCode
            msg("Error code: " + errCode);
        }
        msg("Failed to \"post\" a new representation");
    }
    /* 声明并实现第二个 OcResource.OnPostListener */
    OcResource.OnPostListener onPostListener2 = new OcResource.OnPostListener() {
        /* 当"post"请求成功,会执行下面的事件处理器
         * @param list      list of the header options
         * @param ocRepresentation 资源表示 */
        @Override
        public synchronized void onPostCompleted(List<OcHeaderOption> list,
                                                OcRepresentation ocRepresentation) {
            msg("Second POST request was successful");
            try {
                if (ocRepresentation.hasAttribute(OcResource.CREATED_URI_KEY)) {
                    msg("\tUri of the created resource:" + ocRepresentation.getValue(OcResource.CREATED_URI_KEY));
                } else {
                    mLight.setOcRepresentation(ocRepresentation);
                    msg(mLight.toString());
                }
            } catch (OcException e) {
                Log.e(TAG, e.toString());
            }
            //调用该方法会间接调用 foundLightResource 的 observe //API
            observeFoundLightResource();
        }
        /* 当"post"请求失败,会执行下面的事件处理器
    * @param throwable exception */
        @Override
        public synchronized void onPostFailed(Throwable throwable) {
            if (throwable instanceof OcException) {
                OcException ocEx = (OcException) throwable;
                Log.e(TAG, ocEx.toString());
                ErrorCode errCode = ocEx.getErrorCode();
                //do something based on errorCode
                msg("Error code: " + errCode);
            }
            msg("Failed to \"post\" a new representation");
        }
    };
    /* 开始 Observe 该 light 资源 */
    private void observeFoundLightResource() {
        try {
```

```
        sleep(1);
        mFoundLightResource.observe(ObserveType.OBSERVE, new HashMap<String, String>(), this);
    } catch (OcException e) {
        Log.e(TAG, e.toString());
        msg("Error occurred while invoking \"observe\" API");
    }
}
//保持观察到的当前值
private static int mObserveCount = 0;
/* 当"post"请求成功,会执行下面的事件处理器
 * @param list      list of the header options
 * @param ocRepresentation 资源表示
 * @param sequenceNumber 序列号 */
@Override
public synchronized void onObserveCompleted(List<OcHeaderOption> list,
                                            OcRepresentation ocRepresentation,
                                            int sequenceNumber) {
    if (OcResource.OnObserveListener.REGISTER == sequenceNumber) {
        msg("Observe registration action is successful:");
    } else if (OcResource.OnObserveListener.DEREGISTER == sequenceNumber) {
        msg("Observe De-registration action is successful");
    } else if (OcResource.OnObserveListener.NO_OPTION == sequenceNumber) {
        msg("Observe registration or de-registration action is failed");
    }
    msg("OBSERVE Result:");
    msg("\tSequenceNumber:" + sequenceNumber);
    try {
        mLight.setOcRepresentation(ocRepresentation);
    } catch (OcException e) {
        Log.e(TAG, e.toString());
        msg("Failed to get the attribute values");
    }
    msg(mLight.toString());
    if ((++mObserveCount) == 11) {
        msg("Cancelling Observe...");
        try {
            mFoundLightResource.cancelObserve();
        } catch (OcException e) {
            Log.e(TAG, e.toString());
            msg("Error occurred while invoking \"cancelObserve\" API");
        }
        msg("DONE");
        //准备 SimpleClient 下一次重启
        resetGlobals();
        enableStartButton();
    }
}
/**
 * 当"observe"请求失败,会执行下面的事件处理器
 * @param throwable exception */
@Override
public synchronized void onObserveFailed(Throwable throwable) {
    if (throwable instanceof OcException) {
        OcException ocEx = (OcException) throwable;
```

```
            Log.e(TAG, ocEx.toString());
            ErrorCode errCode = ocEx.getErrorCode();
            //do something based on errorCode
            msg("Error code: " + errCode);
        }
        msg("Observation of the found light resource has failed");
    }
    //OIC 部分的代码结束,设置代码开始
    private final static String TAG = SimpleClient.class.getSimpleName();
    private TextView mConsoleTextView;
    private ScrollView mScrollView;
    protected void onCreate(Bundle savedInstanceState) {
        super.onCreate(savedInstanceState);
        setContentView(R.layout.activity_simple_client);
        mConsoleTextView = (TextView) findViewById(R.id.consoleTextView);
        mConsoleTextView.setMovementMethod(new ScrollingMovementMethod());
        mScrollView = (ScrollView) findViewById(R.id.scrollView);
        mScrollView.fullScroll(View.FOCUS_DOWN);
        final Button button = (Button) findViewById(R.id.button);
         /*修改开始*/
         Button button1 = (Button) findViewById(R.id.button1);
         button1.setOnClickListener(new View.OnClickListener(){
             public void onClick(View v){
             }
         });

        //修改结束
        if (null == savedInstanceState) {
            button.setOnClickListener(new View.OnClickListener() {
                @Override
                public void onClick(View v) {
                    button.setText("Re-start");
                    button.setEnabled(false);
                    new Thread(new Runnable() {
                        public void run() {
                            startSimpleClient();
                        }
                    }).start();
                }
            });
        } else {
            String consoleOutput = savedInstanceState.getString("consoleOutputString");
            mConsoleTextView.setText(consoleOutput);
        }
    }
    @Override
    protected void onSaveInstanceState(Bundle outState) {
        super.onSaveInstanceState(outState);
        outState.putString("consoleOutputString", mConsoleTextView.getText().toString());
    }
    @Override
    protected void onRestoreInstanceState(Bundle savedInstanceState) {
        super.onRestoreInstanceState(savedInstanceState);
        String consoleOutput = savedInstanceState.getString("consoleOutputString");
```

```
        mConsoleTextView.setText(consoleOutput);
    }
    private void enableStartButton() {
        runOnUiThread(new Runnable() {
            public void run() {
                Button button = (Button) findViewById(R.id.button);
                button.setEnabled(true);
            }
        });
    }
    private void sleep(int seconds) {
        try {
            Thread.sleep(seconds * 1000);
        } catch (InterruptedException e) {
            e.printStackTrace();
            Log.e(TAG, e.toString());
        }
    }
    private void msg(final String text) {
        runOnUiThread(new Runnable() {
            public void run() {
                mConsoleTextView.append("\n");
                mConsoleTextView.append(text);
                mScrollView.fullScroll(View.FOCUS_DOWN);
            }
        });
        Log.i(TAG, text);
    }
    private void printLine() {
msg("------------------------------------------------------------------------");
    }
    private synchronized void resetGlobals() {
        mFoundLightResource = null;
        mFoundResources.clear();
        mLight = new Light();
        mObserveCount = 0;
    }
    @Override
    public void onNewIntent(Intent intent) {
        super.onNewIntent(intent);
        Log.d(TAG, "onNewIntent with changes sending broadcast IN ");
        Intent i = new Intent();
        i.setAction(intent.getAction());
        i.putExtra(NfcAdapter.EXTRA_NDEF_MESSAGES,
            intent.getParcelableArrayExtra(NfcAdapter.EXTRA_NDEF_MESSAGES));
        sendBroadcast(i);
        Log.d(TAG, "Initialize Context again resetting");
    }
}
```

2. simpleserver

与 simpleclient 相似，simpleserver 工程中同样包含两个 Java 文件，分别为 Light.java 和 SimpleServer.java。

1）Light.java 文件中的代码

```
package org.iotivity.base.examples;
import android.content.Context;
import android.content.Intent;
import android.util.Log;
import org.iotivity.base.EntityHandlerResult;
import org.iotivity.base.ErrorCode;
import org.iotivity.base.ObservationInfo;
import org.iotivity.base.OcException;
import org.iotivity.base.OcPlatform;
import org.iotivity.base.OcRepresentation;
import org.iotivity.base.OcResource;
import org.iotivity.base.OcResourceHandle;
import org.iotivity.base.OcResourceRequest;
import org.iotivity.base.OcResourceResponse;
import org.iotivity.base.RequestHandlerFlag;
import org.iotivity.base.RequestType;
import org.iotivity.base.ResourceProperty;
import java.util.EnumSet;
import java.util.LinkedList;
import java.util.List;
import java.util.Map;
/* Light,该类表示一个 light 资源 */
public class Light implements OcPlatform.EntityHandler {
    private static final String NAME_KEY = "name";
    private static final String STATE_KEY = "state";
    private static final String POWER_KEY = "power";
    private String mResourceUri;                    //资源 URI
    private String mResourceTypeName;               //资源类型名
    private String mResourceInterface;              //资源接口
    private OcResourceHandle mResourceHandle;
    private String mName;                           //light 名称
    private boolean mState;                         //light 状态
    private int mPower;                             //light 功率
    public Light(String resourceUri, String name, boolean state, int power) {
        mResourceUri = resourceUri;
        mResourceTypeName = "core.light";
        mResourceInterface = OcPlatform.DEFAULT_INTERFACE;
        mResourceHandle = null;                     //资源注册时设置该值
        mName = name;
        mState = state;
        mPower = power;
    }
    public synchronized void registerResource() throws OcException {
        if (null == mResourceHandle) {
            mResourceHandle = OcPlatform.registerResource(
                    mResourceUri,
                    mResourceTypeName,
                    mResourceInterface,
                    this,
                    EnumSet.of(ResourceProperty.DISCOVERABLE, ResourceProperty.OBSERVABLE)
            );
        }
```

```
    }
    /* 注意：这只是事件处理器的一个简单实现
     * @param request
     * @return */
    @Override
    public synchronized EntityHandlerResult handleEntity(final OcResourceRequest request) {
        EntityHandlerResult ehResult = EntityHandlerResult.ERROR;
        if (null == request) {
            msg("Server request is invalid");
            return ehResult;
        }
        //Get the request flags
        EnumSet<RequestHandlerFlag> requestFlags = request.getRequestHandlerFlagSet();
        if (requestFlags.contains(RequestHandlerFlag.INIT)) {
            msg("\t\tRequest Flag: Init");
            ehResult = EntityHandlerResult.OK;
        }
        if (requestFlags.contains(RequestHandlerFlag.REQUEST)) {
            msg("\t\tRequest Flag: Request");
            ehResult = handleRequest(request);
        }
        if (requestFlags.contains(RequestHandlerFlag.OBSERVER)) {
            msg("\t\tRequest Flag: Observer");
            ehResult = handleObserver(request);
        }
        return ehResult;
    }
    private EntityHandlerResult handleRequest(OcResourceRequest request) {
        EntityHandlerResult ehResult = EntityHandlerResult.ERROR;
        //检查查询参数(如果有的话)
        Map<String, String> queries = request.getQueryParameters();
        if (!queries.isEmpty()) {
            msg("Query processing is up to entityHandler");
        } else {
            msg("No query parameters in this request");
        }
        for (Map.Entry<String, String> entry : queries.entrySet()) {
            msg("Query key: " + entry.getKey() + " value: " + entry.getValue());
        }
        //获取请求类型
        RequestType requestType = request.getRequestType();
        switch (requestType) {
            case GET:
                msg("\t\t\tRequest Type is GET");
                ehResult = handleGetRequest(request);
                break;
            case PUT:
                msg("\t\t\tRequest Type is PUT");
                ehResult = handlePutRequest(request);
                break;
            case POST:
                msg("\t\t\tRequest Type is POST");
                ehResult = handlePostRequest(request);
                break;
```

```
            case DELETE:
                msg("\t\t\tRequest Type is DELETE");
                ehResult = handleDeleteRequest();
                break;
        }
        return ehResult;
    }
    private EntityHandlerResult handleGetRequest(final OcResourceRequest request) {
        EntityHandlerResult ehResult;
        OcResourceResponse response = new OcResourceResponse();
        response.setRequestHandle(request.getRequestHandle());
        response.setResourceHandle(request.getResourceHandle());
        if (mIsSlowResponse) {                          //慢响应的情况
            new Thread(new Runnable() {
                public void run() {
                    handleSlowResponse(request);
                }
            }).start();
            ehResult = EntityHandlerResult.SLOW;
        } else {                                        //正常响应的情况
            response.setErrorCode(SUCCESS);
            response.setResponseResult(EntityHandlerResult.OK);
            response.setResourceRepresentation(getOcRepresentation());
            ehResult = sendResponse(response);
        }
        return ehResult;
    }
    private EntityHandlerResult handlePutRequest(OcResourceRequest request) {
        OcResourceResponse response = new OcResourceResponse();
        response.setRequestHandle(request.getRequestHandle());
        response.setResourceHandle(request.getResourceHandle());
        setOcRepresentation(request.getResourceRepresentation());
        response.setResourceRepresentation(getOcRepresentation());
        response.setResponseResult(EntityHandlerResult.OK);
        response.setErrorCode(SUCCESS);
        return sendResponse(response);
    }
    private static int sUriCounter = 1;
    private EntityHandlerResult handlePostRequest(OcResourceRequest request) {
        OcResourceResponse response = new OcResourceResponse();
        response.setRequestHandle(request.getRequestHandle());
        response.setResourceHandle(request.getResourceHandle());
        String newUri = "/a/light" + (++sUriCounter);
        if(null != mContext && mContext instanceof SimpleServer) {
            ((SimpleServer) mContext).createNewLightResource(newUri, "John's light " + sUriCounter);
        }
        OcRepresentation rep_post = getOcRepresentation();
        try {
            rep_post.setValue(OcResource.CREATED_URI_KEY, newUri);
        } catch (OcException e) {
            Log.e(TAG, e.toString());
        }
        response.setResourceRepresentation(rep_post);
        response.setErrorCode(SUCCESS);
```

```
        response.setNewResourceUri(newUri);
        response.setResponseResult(EntityHandlerResult.RESOURCE_CREATED);
        return sendResponse(response);
    }
    private EntityHandlerResult handleDeleteRequest() {
        try {
            this.unregisterResource();
            return EntityHandlerResult.RESOURCE_DELETED;
        } catch (OcException e) {
            Log.e(TAG, e.toString());
            msg("Failed to unregister a light resource");
            return EntityHandlerResult.ERROR;
        }
    }
    private void handleSlowResponse(OcResourceRequest request) {
        sleep(10);
        msg("Sending slow response...");
        OcResourceResponse response = new OcResourceResponse();
        response.setRequestHandle(request.getRequestHandle());
        response.setResourceHandle(request.getResourceHandle());
        response.setErrorCode(SUCCESS);
        response.setResponseResult(EntityHandlerResult.OK);
        response.setResourceRepresentation(getOcRepresentation());
        sendResponse(response);
    }
    private List<Byte> mObservationIds;
    private EntityHandlerResult handleObserver(final OcResourceRequest request) {
        ObservationInfo observationInfo = request.getObservationInfo();
        switch (observationInfo.getObserveAction()) {
            case REGISTER:
                if (null == mObservationIds) {
                    mObservationIds = new LinkedList<>();
                }
                mObservationIds.add(observationInfo.getOcObservationId());
                break;
            case UNREGISTER:
                mObservationIds.remove((Byte)observationInfo.getOcObservationId());
                break;
        }
        //notifyObservers 方法中的一个线程中进行 Observation
        //如果还没有创建线程,则在此处先创建
        if (null == mObserverNotifier) {
            mObserverNotifier = new Thread(new Runnable() {
                public void run() {
                    notifyObservers(request);
                }
            });
            mObserverNotifier.start();
        }
        return EntityHandlerResult.OK;
    }
    private void notifyObservers(OcResourceRequest request) {
        while (true) {
            //每 2s 功率加 10
```

```
            mPower += 10;
            sleep(2);
            msg("Notifying observers...");
            msg(this.toString());
            try {
                if (mIsListOfObservers) {
                    OcResourceResponse response = new OcResourceResponse();
                    response.setErrorCode(SUCCESS);
                    response.setResourceRepresentation(getOcRepresentation());
                    OcPlatform.notifyListOfObservers(
                            mResourceHandle,
                            mObservationIds,
                            response);
                } else {
                    OcPlatform.notifyAllObservers(mResourceHandle);
                }
            } catch (OcException e) {
                ErrorCode errorCode = e.getErrorCode();
                if (ErrorCode.NO_OBSERVERS == errorCode) {
                    msg("No more observers, stopping notifications");
                }
                return;
            }
        }
    }
    private EntityHandlerResult sendResponse(OcResourceResponse response) {
        try {
            OcPlatform.sendResponse(response);
            return EntityHandlerResult.OK;
        } catch (OcException e) {
            Log.e(TAG, e.toString());
            msg("Failed to send response");
            return EntityHandlerResult.ERROR;
        }
    }
    public synchronized void unregisterResource() throws OcException {
        if (null != mResourceHandle) {
            OcPlatform.unregisterResource(mResourceHandle);
        }
    }
    public void setOcRepresentation(OcRepresentation rep) {
        try {
            if (rep.hasAttribute(NAME_KEY)) mName = rep.getValue(NAME_KEY);
            if (rep.hasAttribute(STATE_KEY)) mState = rep.getValue(STATE_KEY);
            if (rep.hasAttribute(POWER_KEY)) mPower = rep.getValue(POWER_KEY);
        } catch (OcException e) {
            Log.e(TAG, e.toString());
            msg("Failed to get representation values");
        }
    }
    public OcRepresentation getOcRepresentation() {
        OcRepresentation rep = new OcRepresentation();
        try {
            rep.setValue(NAME_KEY, mName);
```

```
            rep. setValue(STATE_KEY, mState);
            rep. setValue(POWER_KEY, mPower);
        } catch (OcException e) {
            Log. e(TAG, e. toString());
            msg("Failed to set representation values");
        }
        return rep;
    }
    //OIC 部分代码结束,设置开始
    public void setSlowResponse(boolean isSlowResponse) {
        mIsSlowResponse = isSlowResponse;
    }
    public void useListOfObservers(boolean isListOfObservers) {
        mIsListOfObservers = isListOfObservers;
    }
    public void setContext(Context context) {
        mContext = context;
    }
    @Override
    public String toString() {
        return "\t" + "URI" + ": " + mResourceUri +
                "\n\t" + NAME_KEY + ": " + mName +
                "\n\t" + STATE_KEY + ": " + mState +
                "\n\t" + POWER_KEY + ": " + mPower;
    }
    private void sleep(int seconds) {
        try {
            Thread. sleep(seconds * 1000);
        } catch (InterruptedException e) {
            e. printStackTrace();
            Log. e(TAG, e. toString());
        }
    }
    private void msg(String text) {
        if (null != mContext) {
            Intent intent = new Intent("org. iotivity. base. examples. simpleserver");
            intent. putExtra("message", text);
            mContext. sendBroadcast(intent);
        }
    }
    private final static String TAG = Light. class. getSimpleName();
    private final static int SUCCESS = 200;
    private boolean mIsSlowResponse = false;
    private boolean mIsListOfObservers = false;
    private Thread mObserverNotifier;
    private Context mContext;
}
```

2) SimpleServer. java 文件中的代码

```
package org. iotivity. base. examples;
import android. app. Activity;
import android. content. BroadcastReceiver;
import android. content. Context;
```

```
import android.content.Intent;
import android.content.IntentFilter;
import android.nfc.NfcAdapter;
import android.os.Bundle;
import android.text.method.ScrollingMovementMethod;
import android.util.Log;
import android.view.View;
import android.widget.CompoundButton;
import android.widget.ScrollView;
import android.widget.TextView;
import android.widget.ToggleButton;
import org.iotivity.base.ModeType;
import org.iotivity.base.OcException;
import org.iotivity.base.OcPlatform;
import org.iotivity.base.PlatformConfig;
import org.iotivity.base.QualityOfService;
import org.iotivity.base.ServiceType;
import java.util.LinkedList;
import java.util.List;
/* SimpleServer,SimpleServer 是一个简单的 OIC 服务器端应用.该程序创建了一个灯的资源,并对客户端的呼叫
进行处理 */
public class SimpleServer extends Activity {
  List<Light> lights = new LinkedList<>();
  /* 该方法用于配置和初始化平台,然后创建一个 light 资源 */
  private void startSimpleServer() {
    Context context = this;
    PlatformConfig platformConfig = new PlatformConfig(
          this,
          context,
          ServiceType.IN_PROC,
          ModeType.SERVER,
          "0.0.0.0",                          //通过设置为"0.0.0.0"绑定到所有可用的接口
          0,                                  //使用随机可用的端口
          QualityOfService.LOW
    );
    msg("Configuring platform.");
    OcPlatform.Configure(platformConfig);
    createNewLightResource("/a/light", "John's light");
    msg("Waiting for the requests...");
    printLine();
    enableStartStopButton();
  }
  public void createNewLightResource(String resourceUri, String resourceName){
    msg("Creating a light");
    Light light = new Light(
          resourceUri,                        //URI
    resourceName,                             //名称
    false,                                    //状态
    0                                         //功率
    );
    msg(light.toString());
    light.setContext(this);
    msg("Registering light as a resource");
    try {
```

```
            light.registerResource();
        } catch (OcException e) {
            Log.e(TAG, e.toString());
            msg("Failed to register a light resource");
        }
        lights.add(light);
    }
    private void stopSimpleServer() {
        for (Light light : lights) {
            try {
                light.unregisterResource();
            } catch (OcException e) {
                Log.e(TAG, e.toString());
                msg("Failed to unregister a light resource");
            }
        }
        lights.clear();
        msg("All created resources have been unregistered");
        printLine();
        enableStartStopButton();
    }
    //OIC 部分代码结束,设置开始
    private final static String TAG = SimpleServer.class.getSimpleName();
    private MessageReceiver mMessageReceiver = new MessageReceiver();
    private TextView mConsoleTextView;
    private ScrollView mScrollView;
    @Override
    protected void onCreate(Bundle savedInstanceState) {
        super.onCreate(savedInstanceState);
        setContentView(R.layout.activity_simple_server);
        registerReceiver(mMessageReceiver,
                new IntentFilter("org.iotivity.base.examples.simpleserver"));
        mConsoleTextView = (TextView) findViewById(R.id.consoleTextView);
        mConsoleTextView.setMovementMethod(new ScrollingMovementMethod());
        mScrollView = (ScrollView) findViewById(R.id.scrollView);
        mScrollView.fullScroll(View.FOCUS_DOWN);
        final ToggleButton toggleButton = (ToggleButton) findViewById(R.id.toggleButton);
        if (null == savedInstanceState) {
            toggleButton.setOnCheckedChangeListener(new CompoundButton.OnCheckedChangeListener() {
                public void onCheckedChanged(CompoundButton buttonView, boolean isChecked) {
                    toggleButton.setEnabled(false);
                    if (isChecked) {
                        new Thread(new Runnable() {
                            public void run() {
                                startSimpleServer();
                            }
                        }).start();
                    } else {
                        new Thread(new Runnable() {
                            public void run() {
                                stopSimpleServer();
                            }
                        }).start();
                    }
```

```
            }
        });
    } else {
        String consoleOutput = savedInstanceState.getString("consoleOutputString");
        mConsoleTextView.setText(consoleOutput);
        boolean buttonCheked = savedInstanceState.getBoolean("toggleButtonChecked");
        toggleButton.setChecked(buttonCheked);
    }
}
@Override
public void onDestroy() {
    super.onDestroy();
    onStop();
}
@Override
protected void onStop() {
    //unregisterReceiver(mMessageReceiver);
    super.onStop();
}
@Override
protected void onSaveInstanceState(Bundle outState) {
    super.onSaveInstanceState(outState);
    outState.putString("consoleOutputString", mConsoleTextView.getText().toString());
    ToggleButton toggleButton = (ToggleButton) findViewById(R.id.toggleButton);
    outState.putBoolean("toggleButtonChecked", toggleButton.isChecked());
}
@Override
protected void onRestoreInstanceState(Bundle savedInstanceState) {
    super.onRestoreInstanceState(savedInstanceState);
    String consoleOutput = savedInstanceState.getString("consoleOutputString");
    mConsoleTextView.setText(consoleOutput);
    final ToggleButton toggleButton = (ToggleButton) findViewById(R.id.toggleButton);
    boolean buttonCheked = savedInstanceState.getBoolean("toggleButtonChecked");
    toggleButton.setChecked(buttonCheked);
}
private void msg(final String text) {
    runOnUiThread(new Runnable() {
        public void run() {
            mConsoleTextView.append("\n");
            mConsoleTextView.append(text);
            mScrollView.fullScroll(View.FOCUS_DOWN);
        }
    });
    Log.i(TAG, text);
}
private void printLine() {
msg("------------------------------------------------------------------------");
}
private void sleep(int seconds) {
    try {
        Thread.sleep(seconds * 1000);
    } catch (InterruptedException e) {
        e.printStackTrace();
        Log.e(TAG, e.toString());
```

```
            }
        }
        private void enableStartStopButton() {
            runOnUiThread(new Runnable() {
                public void run() {
                    ToggleButton toggleButton = (ToggleButton) findViewById(R.id.toggleButton);
                    toggleButton.setEnabled(true);
                }
            });
        }
        public class MessageReceiver extends BroadcastReceiver {
            @Override
            public void onReceive(Context context, Intent intent) {
                final String message = intent.getStringExtra("message");
                msg(message);
            }
        }
        @Override
        public void onNewIntent(Intent intent) {
            super.onNewIntent(intent);
            Log.d(TAG, "onNewIntent with changes sending broadcast IN ");
            Intent i = new Intent();
            i.setAction(intent.getAction());
            i.putExtra(NfcAdapter.EXTRA_NDEF_MESSAGES,
                    intent.getParcelableArrayExtra(NfcAdapter.EXTRA_NDEF_MESSAGES));
            sendBroadcast(i);
            Log.d(TAG, "Initialize Context again resetting");
        }
    }
```

12.5 基于 Arduino 的开发方法

本节主要包括配置 Arduino 环境、软件工具的安装、程序编译和实例代码。

12.5.1 配置 Arduino 环境

Iotivity 的 Arduino 代码在 Linux 的交叉环境下编译，环境配置参考 12.3 节。

12.5.2 软件工具的安装

在 iotivity 根目录下运行：

```
$ sudo apt-get install dos2unix
```

12.5.3 程序编译

(1) 编译指令。

在 Iotivity 根目录下运行以下指令：

```
scons resource/csdk/stack/samples/arduino/SimpleClientServer/ \
TARGET_OS=arduino TARGET_ARCH=<target_arch_value> BOARD=<target_board_value> \
```

```
TARGET_TRANSPORT = <target_transport_value> SHIELD = <target_shield_value>
```

(2) 编译选项。

```
Arduino ATMega 2560:
<target_board_value> = mega
<target_arch_value> = avr
<target_transport_value> = BLE | IP
<target_shield_value> = ETH
Arduino Due:
<target_board_value> = arduino_due_x
<target_arch_value> = arm
<target_transport_value> = BLE | IP
<target_shield_value> = ETH
```

(3) 烧写指令。

在 Iotivity 根目录下运行以下指令：

```
scons resource/csdk/stack/samples/arduino/SimpleClientServer/ TARGET_OS = arduino \TARGET_ARCH = <target_
arch_value> BOARD = <target_board_value> \TARGET_TRANSPORT = <target_transport_value> SHIELD = <target_
shield_value> UPLOAD = true
```

烧写操作自动将示例代码编译生成的可执行文件烧写到 Arduino 设备上，代码路径：{IOTIVITY}/resource/csdk/stack/samples/arduino/SimpleClientServer/ocserver/ocserver.cpp。

12.5.4 实例代码

该实例描述作为服务器端的 Arduino，展示基本的资源创建和资源操作。

```
#include "Arduino.h"
#include "logger.h"
#include "ocstack.h"
#include "ocpayload.h"
#include <string.h>
#ifdef ARDUINOWIFI
//Arduino WiFi 扩展板
#include <SPI.h>
#include <WiFi.h>
#include <WiFiUdp.h>
#elif defined ARDUINOETH
//Arduino 以太网扩展板
#include <EthernetServer.h>
#include <Ethernet.h>
#include <Dns.h>
#include <EthernetClient.h>
#include <util.h>
#include <EthernetUdp.h>
#include <Dhcp.h>
#endif
const char *getResult(OCStackResult result);
#define TAG "ArduinoServer"
int gLightUnderObservation = 0;
void createLightResource();
/* 用结构体表示 light 资源 */
```

```
typedef struct LIGHTRESOURCE{
    OCResourceHandle handle;
    bool state;
    int power;
} LightResource;
static LightResource Light;
#ifdef ARDUINOWIFI
//Arduino WiFi 扩展板
//注意：Arduino WiFi 扩展板当前不支持多播，因此，服务器端不能监听 224.0.1.187 多播地址
static const char ARDUINO_WIFI_SHIELD_UDP_FW_VER[] = "1.1.0";
//WiFi 扩展板烧录 Intel 固件补丁
static const char INTEL_WIFI_SHIELD_FW_VER[] = "1.2.0";
//WiFi 网络信息
char ssid[] = "mDNSAP";
char pass[] = "letmein9";
int ConnectToNetwork()
{
    char * fwVersion;
    int status = WL_IDLE_STATUS;
    //检查 WiFi 扩展板的存在
    if (WiFi.status() == WL_NO_SHIELD)
    {
        OIC_LOG(ERROR, TAG, ("WiFi shield not present"));
        return -1;
    }
    //验证 WiFi 扩展板是否正在运行所有 UDP 修复程序的固件
    fwVersion = WiFi.firmwareVersion();
    OIC_LOG_V(INFO, TAG, "WiFi Shield Firmware version %s", fwVersion);
    if ( strncmp(fwVersion, ARDUINO_WIFI_SHIELD_UDP_FW_VER, sizeof(ARDUINO_WIFI_SHIELD_UDP_FW_VER)) != 0 )
    {
        OIC_LOG(DEBUG, TAG, ("!!!!! Upgrade WiFi Shield Firmware version !!!!!!"));
        return -1;
    }
    //尝试连接 WiFi 网络
    while (status != WL_CONNECTED)
    {
        OIC_LOG_V(INFO, TAG, "Attempting to connect to SSID: %s", ssid);
        status = WiFi.begin(ssid,pass);
        //等待 10s 再连接
        delay(10000);
    }
    OIC_LOG(DEBUG, TAG, ("Connected to wifi"));
    IPAddress ip = WiFi.localIP();
    OIC_LOG_V(INFO, TAG, "IP Address: %d.%d.%d.%d", ip[0], ip[1], ip[2], ip[3]);
    return 0;
}
#elif defined ARDUINOETH
//Arduino 以太网扩展板
int ConnectToNetwork()
{
    //注意：使用自己的防火墙 MAC 地址更新此处的 MAC 地址
    uint8_t ETHERNET_MAC[] = {0x90, 0xA2, 0xDA, 0x0E, 0xC4, 0x05};
    uint8_t error = Ethernet.begin(ETHERNET_MAC);
    if (error == 0)
```

```
    {
        OIC_LOG_V(ERROR, TAG, "error is: %d", error);
        return -1;
    }
    IPAddress ip = Ethernet.localIP();
    OIC_LOG_V(INFO, TAG, "IP Address: %d.%d.%d.%d", ip[0], ip[1], ip[2], ip[3]);
    return 0;
}
#endif
//在具有哈佛内存架构的 Arduino Atmel 开发板上,堆栈从顶部向下生长
//堆向上生长.此方法将打印这两者之间的距离(以字节为单位)
/* 详见 http://www.atmel.com/webdoc/AVRLibcReferenceManual/malloc_1malloc_intro.html */
void PrintArduinoMemoryStats()
{
    #ifdef ARDUINO_AVR_MEGA2560
    //此变量在 avr-libc/stdlib/malloc.c 中声明
    //它使得最大的地址没有分配给堆
    extern char *__brkval;
    //tmp 地址提供当前协议栈的边界
    int tmp;
    OIC_LOG_V(INFO, TAG, "Stack: %u Heap: %u", (unsigned int)&tmp, (unsigned int)__brkval);
    OIC_LOG_V(INFO, TAG, "Unallocated Memory between heap and stack: %u",
        ((unsigned int)&tmp - (unsigned int)__brkval));
    #endif
}
//已注册资源的实体处理句柄
//当接收到对资源的请求时被 OCStack 调用
OCEntityHandlerResult OCEntityHandlerCb (OCEntityHandlerFlag flag, OCEntityHandlerRequest *
entityHandlerRequest,void *callbackParam)
{
    OCEntityHandlerResult ehRet = OC_EH_OK;
    OCEntityHandlerResponse response = {0};
    OCRepPayload* payload = OCRepPayloadCreate();
    if(!payload)
    {
        OIC_LOG(ERROR, TAG, ("Failed to allocate Payload"));
        return OC_EH_ERROR;
    }
    if(entityHandlerRequest && (flag & OC_REQUEST_FLAG))
    {
        OIC_LOG (INFO, TAG, ("Flag includes OC_REQUEST_FLAG"));
        if(OC_REST_GET == entityHandlerRequest->method)
        {
            OCRepPayloadSetUri(payload, "/a/light");
            OCRepPayloadSetPropBool(payload, "state", true);
            OCRepPayloadSetPropInt(payload, "power", 10);
        }
        else if(OC_REST_PUT == entityHandlerRequest->method)
        {
            //执行 PUT 操作
            OCRepPayloadSetUri(payload, "/a/light");
            OCRepPayloadSetPropBool(payload, "state", false);
            OCRepPayloadSetPropInt(payload, "power", 0);
        }
```

```
        if (ehRet == OC_EH_OK)
        {
            //生成响应.注意这需要一些与请求消息有关的信息
            response.requestHandle = entityHandlerRequest->requestHandle;
            response.resourceHandle = entityHandlerRequest->resource;
            response.ehResult = ehRet;
            response.payload = (OCPayload*) payload;
            response.numSendVendorSpecificHeaderOptions = 0;
            memset(response.sendVendorSpecificHeaderOptions, 0,
                    sizeof response.sendVendorSpecificHeaderOptions);
            memset(response.resourceUri, 0, sizeof response.resourceUri);
            //表明响应不在缓存中
            response.persistentBufferFlag = 0;
            //发送响应
            if (OCDoResponse(&response) != OC_STACK_OK)
            {
                OIC_LOG(ERROR, TAG, "Error sending response");
                ehRet = OC_EH_ERROR;
            }
        }
    }
    if (entityHandlerRequest && (flag & OC_OBSERVE_FLAG))
    {
        if (OC_OBSERVE_REGISTER == entityHandlerRequest->obsInfo.action)
        {
            OIC_LOG (INFO, TAG, ("Received OC_OBSERVE_REGISTER from client"));
            gLightUnderObservation = 1;
        }
        else if (OC_OBSERVE_DEREGISTER == entityHandlerRequest->obsInfo.action)
        {
            OIC_LOG (INFO, TAG, ("Received OC_OBSERVE_DEREGISTER from client"));
            gLightUnderObservation = 0;
        }
    }
    OCRepPayloadDestroy(payload);
    return ehRet;
}
//此方法用作展示 OC 协议的 Observe 功能
static uint8_t modCounter = 0;
void *ChangeLightRepresentation (void *param)
{
    (void)param;
    OCStackResult result = OC_STACK_ERROR;
    modCounter += 1;
    //匹配 Linux 示例服务器端应用程序用于相同功能的时间
    if(modCounter % 10 == 0)
    {
        Light.power += 5;
        if (gLightUnderObservation)
        {
            OIC_LOG_V(INFO, TAG, " =====> Notifying stack of new power level %d\n", Light.power);
            result = OCNotifyAllObservers (Light.handle, OC_NA_QOS);
            if (OC_STACK_NO_OBSERVERS == result)
            {
```

```
                gLightUnderObservation = 0;
            }
        }
    }
    return NULL;
}
//setup 函数在启动程序时被调用
void setup()
{
    //在此处添加你的初始化代码
    //注意：在此将初始化 Arduino 的串口波特率为 115200
    OIC_LOG_INIT();
    OIC_LOG(DEBUG, TAG, ("OCServer is starting..."));
    //连接以太网或 WiFi 网络
#if defined(ARDUINOWIFI) || defined(ARDUINOETH)
    if (ConnectToNetwork() != 0)
    {
        OIC_LOG(ERROR, TAG, ("Unable to connect to network"));
        return;
    }
#endif
    //在服务器端模式下初始化 OC 协议栈
    if (OCInit(NULL, 0, OC_SERVER) != OC_STACK_OK)
    {
        OIC_LOG(ERROR, TAG, ("OCStack init error"));
        return;
    }
    //声明并创建示例资源 light
    createLightResource();
}
//loop 函数被无休止地调用
void loop()
{
    //在这里保持人为的延迟，以避免 Arduino 微控制器的连续运转
    //根据具体应用需求进行修改
    delay(2000);
    //此调用显示 Arduino 上释放的可用 SRAM 的数量
    PrintArduinoMemoryStats();
    //给 OC 协议栈 CPU 周期来执行发送/接收，以及其他 OC 协议栈的东西
    if (OCProcess() != OC_STACK_OK)
    {
        OIC_LOG(ERROR, TAG, ("OCStack process error"));
        return;
    }
    ChangeLightRepresentation(NULL);
}
void createLightResource()
{
    Light.state = false;
    OCStackResult res = OCCreateResource(&Light.handle,
            "core.light",
            OC_RSRVD_INTERFACE_DEFAULT,
            "/a/light",
            OCEntityHandlerCb,
```

```
            NULL,
            OC_DISCOVERABLE|OC_OBSERVABLE);
    OIC_LOG_V(INFO, TAG, "Created Light resource with result: %s", getResult(res));
}
const char * getResult(OCStackResult result) {
    switch (result) {
    case OC_STACK_OK:
        return "OC_STACK_OK";
    case OC_STACK_INVALID_URI:
        return "OC_STACK_INVALID_URI";
    case OC_STACK_INVALID_QUERY:
        return "OC_STACK_INVALID_QUERY";
    case OC_STACK_INVALID_IP:
        return "OC_STACK_INVALID_IP";
    case OC_STACK_INVALID_PORT:
        return "OC_STACK_INVALID_PORT";
    case OC_STACK_INVALID_CALLBACK:
        return "OC_STACK_INVALID_CALLBACK";
    case OC_STACK_INVALID_METHOD:
        return "OC_STACK_INVALID_METHOD";
    case OC_STACK_NO_MEMORY:
        return "OC_STACK_NO_MEMORY";
    case OC_STACK_COMM_ERROR:
        return "OC_STACK_COMM_ERROR";
    case OC_STACK_INVALID_PARAM:
        return "OC_STACK_INVALID_PARAM";
    case OC_STACK_NOTIMPL:
        return "OC_STACK_NOTIMPL";
    case OC_STACK_NO_RESOURCE:
        return "OC_STACK_NO_RESOURCE";
    case OC_STACK_RESOURCE_ERROR:
        return "OC_STACK_RESOURCE_ERROR";
    case OC_STACK_SLOW_RESOURCE:
        return "OC_STACK_SLOW_RESOURCE";
    case OC_STACK_NO_OBSERVERS:
        return "OC_STACK_NO_OBSERVERS";
    case OC_STACK_ERROR:
        return "OC_STACK_ERROR";
    default:
        return "UNKNOWN";
    }
}
```

12.6 综合实例

本节综合运用 IoTivity 技术，通过 Arduino 实例和 Android 实例，完成综合应用。

12.6.1 Arduino 实例

该实例通过在 Arduino 服务器端创建资源，搭载温湿度传感器 DHT11、光敏传感器和 PM2.5 传感器等，实现采集温湿度值、光强值和 PM2.5 的值。

```
#include "Arduino.h"
#include "logger.h"
#include "ocstack.h"
#include "ocpayload.h"
#include <string.h>
#include <dht11.h>
#ifdef ARDUINOWIFI
//Arduino WiFi 扩展板
#include <SPI.h>
#include <WiFi.h>
#include <WiFiUdp.h>
#elif defined ARDUINOETH
//Arduino 以太网扩展板
#include <EthernetServer.h>
#include <Ethernet.h>
#include <Dns.h>
#include <EthernetClient.h>
#include <util.h>
#include <EthernetUdp.h>
#include <Dhcp.h>
#endif
dht11 DHT11;
#define DHT11PIN 2                                   //数字 2 引脚
int LED_PIN = 30;                                    //LED,引脚 30
int LIGHT_PIN = 0;                                   //模拟引脚 0,连接光敏电阻
int LED_STATE = 0;                                   //LED 当前状态值
int lightval = 0;                                    //光强值
int chk;                                             //DHT11 值
int tempval = 0;                                     //温度值
int humval = 0;                                      //湿度值
int pm25val = 0;
byte Tx[5] = {0xFE,0xA5,0x00,0x00,0xA5};
byte Rx[7] = {0x00,0x00,0x00,0x00,0x00,0x00,0x00};
const char *getResult(OCStackResult result);
#define TAG "ArduinoServer"
//资源接口
#define ARDUINO_RESOURCE_INTERFACE "core.haier.resources"
#define RESOURCE_TYPE "room.haier"
//资源类型
#define TEMPERATURE_RESOURCE_TYPE "room.temperature"
#define LIGHT_RESOURCE_TYPE "ambient.light"
#define LED_RESOURCE_TYPE "platform.led"
//资源 URI
#define TEMPERATURE_RESOURCE_ENDPOINT "/temperature"
#define LIGHT_RESOURCE_ENDPOINT "/ambientlight"
#define LED_RESOURCE_ENDPOINT "/led"
#define PM25_RESOURCE_ENDPOINT "/pm25"
//资源属性
#define TEMPERATURE_RESOURCE_KEY "temperature"
#define LIGHT_RESOURCE_KEY "ambientlight"
#define LED_RESOURCE_KEY "switch"
#define PM25_RESOURCE_KEY "pm25"
#define HUM_RESOURCE_KEY "hum"
int gLightUnderObservation = 0;
```

```
void createLightResource();
void createTemperatureResource();
void createLedResource();
void createPm25Resource();
/*用结构体表示 light 资源*/
typedef struct LIGHTRESOURCE{
    OCResourceHandle lighthandle;
    bool state;
    int power;
    int ambientlight;
} LightResource;
typedef struct TEMPERATURERESOURCE{
    OCResourceHandle temperaturehandle;
    int temperature;
} TemperatureResource;
typedef struct LEDRESOURCE{
    OCResourceHandle ledhandle;
    int state;
} LedResource;
typedef struct PM25RESOURCE{
    OCResourceHandle pm25handle;
} Pm25Resource;
static LightResource Light;
static TemperatureResource Temperature;
static LedResource Led;
static Pm25Resource Pm25;
#ifdef ARDUINOWIFI
//Arduino WiFi 扩展板,注意,Arduino WiFi 扩展板当前不支持多播,因此服务器端不能监听
//224.0.1.187 多播地址
static const char ARDUINO_WIFI_SHIELD_UDP_FW_VER[] = "1.1.0";
//WiFi 扩展板 Intel 固件补丁
static const char INTEL_WIFI_SHIELD_FW_VER[] = "1.2.0";
//WiFi 网络信息
char ssid[] = "mDNSAP";
char pass[] = "letmein9";
int ConnectToNetwork()
{
    char *fwVersion;
    int status = WL_IDLE_STATUS;
    //检查扩展板的存在
    if (WiFi.status() == WL_NO_SHIELD)
    {
        OIC_LOG(ERROR, TAG, ("WiFi shield not present"));
        return -1;
    }
    //验证 WiFi 扩展板是否正在运行所有 UDP 修复程序的固件
    fwVersion = WiFi.firmwareVersion();
    OIC_LOG_V(INFO, TAG, "WiFi Shield Firmware version %s", fwVersion);
    if ( strncmp(fwVersion, ARDUINO_WIFI_SHIELD_UDP_FW_VER, sizeof(ARDUINO_WIFI_SHIELD_UDP_FW_VER)) != 0 )
    {
        OIC_LOG(DEBUG, TAG, ("!!!!! Upgrade WiFi Shield Firmware version !!!!!!"));
        return -1;
    }
    //尝试连接 WiFi 网络
```

```
    while (status != WL_CONNECTED)
    {
        OIC_LOG_V(INFO, TAG, "Attempting to connect to SSID: %s", ssid);
        status = WiFi.begin(ssid,pass);
        //等待 10s 再连接
        delay(10000);
    }
    OIC_LOG(DEBUG, TAG, ("Connected to wifi"));
    IPAddress ip = WiFi.localIP();
    OIC_LOG_V(INFO, TAG, "IP Address: %d.%d.%d.%d", ip[0], ip[1], ip[2], ip[3]);
    return 0;
}
#elif defined ARDUINOETH
//Arduino 以太网扩展板
int ConnectToNetwork()
{
    //注意: 使用您的防火墙的 MAC 地址更新此处的 MAC 地址
    uint8_t ETHERNET_MAC[] = {0x90, 0xA2, 0xDA, 0x0E, 0xC4, 0x05};
    uint8_t error = Ethernet.begin(ETHERNET_MAC);
    if (error == 0)
    {
        OIC_LOG_V(ERROR, TAG, "error is: %d", error);
        return -1;
    }
    IPAddress ip = Ethernet.localIP();
    OIC_LOG_V(INFO, TAG, "IP Address: %d.%d.%d.%d", ip[0], ip[1], ip[2], ip[3]);
    return 0;
}
#endif
//在具有哈佛内存架构的 Arduino Atmel 开发板上,堆栈从顶部向下生长
//堆向上生长.此方法将打印这两者之间的距离(以字节为单位)
/*详见
http://www.atmel.com/webdoc/AVRLibcReferenceManual/malloc_1malloc_intro.html */
void PrintArduinoMemoryStats()
{
    #ifdef ARDUINO_AVR_MEGA2560
    //此变量在 avr-libc/stdlib/malloc.c 中声明
    //它使得最大的地址没有分配给堆
    extern char *__brkval;
    //tmp 地址提供当前协议栈的边界
    int tmp;
    OIC_LOG_V(INFO, TAG, "Stack: %u Heap: %u", (unsigned int)&tmp, (unsigned int)__brkval);
    OIC_LOG_V(INFO, TAG, "Unallocated Memory between heap and stack: %u",
            ((unsigned int)&tmp - (unsigned int)__brkval));
    #endif
}
//已注册 light 资源的实体处理句柄
//当接收到对资源的请求时被 OCStack 调用
OCEntityHandlerResult OCLightEntityHandlerCb (OCEntityHandlerFlag flag, OCEntityHandlerRequest *
entityHandlerRequest,void *callbackParam)
{
    OCEntityHandlerResult ehRet = OC_EH_OK;
    OCEntityHandlerResponse response = {0};
    OCRepPayload* payload = OCRepPayloadCreate();
```

```
if(!payload)
{
    OIC_LOG(ERROR, TAG, ("Failed to allocate Payload"));
    return OC_EH_ERROR;
}
if(entityHandlerRequest && (flag & OC_REQUEST_FLAG))
{
    OIC_LOG (INFO, TAG, ("Flag includes OC_REQUEST_FLAG"));
    if(OC_REST_GET == entityHandlerRequest->method)
    {
        OCRepPayloadSetUri(payload, LIGHT_RESOURCE_ENDPOINT);
        OCRepPayloadSetPropInt(payload, "ambientlight", lightval);
      OCRepPayloadSetPropInt(payload, PM25_RESOURCE_KEY, 20);
    }
    else if(OC_REST_PUT == entityHandlerRequest->method)
    {
        OCRepPayloadSetUri(payload, LIGHT_RESOURCE_ENDPOINT);
        OCRepPayloadSetPropBool(payload, "state", false);
        OCRepPayloadSetPropInt(payload, "power", 10);
    }
    if (ehRet == OC_EH_OK)
    {
        //生成响应.注意这需要一些与请求消息有关的信息
        response.requestHandle = entityHandlerRequest->requestHandle;
        response.resourceHandle = entityHandlerRequest->resource;
        response.ehResult = ehRet;
        response.payload = (OCPayload*) payload;
        response.numSendVendorSpecificHeaderOptions = 0;
        memset(response.sendVendorSpecificHeaderOptions, 0,
        sizeof response.sendVendorSpecificHeaderOptions);
        memset(response.resourceUri, 0, sizeof response.resourceUri);
        //表明响应不在缓存中
        response.persistentBufferFlag = 0;
        //发送响应
        if (OCDoResponse(&response) == OC_STACK_OK)
        {
            OIC_LOG(INFO, TAG, "successful send light response");
            ehRet = OC_EH_OK;
        }
    }
}
if (entityHandlerRequest && (flag & OC_OBSERVE_FLAG))
{
    if (OC_OBSERVE_REGISTER == entityHandlerRequest->obsInfo.action)
    {
        OIC_LOG (INFO, TAG, ("Received OC_OBSERVE_REGISTER from client"));
        gLightUnderObservation = 1;
    }
    else if (OC_OBSERVE_DEREGISTER == entityHandlerRequest->obsInfo.action)
    {
        OIC_LOG (INFO, TAG, ("Received OC_OBSERVE_DEREGISTER from client"));
        gLightUnderObservation = 0;
    }
}
```

```
    OCRepPayloadDestroy(payload);
    return ehRet;
}
//已注册 Temperature 资源的实体处理句柄
//当接收到对资源的请求时被 OCStack 调用
OCEntityHandlerResult OCTemperatureEntityHandlerCb(OCEntityHandlerFlag flag, OCEntityHandlerRequest *
entityHandlerRequest,void * callbackParam)
{
    OCEntityHandlerResult ehRet = OC_EH_OK;
    OCEntityHandlerResponse response = {0};
    OCRepPayload * payload = OCRepPayloadCreate();
    if(!payload)
    {
        OIC_LOG(ERROR, TAG, ("Failed to allocate Payload"));
        return OC_EH_ERROR;
    }
    if(entityHandlerRequest && (flag & OC_REQUEST_FLAG))
    {
        OIC_LOG (INFO, TAG, ("Flag includes OC_REQUEST_FLAG"));
        if(OC_REST_GET == entityHandlerRequest->method)
        {
        OCRepPayloadSetUri(payload, TEMPERATURE_RESOURCE_KEY);
        OCRepPayloadSetPropInt(payload, TEMPERATURE_RESOURCE_KEY, tempval);
         OCRepPayloadSetPropInt(payload, PM25_RESOURCE_KEY, pm25val);
         OCRepPayloadSetPropInt(payload, HUM_RESOURCE_KEY, humval);
   if(pm25val>200)
         digitalWrite(4,HIGH);
        else
         digitalWrite(4,LOW);
        }
        if (ehRet == OC_EH_OK)
        {
            //生成响应.注意这需要一些与请求消息有关的信息
            response.requestHandle = entityHandlerRequest->requestHandle;
            response.resourceHandle = entityHandlerRequest->resource;
            response.ehResult = ehRet;
            response.payload = (OCPayload *) payload;
            response.numSendVendorSpecificHeaderOptions = 0;
            memset(response.sendVendorSpecificHeaderOptions, 0,
            sizeof response.sendVendorSpecificHeaderOptions);
            memset(response.resourceUri, 0, sizeof response.resourceUri);
            //表明响应不在缓存中
            response.persistentBufferFlag = 0;
            //发送响应
       if (OCDoResponse(&response) == OC_STACK_OK)
            {
                OIC_LOG(INFO, TAG, "successful send temp response");
                ehRet = OC_EH_OK;
            }
        }
    }
    if (entityHandlerRequest && (flag & OC_OBSERVE_FLAG))
    {
        if (OC_OBSERVE_REGISTER == entityHandlerRequest->obsInfo.action)
```

```
        {
            OIC_LOG (INFO, TAG, ("Received OC_OBSERVE_REGISTER from client"));
            gLightUnderObservation = 1;
        }
        else if (OC_OBSERVE_DEREGISTER == entityHandlerRequest->obsInfo.action)
        {
            OIC_LOG (INFO, TAG, ("Received OC_OBSERVE_DEREGISTER from client"));
            gLightUnderObservation = 0;
        }
    }
    OCRepPayloadDestroy(payload);
    return ehRet;
}
//已注册的 LED 资源的实体处理句柄
//当接收到对资源的请求时被 OCStack 调用
OCEntityHandlerResult OCLedEntityHandlerCb ( OCEntityHandlerFlag flag, OCEntityHandlerRequest *
entityHandlerRequest,void * callbackParam)
{
    OCEntityHandlerResult ehRet = OC_EH_OK;
    OCEntityHandlerResponse response = {0};
    OCRepPayload* payload = OCRepPayloadCreate();
    if(!payload)
    {
        OIC_LOG(ERROR, TAG, ("Failed to allocate Payload"));
        return OC_EH_ERROR;
    }
    if(entityHandlerRequest && (flag & OC_REQUEST_FLAG))
    {
        OIC_LOG (INFO, TAG, ("Flag includes OC_REQUEST_FLAG"));
        if(OC_REST_PUT == entityHandlerRequest->method)
        {
            //执行 PUT 操作
            OCRepPayload* input = reinterpret_cast<OCRepPayload*>(entityHandlerRequest->payload);
              int64_t Lightstate;
              if(OCRepPayloadGetPropInt(input, LED_RESOURCE_KEY, &Lightstate))
              {
                  OIC_LOG(INFO, TAG, "successful Get led switch");
                  if(Lightstate == 1)
                  {
                      digitalWrite(LED_PIN,HIGH);
                      LED_STATE = 1;
                  }
                  else
                  {
                      digitalWrite(LED_PIN,LOW);
                      LED_STATE = 0;
                  }
              OCRepPayloadSetUri(payload, LED_RESOURCE_ENDPOINT);
              OCRepPayloadSetPropInt(payload, LED_RESOURCE_KEY, Lightstate);
              }
        }
    else if(OC_REST_GET == entityHandlerRequest->method)
        {
            OCRepPayloadSetUri(payload, LED_RESOURCE_KEY);
```

```
            OCRepPayloadSetPropInt(payload, LED_RESOURCE_KEY, LED_STATE);
        }
        if (ehRet == OC_EH_OK)
        {
            //生成响应.注意这需要一些与请求消息有关的信息
            response.requestHandle = entityHandlerRequest->requestHandle;
            response.resourceHandle = entityHandlerRequest->resource;
            response.ehResult = ehRet;
            response.payload = (OCPayload*) payload;
            response.numSendVendorSpecificHeaderOptions = 0;
            memset(response.sendVendorSpecificHeaderOptions, 0,sizeof response.sendVendorSpecificHeaderOptions);
            memset(response.resourceUri, 0, sizeof response.resourceUri);
            //表明响应不在持续的缓存中
            response.persistentBufferFlag = 0;
            //发送响应
                if (OCDoResponse(&response) == OC_STACK_OK)
            {
                OIC_LOG(INFO, TAG, "successful send led response");
                ehRet = OC_EH_OK;
            }
        }
    }
    if (entityHandlerRequest && (flag & OC_OBSERVE_FLAG))
    {
        if (OC_OBSERVE_REGISTER == entityHandlerRequest->obsInfo.action)
        {
            OIC_LOG (INFO, TAG, ("Received OC_OBSERVE_REGISTER from client"));
            gLightUnderObservation = 1;
        }
        else if (OC_OBSERVE_DEREGISTER == entityHandlerRequest->obsInfo.action)
        {
            OIC_LOG (INFO, TAG, ("Received OC_OBSERVE_DEREGISTER from client"));
            gLightUnderObservation = 0;
        }
    }
    OCRepPayloadDestroy(payload);
    return ehRet;
}
//已注册 PM2.5 资源的实体处理句柄
//当接收到对资源的请求时被 OCStack 调用
OCEntityHandlerResult OCPm25EntityHandlerCb (OCEntityHandlerFlag flag, OCEntityHandlerRequest *
entityHandlerRequest,void *callbackParam)
{
    OCEntityHandlerResult ehRet = OC_EH_OK;
    OCEntityHandlerResponse response = {0};
    OCRepPayload* payload = OCRepPayloadCreate();
    if(!payload)
    {
        OIC_LOG(ERROR, TAG, ("Failed to allocate Payload"));
        return OC_EH_ERROR;
    }
    if(entityHandlerRequest && (flag & OC_REQUEST_FLAG))
    {
        OIC_LOG (INFO, TAG, ("Flag includes OC_REQUEST_FLAG"));
```

```
        if(OC_REST_GET == entityHandlerRequest->method)
        {
            OCRepPayloadSetUri(payload, LIGHT_RESOURCE_ENDPOINT);
            OCRepPayloadSetPropBool(payload, "state", true);
            OCRepPayloadSetPropInt(payload, "ambientlight", 10);
        }
        else if(OC_REST_PUT == entityHandlerRequest->method)
        {
            //执行 PUT 操作
            OCRepPayloadSetUri(payload, LIGHT_RESOURCE_ENDPOINT);
            OCRepPayloadSetPropBool(payload, "state", false);
            OCRepPayloadSetPropInt(payload, "power", 10);
        }
        if (ehRet == OC_EH_OK)
        {
            //生成响应,注意这需要一些与请求消息有关的信息
            response.requestHandle = entityHandlerRequest->requestHandle;
            response.resourceHandle = entityHandlerRequest->resource;
            response.ehResult = ehRet;
            response.payload = (OCPayload*) payload;
            response.numSendVendorSpecificHeaderOptions = 0;
            memset(response.sendVendorSpecificHeaderOptions, 0,
                   sizeof response.sendVendorSpecificHeaderOptions);
            memset(response.resourceUri, 0, sizeof response.resourceUri);
            //表明响应不在缓存中
            response.persistentBufferFlag = 0;
            //发送响应
               if (OCDoResponse(&response) == OC_STACK_OK)
            {
               OIC_LOG(INFO, TAG, "successful send response");
               ehRet = OC_EH_OK;
            }
        }
    }
    if (entityHandlerRequest && (flag & OC_OBSERVE_FLAG))
    {
        if (OC_OBSERVE_REGISTER == entityHandlerRequest->obsInfo.action)
        {
            OIC_LOG (INFO, TAG, ("Received OC_OBSERVE_REGISTER from client"));
            gLightUnderObservation = 1;
        }
        else if (OC_OBSERVE_DEREGISTER == entityHandlerRequest->obsInfo.action)
        {
            OIC_LOG (INFO, TAG, ("Received OC_OBSERVE_DEREGISTER from client"));
            gLightUnderObservation = 0;
        }
    }
    OCRepPayloadDestroy(payload);
    return ehRet;
}
//setup 函数在启动程序时被调用
void setup()
{
    //在此处添加初始化代码
```

```
    //注意：在此将初始化 Arduino 的串口波特率为 115200
    OIC_LOG_INIT();
    OIC_LOG(DEBUG, TAG, ("OCServer is starting..."));
    pinMode(LED_PIN,OUTPUT);
    pinMode(4,OUTPUT);
    digitalWrite(LED_PIN,LOW);
    digitalWrite(4,LOW);
     LED_STATE = 0;
    Serial2.begin(1200);                          //串口 2,引脚 16,17
    //连接以太网或 WiFi 网络
#if defined(ARDUINOWIFI) || defined(ARDUINOETH)
    if (ConnectToNetwork() != 0)
    {
        OIC_LOG(ERROR, TAG, ("Unable to connect to network"));
        return;
    }
#endif
    //在服务器端模式下初始化 OC 协议栈
    if (OCInit(NULL, 0, OC_SERVER) != OC_STACK_OK)
    {
        OIC_LOG(ERROR, TAG, ("OCStack init error"));
        return;
    }
      createLightResource();
    OIC_LOG(DEBUG, TAG, ("createLightResource..."));
    createTemperatureResource();
    OIC_LOG(DEBUG, TAG, ("createTemperatureResource..."));
    createLedResource();
    OIC_LOG(DEBUG, TAG, ("createLedResource..."));
}
//loop 函数被无休止地调用
void loop()
{
    //在这里保持人为的延迟,以避免 Arduino 微控制器的连续旋转
    //根据具体应用需求进行修改
    delay(200);
    int data;
    chk = DHT11.read(DHT11PIN);
    lightval = analogRead(LIGHT_PIN);
    tempval = DHT11.temperature;
    humval = DHT11.humidity;
    Serial.print("the temp is:");
    Serial.println(tempval);
    Serial.print("the hum is:");
    Serial.println(humval);
    Serial.print("the LED state is:");
    Serial.println(LED_STATE);
    //向 PM2.5 传感器发送串口请求
    Serial2.write(Tx,5);
      int i = 0;
    while(Serial2.available())
    {
      Rx[i] = byte(Serial2.read());
       i++;
```

```
            delay(2);
        }
        Serial.println(Rx[4]);
        Serial.println(Rx[5]);
        pm25val = Rx[4] * 256 + Rx[5];
        Serial.print("The PM25 is: ");
        Serial.println(pm25val);
        Serial.println("####################");
    //此调用显示 Arduino 上释放的可用 SRAM 的数量
    PrintArduinoMemoryStats();
    //给 OC 协议栈 CPU 周期来执行发送/接收,以及其他 OC 协议栈的东西
    if (OCProcess() != OC_STACK_OK)
    {
        OIC_LOG(ERROR, TAG, ("OCStack process error"));
        return;
    }
}
void createLightResource()
{
    OCStackResult res = OCCreateResource(&Light.lighthandle,
            RESOURCE_TYPE,
            ARDUINO_RESOURCE_INTERFACE,
            LIGHT_RESOURCE_ENDPOINT,
            OCLightEntityHandlerCb,
            NULL,
            OC_DISCOVERABLE|OC_OBSERVABLE);
    OIC_LOG_V(DEBUG, TAG, "Created Light resource with result: %s", getResult(res));
}
void createTemperatureResource()
{
    OCStackResult res = OCCreateResource(&Temperature.temperaturehandle,
            RESOURCE_TYPE,
            ARDUINO_RESOURCE_INTERFACE,
            TEMPERATURE_RESOURCE_ENDPOINT,
            OCTemperatureEntityHandlerCb,
            NULL,
            OC_DISCOVERABLE|OC_OBSERVABLE);
    OIC_LOG_V(DEBUG, TAG, "Created Temperature resource with result: %s", getResult(res));
}
void createLedResource()
{
    OCStackResult res = OCCreateResource(&Led.ledhandle,
            RESOURCE_TYPE,
            ARDUINO_RESOURCE_INTERFACE,
            LED_RESOURCE_ENDPOINT,
            OCLedEntityHandlerCb,
            NULL,
            OC_DISCOVERABLE|OC_OBSERVABLE);
    OIC_LOG_V(DEBUG, TAG, "Created Led resource with result: %s", getResult(res));
}
void createPm25Resource()
{
    OCStackResult res = OCCreateResource(&Pm25.pm25handle,
            RESOURCE_TYPE,
```

```
            ARDUINO_RESOURCE_INTERFACE,
            PM25_RESOURCE_ENDPOINT,
            OCPm25EntityHandlerCb,
            NULL,
            OC_DISCOVERABLE|OC_OBSERVABLE);
    OIC_LOG_V(DEBUG, TAG, "Created Led resource with result: %s", getResult(res));
}
const char * getResult(OCStackResult result) {
    switch (result) {
    case OC_STACK_OK:
        return "OC_STACK_OK";
    case OC_STACK_INVALID_URI:
        return "OC_STACK_INVALID_URI";
    case OC_STACK_INVALID_QUERY:
        return "OC_STACK_INVALID_QUERY";
    case OC_STACK_INVALID_IP:
        return "OC_STACK_INVALID_IP";
    case OC_STACK_INVALID_PORT:
        return "OC_STACK_INVALID_PORT";
    case OC_STACK_INVALID_CALLBACK:
        return "OC_STACK_INVALID_CALLBACK";
    case OC_STACK_INVALID_METHOD:
        return "OC_STACK_INVALID_METHOD";
    case OC_STACK_NO_MEMORY:
        return "OC_STACK_NO_MEMORY";
    case OC_STACK_COMM_ERROR:
        return "OC_STACK_COMM_ERROR";
    case OC_STACK_INVALID_PARAM:
        return "OC_STACK_INVALID_PARAM";
    case OC_STACK_NOTIMPL:
        return "OC_STACK_NOTIMPL";
    case OC_STACK_NO_RESOURCE:
        return "OC_STACK_NO_RESOURCE";
    case OC_STACK_RESOURCE_ERROR:
        return "OC_STACK_RESOURCE_ERROR";
    case OC_STACK_SLOW_RESOURCE:
        return "OC_STACK_SLOW_RESOURCE";
    case OC_STACK_NO_OBSERVERS:
        return "OC_STACK_NO_OBSERVERS";
    case OC_STACK_ERROR:
        return "OC_STACK_ERROR";
    default:
        return "UNKNOWN";
    }
}
```

12.6.2 Android实例

Android端应用程序作为OCF客户端运行，实现手机实时监测Arduino端传感器的值，并实现开关灯等功能。Android端应用程序包括Constant.java、LEDResource.java、LightResource.java、TempResource.java和SimpleClient.java五部分。

1. Constant. java

```
package org.iotivity.base.examples;
public class Constants {
    public static String ARDUINO_RESOURCE_INTERFACE = "core.haier.resources";
    public static String RESOURCE_TYPE = "room.haier";
    //资源 URI
    public static String TEMPERATURE_RESOURCE_ENDPOINT = "/temperature";
    public static String LIGHT_RESOURCE_ENDPOINT = "/ambientlight";
    public static String LED_RESOURCE_ENDPOINT = "/led";
    public static String PM25_RESOURCE_ENDPOINT = "/pm2.5";
    //属性
    public static String TEMPERATURE_RESOURCE_KEY = "temperature";
    public static String LIGHT_RESOURCE_KEY = "ambientlight";
    public static String LED_RESOURCE_KEY = "switch";
    public static String PM25_RESOURCE_KEY = "pm2.5";
    public static String HUM_RESOURCE_KEY = "hum";
```

2. LEDResource. java

```
package org.iotivity.base.examples;
import org.iotivity.base.OcException;
import org.iotivity.base.OcRepresentation;
public class LEDResource {
    public static final String STATE_KEY = Constants.LED_RESOURCE_KEY;
    private int mState;
    public LEDResource(){
        mState = 0;
    }
    public void setOcRepresentation(OcRepresentation rep) throws OcException {
        mState = rep.getValue(LEDResource.STATE_KEY);
    }
    public OcRepresentation getOcRepresentation() throws OcException {
        OcRepresentation rep = new OcRepresentation();
        rep.setValue(STATE_KEY, mState);
        return rep;
    }
    public int getState() {
        return mState;
    }
    public void setState(int state) {
        this.mState = state;
    }
    public String toString() {
        return "\t" + STATE_KEY + ": " + mState;
    }
}
```

3. LightResource. java

```
package org.iotivity.base.examples;
import org.iotivity.base.OcException;
import org.iotivity.base.OcRepresentation;
public class LightResource {
    public static final String LIGHT_KEY = Constants.LIGHT_RESOURCE_KEY;
```

```
    private int light;
    public LightResource(){
        light = 0;
    }
    public void setOcRepresentation(OcRepresentation rep) throws OcException {
        light = rep.getValue(LightResource.LIGHT_KEY);
    }
    public OcRepresentation getOcRepresentation() throws OcException {
        OcRepresentation rep = new OcRepresentation();
        rep.setValue(LIGHT_KEY, light);
        return rep;
    }
    public int getLight() {
        return light;
    }
    public void setLight(int light) {
        this.light = light;
    }
    public String toString() {
        return "\t" + LIGHT_KEY + ": " + light;
    }
}
```

4. TempResource.java

```
package org.iotivity.base.examples;
import org.iotivity.base.OcException;
import org.iotivity.base.OcRepresentation;
public class TempResource {
    public static final String TEMP_KEY = Constants.TEMPERATURE_RESOURCE_KEY;
    public static final String HUMI_KEY = Constants.HUM_RESOURCE_KEY;
    public static final String PM25_KEY = Constants.PM25_RESOURCE_KEY;
    private int temp;
    private int humi;
    private int pm25;
    public TempResource(){
        temp = 0;
        humi = 0;
        pm25 = 0;
    }
    public void setOcRepresentation(OcRepresentation rep) throws OcException {
        temp = rep.getValue(TempResource.TEMP_KEY);
        humi = rep.getValue(TempResource.HUMI_KEY);
        pm25 = rep.getValue(TempResource.PM25_KEY);
    }
    public OcRepresentation getOcRepresentation() throws OcException {
        OcRepresentation rep = new OcRepresentation();
        rep.setValue(TEMP_KEY, temp);
        rep.setValue(HUMI_KEY, humi);
        rep.setValue(PM25_KEY, pm25);
        return rep;
    }
    public int getTemp() {
        return temp;
```

```
    }
    public void setTemp(int temp) {
        this.temp = temp;
    }
    public int getHumi() {
        return humi;
    }
    public void setHumi(int humi) {
        this.humi = humi;
    }

    public int getPm25() {
        return pm25;
    }
    public void setPm25(int pm25) {
        this.pm25 = pm25;
    }

    @Override
    public String toString() {
        return "\t" + TEMP_KEY + ": " + temp +
                "\n\t" + HUMI_KEY + ": " + humi +
                "\n\t" + PM25_KEY + ": " + pm25;
    }
}
```

5. SimpleClient.java

```
package org.iotivity.base.examples;
import android.app.Activity;
import android.content.Context;
import android.content.Intent;
import android.nfc.NfcAdapter;
import android.os.Bundle;
import android.text.method.ScrollingMovementMethod;
import android.util.Log;
import android.widget.Button;
import android.widget.ScrollView;
import android.widget.TextView;
import android.view.LayoutInflater;
import android.view.View;
import android.view.ViewGroup;
import android.widget.BaseAdapter;
import android.widget.CompoundButton;
import android.widget.ImageView;
import android.widget.ListView;
import android.widget.Switch;
import android.app.Dialog;
import android.content.DialogInterface;
import android.app.AlertDialog;
import android.os.Handler;
import android.os.Message;
import android.widget.Toast;
import android.widget.RelativeLayout;
```

```
import android.widget.LinearLayout;
import org.iotivity.base.ErrorCode;
import org.iotivity.base.ModeType;
import org.iotivity.base.ObserveType;
import org.iotivity.base.OcConnectivityType;
import org.iotivity.base.OcException;
import org.iotivity.base.OcHeaderOption;
import org.iotivity.base.OcPlatform;
import org.iotivity.base.OcRepresentation;
import org.iotivity.base.OcResource;
import org.iotivity.base.OcResourceIdentifier;
import org.iotivity.base.PlatformConfig;
import org.iotivity.base.QualityOfService;
import org.iotivity.base.ServiceType;
import java.util.ArrayList;
import java.util.EnumSet;
import java.util.HashMap;
import java.util.List;
import java.util.Map;
import java.util.Timer;
import java.util.TimerTask;
public class SimpleClient extends Activity implements
        OcPlatform.OnResourceFoundListener,
        OcResource.OnGetListener,
        OcResource.OnPutListener,
        OcResource.OnObserveListener {
    private Map<OcResourceIdentifier, OcResource> mFoundResources = new HashMap<>();
    private OcResource mFoundLightResource = null;
    private OcResource mFoundLedResource = null;
    private OcResource mFoundTempResource = null;
    //local representation of a server's light resource
    private LightResource mLight = new LightResource();
    private LEDResource mLED = new LEDResource();
    private TempResource mTemp = new TempResource();
    /*该方法配置并初始化平台,然后寻找资源类型为"room.haier"的资源*/
    private void startSimpleClient() {
        Context context = this;
        PlatformConfig platformConfig = new PlatformConfig(
                this,
                context,
                ServiceType.IN_PROC,
                ModeType.CLIENT,
                "0.0.0.0",                        //通过设置为"0.0.0.0"绑定到所有可用的接口
                0,                                //使用随机可用的端口
                QualityOfService.LOW
        );
        msg("Configuring platform.");
        OcPlatform.Configure(platformConfig);
        try {
            msg("Finding all resources of type \"room.haier\".");
            String requestUri = OcPlatform.WELL_KNOWN_QUERY + "?rt=" + Constants.RESOURCE_TYPE;
            OcPlatform.findResource("",
                    requestUri,
                    EnumSet.of(OcConnectivityType.CT_DEFAULT),
```

```
                this
            );
            sleep(1);
        } catch (OcException e) {
            Log.e(TAG, e.toString());
            msg("Failed to invoke find resource API");
        }
        printLine();
    }
    /* 当一个"findResource"请求成功完成时,执行该事件处理器
     * @param ocResource found resource */
    @Override
    public synchronized void onResourceFound(OcResource ocResource) {
        if (null == ocResource) {
            Message msg = new Message();
            msg.what = UNCONNECT;
            handler.sendMessage(msg);
            msg("Found resource is invalid");
            return;
        }
        //获取资源 URI
        String resourceUri = ocResource.getUri();
        //获取资源主机地址
        String hostAddress = ocResource.getHost();
        msg("\tURI of the resource: " + resourceUri);
        msg("\tHost address of the resource: " + hostAddress);
        //获取资源类型
        msg("\tList of resource types: ");
        for (String resourceType : ocResource.getResourceTypes()) {
            msg("\t\t" + resourceType);
        }
        msg("\tList of resource interfaces:");
        for (String resourceInterface : ocResource.getResourceInterfaces()) {
            msg("\t\t" + resourceInterface);
        }
        msg("\tList of resource connectivity types:");
        for (OcConnectivityType connectivityType : ocResource.getConnectivityTypeSet()) {
            msg("\t\t" + connectivityType);
        }
        printLine();
        Message msg = new Message();
        msg.what = CONNECT;
        handler.sendMessage(msg);
        if (resourceUri.equals(Constants.LED_RESOURCE_ENDPOINT)) {
            mFoundLedResource = ocResource;
            getLedResourceRepresentation();
        }else if(resourceUri.equals(Constants.LIGHT_RESOURCE_ENDPOINT)){
            mFoundLightResource = ocResource;
            getLightResourceRepresentation();
        }else if(resourceUri.equals(Constants.TEMPERATURE_RESOURCE_ENDPOINT)){
            mFoundTempResource = ocResource;
            getTempResourceRepresentation();
        }
    }
}
```

```
private void getTempResourceRepresentation(){
    msg("Getting Temp Representation...");
    Map<String, String> queryParams = new HashMap<>();
    try {
        sleep(1);
        mFoundTempResource.get(queryParams, this);
    } catch (OcException e) {
        Log.e(TAG, e.toString());
        msg("Error occurred while invoking \"get\" API of Led");
    }
}
private void getLedResourceRepresentation(){
    msg("Getting Led Representation...");
    Map<String, String> queryParams = new HashMap<>();
    try {
        sleep(1);
        mFoundLedResource.get(queryParams, this);
    } catch (OcException e) {
        Log.e(TAG, e.toString());
        msg("Error occurred while invoking \"get\" API of Led");
    }
}
/*本地方法,获取一个已发现的light资源的表示*/
private void getLightResourceRepresentation() {
    msg("Getting Light Representation...");
    Map<String, String> queryParams = new HashMap<>();
    try {
        sleep(1);
        mFoundLightResource.get(queryParams, this);
    } catch (OcException e) {
        Log.e(TAG, e.toString());
        msg("Error occurred while invoking \"get\" API");
    }
}
/*当"get"请求成功,执行该事件处理器*/
@Override
public synchronized void onGetCompleted(List<OcHeaderOption> list,OcRepresentation ocRepresentation) {
    msg("GET request was successful");
    msg("Resource URI: " + ocRepresentation.getUri());
    String resourceURI = ocRepresentation.getUri();
    try {
        if(resourceURI.equals(Constants.LED_RESOURCE_ENDPOINT)){
            mLED.setOcRepresentation(ocRepresentation);
            msg("Led attributes: ");
            msg(mLED.toString());
            printLine();
        }
        else if(resourceURI.equals(Constants.LIGHT_RESOURCE_ENDPOINT)){
            //将属性值读取到light的本地表示中
            mLight.setOcRepresentation(ocRepresentation);
            msg("Light attributes: ");
            msg(mLight.toString());
            printLine();
        }else if(resourceURI.equals(Constants.TEMPERATURE_RESOURCE_ENDPOINT)){
```

```
            mTemp.setOcRepresentation(ocRepresentation);
            msg("Light attributes: ");
            msg(mTemp.toString());
            printLine();
        }
        Message msg = new Message();
        msg.what = REFRESH;
        handler.sendMessage(msg);
    } catch (OcException e) {
        Log.e(TAG, e.toString());
        msg("Failed to read the attributes of a light resource");
    }
}
/*"get"请求失败时应该执行的操作*/
@Override
public synchronized void onGetFailed(Throwable throwable) {
    if (throwable instanceof OcException) {
        OcException ocEx = (OcException) throwable;
        Log.e(TAG, ocEx.toString());
        ErrorCode errCode = ocEx.getErrorCode();
        //根据错误代码进行一些处理
        msg("Error code: " + errCode);
    }
    msg("Failed to get representation of a found light resource");
}
private void putLedOnRepersentation(){
    mLED.setState(1);
    msg("Putting led representation...");
    OcRepresentation representation = null;
    try {
        representation = mLED.getOcRepresentation();
    } catch (OcException e) {
        Log.e(TAG, e.toString());
        msg("Failed to get OcRepresentation from a led");
    }
    Map<String, String> queryParams = new HashMap<>();
    try {
        sleep(1);
        mFoundLedResource.put(representation, queryParams, this);
    } catch (OcException e) {
        Log.e(TAG, e.toString());
        msg("Error occurred while invoking \"put\" API");
    }
}
private void putLedOffRepersentation(){
    mLED.setState(0);
    msg("Putting led representation...");
    OcRepresentation representation = null;
    try {
        representation = mLED.getOcRepresentation();
    } catch (OcException e) {
        Log.e(TAG, e.toString());
        msg("Failed to get OcRepresentation from a led");
    }
```

```
            Map<String, String> queryParams = new HashMap<>();
            try {
                sleep(1);
                mFoundLedResource.put(representation, queryParams, this);
            } catch (OcException e) {
                Log.e(TAG, e.toString());
                msg("Error occurred while invoking \"put\" API");
            }
        }
        /* "put"操作成功时进行的操作 */
        @Override
        public synchronized void onPutCompleted(List<OcHeaderOption> list, OcRepresentation ocRepresentation) {
            msg("PUT request was successful");
            String resourceURI = ocRepresentation.getUri();
            if (resourceURI.equals(Constants.LED_RESOURCE_ENDPOINT)) {
                msg("Led attributes: ");
                msg(mLED.toString());
                printLine();
            }
        }
        /* "put"请求失败时进行的操作 */
        @Override
        public synchronized void onPutFailed(Throwable throwable) {
            if (throwable instanceof OcException) {
                OcException ocEx = (OcException) throwable;
                Log.e(TAG, ocEx.toString());
                ErrorCode errCode = ocEx.getErrorCode();
                //do something based on errorCode
                msg("Error code: " + errCode);
            }
            msg("Failed to \"put\" a new representation");
        }
        /* 该方法开始观察 light 资源 */
        private void observeFoundLightResource() {
            try {
                sleep(1);
                mFoundLightResource.observe(ObserveType.OBSERVE, new HashMap<String, String>(), this);
            } catch (OcException e) {
                Log.e(TAG, e.toString());
                msg("Error occurred while invoking \"observe\" API");
            }
    }
        private static int mObserveCount = 0;
        /* "post"请求成功时进行的操作 */
        @Override
        public synchronized void onObserveCompleted(List<OcHeaderOption> list,
                                                    OcRepresentation ocRepresentation,
                                                    int sequenceNumber) {
            if (OcResource.OnObserveListener.REGISTER == sequenceNumber) {
                msg("Observe registration action is successful:");
            } else if (OcResource.OnObserveListener.DEREGISTER == sequenceNumber) {
                msg("Observe De-registration action is successful");
            } else if (OcResource.OnObserveListener.NO_OPTION == sequenceNumber) {
                msg("Observe registration or de-registration action is failed");
```

```
        }
        msg("OBSERVE Result:");
        msg("\tSequenceNumber:" + sequenceNumber);
        try {
            mLight.setOcRepresentation(ocRepresentation);
        } catch (OcException e) {
            Log.e(TAG, e.toString());
            msg("Failed to get the attribute values");
        }
        msg(mLight.toString());
        if ((++mObserveCount) == 11) {
            msg("Cancelling Observe...");
            try {
                mFoundLightResource.cancelObserve();
            } catch (OcException e) {
                Log.e(TAG, e.toString());
                msg("Error occurred while invoking \"cancelObserve\" API");
            }
            msg("DONE");
            //prepare for the next restart of the SimpleClient
            resetGlobals();
            enableStartButton();
        }
    }
    /* "observe"请求失败时执行的操作 */
    @Override
    public synchronized void onObserveFailed(Throwable throwable) {
        if (throwable instanceof OcException) {
            OcException ocEx = (OcException) throwable;
            Log.e(TAG, ocEx.toString());
            ErrorCode errCode = ocEx.getErrorCode();
            msg("Error code: " + errCode);
        }
        msg("Observation of the found light resource has failed");
    }
    //OCF 部分的代码结束,开始设置
    private TextView pm25Value, tempValue, humiValue, lightValue;
    private Switch ledSwitch;
    private AlertDialog.Builder builder;
    private Button refreshButton;
    private static final int CONNECT = 0;
    private static final int REFRESH = 1;
    private static final int UNCONNECT = 2;
    private static final int OVERPM25 = 3;
    private static final int BELOWPM25 = 4;
    private boolean isLedOpen = false;
    private static boolean isConnected = false;
    private static boolean isAirContionerOpen = false;
    private final static String TAG = SimpleClient.class.getSimpleName();
    private TextView mConsoleTextView;
    private ScrollView mScrollView;
    private Handler handler = new Handler() {
        @Override
        public void handleMessage(Message msg) {
```

```
            if (msg.what == CONNECT && !isConnected) {
                if(builder != null){
                    builder.show();
                    isConnected = true;
                    setData();
                    isLedOpen = mLED.getState() == 0 ? false : true;
                    ledSwitch.setChecked(isLedOpen);
                }
            }
            if(msg.what == UNCONNECT && isConnected){
                if(builder != null){
                    builder.setMessage("Can not find resources");
                    builder.show();
                    isConnected = false;
                }
            }
            if(msg.what == OVERPM25 && !isAirContionerOpen){
                Toast toast = Toast.makeText(getApplicationContext(), "Haier air conditioner is opening.",
Toast.LENGTH_SHORT);
                LinearLayout linearLayout = (LinearLayout) toast.getView();
                TextView messageTextView = (TextView) linearLayout.getChildAt(0);
                messageTextView.setTextSize(20);
                toast.show();
                isAirContionerOpen = true;
            }
            if(msg.what == BELOWPM25 && isAirContionerOpen){
                Toast toast = Toast.makeText(getApplicationContext(), "Haier air conditioner is closed.",
Toast.LENGTH_SHORT);
                LinearLayout linearLayout = (LinearLayout) toast.getView();
                TextView messageTextView = (TextView) linearLayout.getChildAt(0);
                messageTextView.setTextSize(20);
                toast.show();
                isAirContionerOpen = false;
            }
            if(msg.what == REFRESH){
                setData();
                ledSwitch.setChecked(isLedOpen);
            }
        }
    };
    @Override
    protected void onCreate(Bundle savedInstanceState) {
        super.onCreate(savedInstanceState);
        setContentView(R.layout.activity_simple_client);
        builder = new AlertDialog.Builder(this);
        builder.setTitle("Notification").
                setTitle("Notification").
                setMessage("Find LED Resource. \n Find Environment Resource. \n Find pm25 Resource. \n Connect
Success!").
                setPositiveButton("Confirm", new DialogInterface.OnClickListener() {
                    @Override
                    public void onClick(DialogInterface dialog, int which) {
                    }
                }).create();
```

```
        initGUI();
        mConsoleTextView = (TextView) findViewById(R.id.consoleTextView);
        mConsoleTextView.setMovementMethod(new ScrollingMovementMethod());
        mScrollView = (ScrollView) findViewById(R.id.scrollView);
        mScrollView.fullScroll(View.FOCUS_DOWN);
    }
    private void initGUI() {
        drawGUI();
        new Thread(new Runnable() {
            public void run() {
                startSimpleClient();
            }
        }).start();
        new Thread(new Runnable() {
            public void run() {
                while (true) {
                    if (isConnected) {
                        if (mFoundTempResource != null) {
                            getTempResourceRepresentation();
                        }
                        if (mFoundLightResource != null) {
                            getLightResourceRepresentation();
                        }
                        if (mFoundLedResource != null) {
                            getLedResourceRepresentation();
                        }
                    }
                    sleep(6);
                }
            }
        }).start();
        refreshButton.setOnClickListener(new View.OnClickListener() {
            @Override
            public void onClick(View v) {
                new Thread(new Runnable() {
                    public void run() {
                            startSimpleClient();
                            if (mFoundTempResource != null) {
                                getTempResourceRepresentation();
                            }
                            if (mFoundLightResource != null) {
                                getLightResourceRepresentation();
                            }
                            if (mFoundLedResource != null) {
                                getLedResourceRepresentation();
                            }
                    }
                }).start();
            }
        });
        ledSwitch.setChecked(isLedOpen);
        ledSwitch.setOnCheckedChangeListener(new CompoundButton.OnCheckedChangeListener() {
            @Override
            public void onCheckedChanged(CompoundButton buttonView, boolean isChecked) {
```

```
                if(isChecked){
                    isLedOpen = true;
                    new Thread(new Runnable() {
                        public void run() {
                            if(mFoundLedResource != null){
                                putLedOnRepersentation();
                            }else{
                                Message msg = new Message();
                                msg.what = UNCONNECT;
                                handler.sendMessage(msg);
                            }
                        }
                    }).start();
                }else{
                    isLedOpen = false;
                    new Thread(new Runnable() {
                        public void run() {
                            if(mFoundLedResource != null){
                                putLedOffRepersentation();
                            }else{
                                Message msg = new Message();
                                msg.what = UNCONNECT;
                                handler.sendMessage(msg);
                            }
                        }
                    }).start();
                }
            }
        });
    }
    private void drawGUI(){
        ImageView pm25Image;
        TextView pm25Title;
        RelativeLayout pm25Layout = (RelativeLayout) findViewById(R.id.pm25_layout);
        if (pm25Layout != null) {
            pm25Layout.setBackgroundColor(getResources().getColor(R.color.blue));
            pm25Image = (ImageView) pm25Layout.findViewById(R.id.image);
            pm25Image.setImageDrawable(getResources().getDrawable(R.drawable.pm25));
            pm25Title = (TextView) pm25Layout.findViewById(R.id.title);
            pm25Title.setText("PM25");
            pm25Value = (TextView) pm25Layout.findViewById(R.id.value);
            pm25Value.setText("22 ug");
        }
        ImageView tempImage;
        TextView tempTitle;
        RelativeLayout tempLayout = (RelativeLayout) findViewById(R.id.temp_layout);
        if(tempLayout != null){
            tempLayout.setBackgroundColor(getResources().getColor(R.color.green));
            tempImage = (ImageView) tempLayout.findViewById(R.id.image);
            tempImage.setImageDrawable(getResources().getDrawable(R.drawable.temp));
            tempTitle = (TextView) tempLayout.findViewById(R.id.title);
            tempTitle.setText("Temperature");
            tempValue = (TextView) tempLayout.findViewById(R.id.value);
            tempValue.setText("30 ℃");
```

```
        }
        ImageView humiImage;
        TextView humiTitle;
        RelativeLayout humiLayout = (RelativeLayout) findViewById(R.id.humi_layout);
        if(humiLayout != null){
            humiLayout.setBackgroundColor(getResources().getColor(R.color.orange));
            humiImage = (ImageView) humiLayout.findViewById(R.id.image);
            humiImage.setImageDrawable(getResources().getDrawable(R.drawable.humi));
            humiTitle = (TextView) humiLayout.findViewById(R.id.title);
            humiTitle.setText("Humiditity");
            humiValue = (TextView) humiLayout.findViewById(R.id.value);
            humiValue.setText("50 %");
        }
        ImageView lightImage;
        TextView lightTitle;
        RelativeLayout lightLayout = (RelativeLayout) findViewById(R.id.light_layout);
        if(lightLayout != null){
            lightLayout.setBackgroundColor(getResources().getColor(R.color.gray));
            lightImage = (ImageView) lightLayout.findViewById(R.id.image);
            lightImage.setImageDrawable(getResources().getDrawable(R.drawable.ambientlight));
            lightTitle = (TextView) lightLayout.findViewById(R.id.title);
            lightTitle.setText("Light");
            lightValue = (TextView) lightLayout.findViewById(R.id.value);
            lightValue.setText("100 Lux");
        }
        RelativeLayout ledLayout = (RelativeLayout) findViewById(R.id.led_layout);
        ledSwitch = (Switch) ledLayout.findViewById(R.id.switch_icon);
        refreshButton = (Button) findViewById(R.id.refresh_button);
    }
    private void setData(){
        pm25Value.setText(mTemp.getPm25() + " ug");
        tempValue.setText(mTemp.getTemp() + " ℃");
        humiValue.setText(mTemp.getHumi() + " %");
        lightValue.setText(mLight.getLight() + " Lux");
        if(mTemp.getPm25() > 200){
            Message msg = new Message();
            msg.what = OVERPM25;
            handler.sendMessage(msg);
        }
        if(mTemp.getPm25() < 200){
            Message msg = new Message();
            msg.what = BELOWPM25;
            handler.sendMessage(msg);
        }
    }
    @Override
    protected void onSaveInstanceState(Bundle outState) {
        super.onSaveInstanceState(outState);
        outState.putString("consoleOutputString", mConsoleTextView.getText().toString());
    }
    @Override
    protected void onRestoreInstanceState(Bundle savedInstanceState) {
        super.onRestoreInstanceState(savedInstanceState);
        String consoleOutput = savedInstanceState.getString("consoleOutputString");
```

```
        mConsoleTextView.setText(consoleOutput);
    }
    private void enableStartButton() {
        runOnUiThread(new Runnable() {
            public void run() {
            }
        });
    }
    private void sleep(int seconds) {
        try {
            Thread.sleep(seconds * 500);
        } catch (InterruptedException e) {
            e.printStackTrace();
            Log.e(TAG, e.toString());
        }
    }
    private void msg(final String text) {
        runOnUiThread(new Runnable() {
            public void run() {
                mConsoleTextView.append("\n");
                mConsoleTextView.append(text);
                mScrollView.fullScroll(View.FOCUS_DOWN);
            }
        });
        Log.i(TAG, text);
    }
    private void printLine() {
msg("----------------------------------------------------------------------------");
    }
    private synchronized void resetGlobals() {
        mFoundLightResource = null;
        mFoundResources.clear();
        mLight = new LightResource();
        mLED = new LEDResource();
        mTemp = new TempResource();
        mObserveCount = 0;
    }
    @Override
    public void onNewIntent(Intent intent) {
        super.onNewIntent(intent);
        Log.d(TAG, "onNewIntent with changes sending broadcast IN ");
        Intent i = new Intent();
        i.setAction(intent.getAction());
        i.putExtra(NfcAdapter.EXTRA_NDEF_MESSAGES,
                intent.getParcelableArrayExtra(NfcAdapter.EXTRA_NDEF_MESSAGES));
        sendBroadcast(i);
        Log.d(TAG, "Initialize Context again resetting");
    }
}
```